Astrophysical Spectropolarimetry

The polarization of light is the key to obtaining a wealth of essential information that lies encoded in the electromagnetic radiation reaching us from cosmic objects. Spectropolarimetry and imaging polarimetry provide powerful diagnostics of the physical conditions in astrophysical plasmas, for instance, concerning magnetic fields, which cannot be obtained via conventional spectroscopy. Spectropolarimetry is being used with great success in solar physics. Yet, its application to other fields of astrophysics is still in an early stage of development.

This book on Astrophysical Spectropolarimetry comes at a time of growing awareness of the new possibilities offered by this field. This is mainly due to the observational opportunities opened up by the new generation of telescopes, both ground-based and space-borne, and their associated instrumentation as well as to recent advances in the theory and numerical modelling of the generation and transfer of polarized radiation. The book contains the lectures delivered at the XII Canary Islands Winter School of Astrophysics on the following topics: the physics of polarization, polarized radiation diagnostics of solar magnetic fields, stellar magnetic fields, polarization insights for active galactic nuclei, compact objects and accretion disks, astronomical masers and their polarization, interstellar magnetic fields and infrared-submillimeter spectropolarimetry, and instrumentation for astrophysical spectropolarimetry. They are written by prestigious researchers working in several areas of astrophysics, all of them sharing an active interest in theoretical and observational spectropolarimetry.

This timely volume provides graduate students and researchers with an unprecedented introduction to Astrophysical Spectropolarimetry.

Astrophysical Spectropolarimetry

Proceedings of the XII Canary Islands Winter School of Astrophysics
Puerto de la Cruz, Tenerife, Spain
November 13–24, 2000

Edited by

J. TRUJILLO-BUENO
Instituto de Astrofísica de Canarias

F. MORENO-INSERTIS
Instituto de Astrofísica de Canarias

F. SÁNCHEZ
Instituto de Astrofísica de Canarias

CAMBRIDGE
UNIVERSITY PRESS

PUBLISHED BY THE PRESS SYNDICATE OF THE UNIVERSITY OF CAMBRIDGE
The Pitt Building, Trumpington Street, Cambridge, United Kingdom

CAMBRIDGE UNIVERSITY PRESS
The Edinburgh Building, Cambridge CB2 2RU, UK
40 West 20th Street, New York, NY 10011–4211, USA
477 Williamstown Road, Port Melbourne, VIC 3207, Australia
Ruiz de Alarcón 13, 28014 Madrid, Spain
Dock House, The Waterfront, Cape Town 8001, South Africa

http://www.cambridge.org

First published 2002

Printed in the United Kingdom at the University Press, Cambridge

A catalogue record for this book is available from the British Library

ISBN 0 521 80998 3 hardback

Contents

The Physics of Polarization

E. Landi Degl'Innocenti

Polarized Radiation Diagnostics of Solar Magnetic Fields

J. O. Stenflo

Polarized Radiation Diagnostics of Stellar Magnetic Fields

G. Mathys

Polarization Insights for Active Galactic Nuclei

R. Antonucci

Compact Objects and Accretion Disks

R. Blandford, E. Agol, A. Broderick, J. Heyl, L. Koopmans, H.-W. Lee

Astronomical Masers and their Polarization

M. Elitzur

Interstellar magnetic fields and infrared-submillimeter spectropolarimetry

R. H. Hildebrand

Instrumentation for Astrophysical Spectropolarimetry

C. U. Keller

Participants

Ageorges, Nancy	European Southern Observatory (Chile)
Antonucci, Robert	UC Santa Barbara (USA)
Aryal, Binil	University of Innsbruck (Austria)
Asensio, Andrés	Instituto de Astrofísica de Canarias (Spain)
Behlke, Rico	Swedish Institute for Space Physics (Sweden)
Berdyugina, Svetlana	University of Oulu (Finland)
Bianda, Michele	Istituto Ricerche Solari Locarno (Switzerland)
Blandford, Roger	California Institute of Technology (USA)
Brogan, Crystal	National Radio Astronomy Obs. (USA)
Carroll, Thorsten Anthony	Astrophysikalisches Institut Postdam (Germany)
Collados Vera, Manuel	Instituto de Astrofísica de Canarias (Spain)
Deetjen, Jochen	Int. fur Astronomie und Astrophysik (Germany)
Dittmann, Olaf	Instituto de Astrofísica de Canarias (Spain)
Elitzur, Moshe	University of Kentucky (USA)
Etoka, Sandra	Jodrell Bank Observatory (UK)
Feautrier, Nicole	Observatoire de Paris (France)
Fluri, Dominique	ETH Zurich, Institute for Astronomy (Switzerland)
Fuhiyoshi, Takuya	Cavendish Laboratory (UK)
Gálvez Ortiz, Mª Cruz	Universidad Complutense de Madrid (Spain)
Gandofer, Achim	ETH Zurich, Institute for Astronomy (Switzerland)
García Álvarez, David	Armagh Observatory (Ireland)
García Lorenzo, Begoña	Isaac Newton Group (Spain)
Gisler, Daniel	ETH Zurich, Institute for Astronomy (Switzerland)
González Alfonso, Eduardo	I.E.M – CSIC (Spain)
González Delgado, David	Stockholm Observatory (Sweden)
González Hernández, José J.	Universidad de La Laguna (Spain)
Gupta, Alok Chandra	Physical Research Laboratory (India)
Herpin, Fabrice	I.E.M – CSIC (Spain)
Hildebrand, Robert	University of Chicago (USA)
Hoffman, Jennifer L.	University of Wisconsin-Madison (USA)
Holloway, Richard	University of Hertfordshire (UK)
Ikeda, Yuji	University of Tohoku (Japan)
Ilyin, Ilya	University of Oulu (Finland)
Janssen, Katja	Georg-August Universitat Gottingen (Germany)
Keller, Christoph	National Solar Observatory (USA)
Kelz, Andreas	Astrophysikalisches Institut Postdam (Germany)
Khomenko, Elena	Main Astronomical Observatory (Ukraine)
Kilbinger, Martin	Universitate Bonn (Germany)
Klement, Joachim	ETH Zurich, Institute for Astronomy (Switzerland)
Kochukhov, Oleg	Uppasala University (Sweden)
Kokkonen, Krista K.	Tuorla Observatory (Finland)
Korhonen, Heidi	University of Oulu (Finland)
Landi Degl'Innocenti, Egidio	Università di Firenze (Italy)
López Santiago, Javier	Universidad Complutense de Madrid (Spain)
Manso Sainz, Rafael	Instituto de Astrofísica de Canarias (Spain)
Martín, Fabiola	Universidad de La Laguna (Spain)
Mathys, Gautier	European Southern Observatory (Chile)
Modigliani, Andrea	European Southern Observatory (Germany)

Moreno-Insertis, Fernando	Instituto de Astrofísica de Canarias (Spain)
Muglach, Karin	Astrophysikalisches Institut Postdam (Germany)
Müller, Daniel	Kiepenheuer-Institut für Sonnenphysik (Germany)
Neiner, Coralie	Observatoire de Meudon (France)
Nielsen, Krister	Lund University (Sweden)
O'Connor, Padraig	National University of Ireland (Ireland)
O'Shea, Eoghan	ESTEC / Solar System Division (The Netherlands)
Packham, Chris	University of Florida (USA)
Preuss, Oliver	Max-Planck-Institut fur Aeronomie (Germany)
Romoli, Marco	Università di Firenze (Italy)
Sainz Dalda, Alberto	THEMIS (Spain)
Schlichenmaier, Rolf	Kiepenheuer-Institut für Sonnenphysik (Germany)
Skender, Marina	Institute Rudjer Boskovic (Croatia)
Stenflo, Jan Olof	Institute of Astronomy (Switzerland)
Suetterlin, Peter	Sterrenkundig Instituut Utrecht (The Netherlands)
Telfer, Deborah	University of Glasgow (UK)
Thum, Clemens	IRAM (France)
Trujillo-Bueno, Javier	Instituto de Astrofísica de Canarias (Spain)
Tziotziou, Konstantinos	Observatoire de Paris (France)
Vlemmings, Wouter	Sterrewacht Leiden (The Netherlands)
Vogt, Etienne	Osservatorio Astronomico di Capodimonte (Italy)
Weber, Michael	University of Vienna (Austria)
Wenzler, Thomas	ETH Zurich, Institute for Astronomy (Switzerland)

Group Photograph

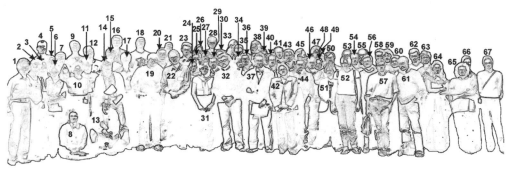

1	Robert Hildebrand	24	Katja Janssen	47	Michael Weber
2	Svetlana Berdyugina	25	Joachim Klement	48	Coralie Neiner
3	Crystal Brogan	26	Christoper Keller	49	Marina Skender
4	Thorsten Anthony Carroll	27	Clemens Thum	50	Ilya Ilyin
5	Thomas Wenzler	28	Moshe Elitzur	51	Nicole Feautrier
6	Heidi Korhonen	29	Dominique Fluri	52	Deborah Telfer
7	Binil Aryal	30	Krister Nielsen	53	Wouter Vlemmings
8	Krista K. Kokkonen	31	Nieves Villoslada Dionis	54	Unknown
9	Takuya Fuhiyoshi	32	Javier Trujillo-Bueno	55	Oliver Preuss
10	Jennifer L. Hoffman	33	Daniel Müller	56	Alberto Sainz Dalda
11	Daniel Gisler	34	Yuji Ikeda	57	Lucio Crivellari
12	Andrea Modigliani	35	Konstantinos Tziotziou	58	José J. González Hernández
13	Karin Muglach	36	Oleg Kochukhov	59	Andrés Asensio
14	Manuel Collados Vera	37	Fernando Moreno-Insertis	60	Begoña García Lorenzo
15	Javier López Santiago	38	Rolf Schlichenmaier	61	Egidio Landi Degl'Innocenti
16	Richard Holloway	39	Peter Suetterlin	62	David García Álvarez
17	Alok Chandra Gupta	40	Andreas Kelz	63	David González Delgado
18	Marco Romoli	41	Gautier Mathys	64	Fabiola Martín
19	Roger Blandford	42	Lourdes González	65	Sandra Etoka
20	Jochen Deetjen	43	Achim Gandofer	66	Mª Cruz Gálvez Ortiz
21	Eoghan O'Shea	44	Padraig O'Connor	67	Elena Khomenko
22	Jan Olof Stenflo	45	Fabrice Herpin		
23	Rico Behlke	46	Michele Bianda		

Preface

Most observational work in astrophysics has so far been carried out mainly on the basis of the intensity of the radiation received from the object observed as a function of wavelength. However, an important and frequently overlooked aspect of electromagnetic radiation is its state of polarization, which is related to the orientation of the electric field of the wave. The state of polarization can be conveniently characterized in terms of four quantities that can be measured by furnishing our telescopes with a polarimeter. These observables are the four Stokes parameters (I, Q, U, V) which were formulated by Sir George Stokes in 1852 and introduced into astrophysics by the Nobel laureate Subrahmanyan Chandrasekhar in 1946. A quick, intuitive definition of the meaning of these four parameters can be obtained from Figure 1 of the chapter by Prof. Landi Degl'Innocenti in this book, which we borrowed for the poster announcing the Twelfth Canary Islands Winter School on Astrophysical Spectropolarimetry.

In physics laboratory experiments, where the magnetic field is known beforehand, the observed polarization signals are used to obtain information on the atomic and molecular structure of the system under study. In astrophysics we have the inverse problem, the magnetic field being the unknown quantity. To obtain information about cosmic magnetic fields, therefore, we have to learn how to interpret spectropolarimetric observations correctly by resorting to our knowledge of atomic and molecular physics.

The importance of the information contained in the polarization of electromagnetic radiation has been recognized for decades in solar physics, where the spatial resolution and signal-to-noise ratio in spectroscopic observations are much more favourable than in night-time observation. In recent years, developments in theoretical astrophysics and astronomical instrumentation (telescopes with large light-collecting areas and innovative spectropolarimeters) are leading an ever-growing number of astrophysicists to learn to appreciate the enormous diagnostic potential offered by spectropolarimetry. The polarization of light is the key to unlocking new discoveries and obtaining the information we need to understand the physics of many phenomena occurring in the Universe. Particularly relevant examples, besides the magnetized plasmas of the Sun and peculiar A- and B-type stars, are young stellar objects and their surrounding discs, Herbig-Haro objects, symbiotic stars, hot stellar winds, active galactic nuclei, radio galaxies, black holes, the interstellar medium, the cosmic microwave background radiation and its cosmological implications, etc.

This book contains the lectures delivered at the XII Canary Islands Winter School of Astrophysics, organized by the Instituto de Astrofísica de Canarias (IAC), on Astrophysical Spectropolarimetry. The time is clearly ripe for such a book on the subject. There is increased awareness of the relevance of spectropolarimetry for astrophysics thanks, in part, to the new generation of 10-metre-class telescopes. Some of these large telescopes, such as the GTC at the Observatorio del Roque de los Muchachos (La Palma), are right now under construction while others are already in operation at observatories worldwide. Because of the large collecting surface of this class of telescope, the development of spectropolarimeters capable of quantifying with high precision the state of polarization of the light, and the recent unprecedented advances in the field of theoretical and numerical astrophysics, spectropolarimetry is gradually emerging as a powerful new diagnostic tool for probing the physical conditions and the magnetic fields of the Universe.

The application of spectropolarimetry in astrophysics is still at an early stage of development. This makes it especially attractive for young researchers eager to contribute to the advance of astrophysics. In this field, theoretical and observational astrophysics, numerical simulations and instrumental developments are frequently called upon. This new window on the Universe offers the opportunity to make new discoveries through the rigorous physical interpretation of spectropolarimetric observations, which provide information that it is impossible to acquire through conventional spectroscopy.

We have edited this book in the desire to present an introduction to the field of Astrophysical Spectropolarimetry, with a view to encouraging young researchers to investigate rigorously and in depth the "polarized Universe". We are convinced that the achievements of spectropolarimetry in solar physics will soon be possible in other areas of astrophysics as well.

Javier Trujillo-Bueno & Fernando Moreno-Insertis
Instituto de Astrofísica de Canarias
January 2001

Foreword

After twelve years, the Canary Islands Winter School continues to provide a unique opportunity for the participants to broaden their knowledge in a key field of astrophysics. The idea works because promising young scientists and invited lecturers interact, learn and enjoy science in the pleasant environment of the Canary Islands.

The XII edition of the Canary Islands Winter School looked at the Universe from a relatively unexploited viewpoint, namely that fostered by a multidisciplinary branch of science which has a great future in store: spectropolarimetry. Thanks to theoretical and observational spectropolarimetry we will be able to explore new facets of the Universe while unveiling new discoveries still hidden in the electromagnetic radiation we receive. The large telescopes of the future – among them the 10.4 m Gran Telescopio Canarias – and advanced postfocus instrumentation should be designed with a view to rendering feasible high precision spectropolarimetric observations. The theoretical interpretation of observed polarization signals will allow new fundamental advances in our knowledge of cosmic magnetic fields. Spectropolarimetry could well be a revolutionary technique in the astrophysics of the XXI century. That is why the XII Canary Islands Winter School of Astrophysics has been devoted to this promising and exciting field.

Francisco Sánchez
Director of the Instituto de Astrofísica de Canarias

Acknowledgements

In staging the XII Canary Islands Winter School the editors have received help from many people and institutions. We take this opportunity to express our gratitude to them all. Firstly, we thank the eight lecturers for their efforts in preparing their classes and writing the chapters of this book. We also thank all the participants for their enthusiasm and for creating a truly scientific atmosphere during the School. Our thanks also go to the following institutions for their generous support: The European Commission (DG XII), the Spanish Ministry for Science and Technology, IBERIA, the European Solar Magnetometry Network, and the local governments (Cabildos) of the Islands of La Palma and Tenerife. Last but not least, we are grateful to the following IAC staff members whose efficient work was essential for the success of the School: Lourdes González, Nieves Villoslada, Monica Murphy, Carmen del Puerto, Begoña López Betancor, Jesús Burgos, Terry Mahoney and Gabriel Pérez. We also wish to acknowledge the work done by Dr. Gabriel Gómez in preparing the camera-ready copy for Cambridge University Press.

THE PHYSICS OF POLARIZATION

By EGIDIO LANDI DEGL'INNOCENTI

Dipartimento di Astronomia e Scienza dello Spazio, Università di Firenze, Largo E. Fermi 5, 50125 Firenze, Italy

This course is intended to give a description of the basic physical concepts which underlie the study and the interpretation of polarization phenomena. Apart from a brief historical introduction (Sect. 1), the course is organized in three parts. A first part (Sects. 2-6) covers the most relevant facts about the polarization phenomena that are typically encountered in laboratory applications and in everyday life. In Sect. 2, the modern description of polarization in terms of the Stokes parameters is recalled, whereas Sect. 3 is devoted to introduce the basic tools of laboratory polarimetry, such as the Jones calculus and the Mueller matrices. The polarization phenomena which are met in the reflection and refraction of a beam of radiation at the separation surface between two dielectrics, or between a dielectric and a metal, are recalled in Sect. 4. Finally, Sect. 5 gives an introduction to the phenomena of dichroism and of anomalous dispersion and Sect. 6 summarizes the polarization phenomena that are commonly encountered in everyday life. The second part of this course (Sects. 7-14) deals with the description, within the formalism of classical physics, of the spectro-polarimetric properties of the radiation emitted by accelerated charges. Such properties are derived by taking as starting point the Liénard and Wiechert equations that are recalled and discussed in Sect. 7 both in the general case and in the non-relativistic approximation. The results are developed to find the percentage polarization, the radiation diagram, the cross-section and the spectral characteristics of the radiation emitted in different phenomena particularly relevant from the astrophysical point of view. The emission of a linear antenna is derived in Sect. 8. The other Sections are devoted to Thomson scattering (Sect. 9), Rayleigh scattering (Sect. 10), Mie scattering (Sect. 11), bremsstrahlung radiation (Sect. 12), cyclotron radiation (sect. 13), and synchrotron radiation (Sect. 14). Finally, the third part (Sects. 15-19) is devoted to give a sketch of the theory of the generation and transfer of polarized radiation in spectral lines. After a general introduction to the argument (Sect. 15), the concepts of density-matrix and of atomic polarization are illustrated in Sect. 16. In Sect. 17, a parallelism is established, within the framework of the theory of stellar atmospheres, between the usual formalism, which neglects polarization phenomena, and the more involved formalism needed for the interpretation of spectro-polarimetric observations. Some consequences of the radiative transfer equations for polarized radiation, pointing to the importance of dichroism phenomena in establishing the amplification condition via stimulated emission, are discussed in Sect. 18. The last section (Sect. 19) is devoted to introduce the problem of finding a self-consistent solution of the radiative transfer equations for polarized radiation and of the statistical equilibrium equations for the density matrix (non-LTE of the 2nd kind).

---

1. Introduction

Polarization is an important physical property of electromagnetic waves which is connected with the transversality character, with respect to the direction of propagation, of the electric and magnetic field vectors. Under this respect, the phenomenon of polarization is not restricted to electromagnetic waves, but could in principle be defined for any wave having a transverse character, such as, for instance, transverse elastic waves propagating in a solid, transverse seismic waves, waves in a guitar string, and so on. On the contrary, polarization phenomena are obviously inexistent for longitudinal waves, such as the usual acoustic waves propagating in a gas or in a liquid.

From an historical perspective (see Swyndell, 1975, for a more exhaustive treatment of the argument), the study of the polarization characteristics of electromagnetic waves started as early as the 17th century with an interesting treatise by the Dutch physicist Erasmus Bartholinus entitled "*Experimenta crystalli islandici disdiaclastici, quibus mira et insolita refractio detegitur*" ("*Experiments on double-refracting Icelandic crystals, showing amazing and unusual refraction*", 1670 –detailed references to this work, as well as to the other papers or books appeared earlier than 1850, can be found in Swyndell, 1975). In this work, one can find the earliest account of a phenomenon, double refraction in a crystal, which is intimately connected with the polarization characteristics of light. We now know that the two rays resulting from the refraction inside a crystal such as an Iceland spar have different polarization characteristics, a fact that was however ignored by Bartholinus.

Reflecting on Bartholinus' experiments, first Christian Huyghens in his "*Treatise on Light*" (1690), and later Isaac Newton in his "*Optiks*" (1730), though working in the framework of two competing theories of light, arrived to the conclusion that light should have some "transversality" property, a property, however, that was not yet called "polarization". Newton, for instance, refers to the phenomenon of polarization by saying that a ray of light has "sides".

After many years from Huyghens and Newton, the French physicist Etienne Louis Malus introduces in the scientific literature the word "*Polarization*" and brings several significant contributions to the establishment of the concept of polarization in modern terms. In his paper "*Sur une propriété de la lumière réfléchie*" (1809) Malus proves that polarization is an intrinsic property of light (and not a property "induced" in the light by crossing an Iceland spar), he demonstrates that polarization can be easily produced through the phenomena of reflection and refraction, and he also proves the famous $\cos^2 \theta$ law (giving the fraction of the intensity transmitted by two polarizers crossed at an angle θ), nowadays known as Malus law. This work opens the way to the achievements of another physicist, probably the most renowned optician of all times, Augustin Fresnel, who definitely proves the transversality of light despite the widespread belief of the times according to which, the ether being a fluid, the light should be composed of longitudinal waves. Around 1830, in his paper "*Mémoires sur la réflexion de la lumière polarisée*", Fresnel proves his famous laws concerning the relationships among the polarization properties of the incident beam and the same properties of the beams reflected and refracted at the surface of a dielectric. Despite the fact that the electromagnetic nature of light was not yet known, Fresnel's laws are correct and are still in use today.

The story of polarization continues in the 18th century with several significant contributions by François Arago and Jean-Baptiste Biot (who discover the phenomenon of *Optical Activity* in crystals and in solutions, respectively), David Brewster (nowadays known for the "Brewster angle"), William Nicol (who builds the first polarizer, the so-called Nicol prism), and Michael Faraday (who discovers an effect today known as the "Faraday effect"). However, it is only with the fundamental work of George Stokes, "*On the Composition and Resolution of Streams of Polarized Light from Different Sorces*" (1852), that the description of polarized radiation becomes fully consistent. This is achieved by giving an operational definition of four quantities, the so-called Stokes parameters, and by introducing a statistical description of the polarization property of radiation, as we will see in the next Section.

At the middle of the 18th century, the phenomenon of polarization is thus fairly well understood but it is necessary to wait almost 60 years before assisting to the first application of polarimetry to astronomy. In 1908, George Ellery Hale, has the brilliant idea of observing the solar spectrum with the help of some polarizing devices (Hale, 1908). By means of a Fresnel rhomb (acting as a quarter-wave plate) and a Nicol prism (acting as a polarizer), Hale succeeds in observing the spectrum of a sunspot in two opposite directions of circular polarization and, from the observed shift of spectral lines, induces for the first time the existence of magnetic fields in an astronomical object. Since its birth, astronomical polarimetry has evolved through the years and has given a relevant contribution to our present understanding of the physical Universe. Among the various astronomical discoveries that have relied on the use of polarimetric techniques it is just enough to quote here the discovery of the first magnetic star (Babcock, 1947) and the discovery of the existence of magnetic white dwarfs (Kemp *et al.*, 1970).

Notwithstanding these remarkable successes, polarimetry has remained for a long time a secondary discipline in astronomy. However, mostly in the last ten years, the situation has rapidly evolved and we are now undoubtedly assisting to a revival of this discipline that seems capable of capturing the scientific interests of a large community of persons and a non negligible fraction of the funds allocated to astronomical research (the organization of the present Winter School is a clear example of this trend). Probably, this is far from being an accidental event. Now that all the possible "windows" of the electromagnetic spectrum have been opened (from γ-rays to radio-waves), the possibility of new discoveries –including the serendipitous ones– relies on the development of new technologies aimed to increase the accuracy of older instrumentation (better angular, temporal, or spectral resolution, better photometric accuracy, and so on). Polarimetry perfectly fits into this trend also because, for almost a century, it has generally trailed behind the other disciplines as a possible target of novel technologies.

Apart from these historical notes, I feel necessary to spend some more introductory words about polarimetry in the astronomical context. The first thing to be remarked is that polarization is an invaluable source of information about *the geometry* of the astronomical object observed, or about any physical agent (like for instance a magnetic field) that is capable of altering, to some extent, the geometrical scenario of the same object. In polarimetry, more than in any other discipline of astronomy, the words of Galileo about geometry and the physical world still stand, after almost four centuries, as a must: *"Egli (l'Universo) è scritto in lingua matematica, e i caratteri son triangoli, cerchi, ed altre figure geometriche, senza i quali mezzi è impossibilie a intendere umanamante parola..."* (*"The Universe is written in mathematical language, and its characters are triangles, circles, and other geometrical figures, without which it is humanly impossible to understand a single word..."*).

Only a perfectly symmetric object, devoided of any physical agent capable of introducing the minimum dissimmetry in its geometrical scenario, is capable of emitting a completely unpolarized beam of radiation. An ideal black-body could provide an example of such an object, but, as we all know, ideal objects do not exist in real life and we have then to expect that some polarization signal, even if exceedingly small, may always be present in no matter which astronomical object.

The real challenge for the future of astronomical polarimetry is to increase the sensitivity of the present polarimeters operating in the different regions of the electromagnetic spectrum. Quite recently, solar physicists have succeeded in lowering the sensitivity of their polarimeters, operating in the visible range of the electromagnetic spectrum, below the limit of 10^{-4}, thus discovering a wealth of new and unexpected phenomena that are taking place in the higher layers of the solar atmosphere and that are stimulating novel theoretical approaches for their interpretation. It is my impression that, quite similarly, new exciting discoveries may be obtained for any spectral domain and any discipline of astronomy once the major effort of building a new-technology polarimeter has reached the ultimate goal of lowering the sensitivity of presently available instruments.

2. Description of Polarized Radiation

Consider an electromagnetic, monochromatic plane wave of angular frequency ω that is propagating in vacuum along a direction that we assume as the z-axis of a right-handed reference system. In a given point of space, the electric and magnetic field vectors of the wave oscillate in the x-y plane according to equations of the form

$$E_x(t) = E_1 \cos(\omega t - \phi_1) \,, \qquad E_y(t) = E_2 \cos(\omega t - \phi_2) \quad,$$

where E_1, E_2, ϕ_1, and ϕ_2 are constants. The same oscillation can also be described in terms of complex quantities by writing

$$E_x(t) = \mathrm{Re}(\mathcal{E}_1 e^{-i\omega t}) \,, \qquad E_y(t) = \mathrm{Re}(\mathcal{E}_2 e^{-i\omega t}) \quad,$$

where $\mathcal{E}_1$ and $\mathcal{E}_2$ are given by

$$\mathcal{E}_1 = E_1 e^{i\phi_1} \,, \qquad \mathcal{E}_2 = E_2 e^{i\phi_2} \quad.$$

As is well known, the composition of two orthogonal oscillations of the same frequency gives rise to an ellipse. The tip of the electric field vector thus describes an ellipse at the angular frequency ω, and, when trying to recover the geometrical parameters of the ellipse from the quantities previously introduced, one finds that the following four combinations,

$$P_I = E_1^2 + E_2^2 = \mathcal{E}_1^* \mathcal{E}_1 + \mathcal{E}_2^* \mathcal{E}_2 \,, \quad P_Q = E_1^2 - E_2^2 = \mathcal{E}_1^* \mathcal{E}_1 - \mathcal{E}_2^* \mathcal{E}_2 \,,$$

$$P_U = 2E_1 E_2 \cos(\phi_1 - \phi_2) = \mathcal{E}_1^* \mathcal{E}_2 + \mathcal{E}_2^* \mathcal{E}_1 \,, \quad P_V = 2E_1 E_2 \sin(\phi_1 - \phi_2) = i\,(\mathcal{E}_1^* \mathcal{E}_2 - \mathcal{E}_2^* \mathcal{E}_1) \,,$$

come naturally into play. The ratio between the minor and major axes of the ellipse, for instance, is given by

$$\frac{b}{a} = \frac{\left| \sqrt{P_I - P_V} - \sqrt{P_I + P_V} \right|}{\sqrt{P_I - P_V} + \sqrt{P_I + P_V}} \quad,$$

whereas the angle χ that the major axis of the ellipse forms with the x-axis can be found through the equation

$$\tan(2\chi) = \frac{P_U}{P_Q} \quad.$$

The quantities P_I, P_Q, P_U, and P_V now introduced are not independent. Indeed they obey the relationship

$$P_I^2 = P_Q^2 + P_U^2 + P_V^2 \quad,$$

and, varying their values, any kind of polarization ellipse can be described. Circular polarization is obtained by setting $P_Q = P_U = 0$, and one speaks about positive (or right-handed) circular polarization if $P_V = P_I$ and of negative (or left-handed) circular polarization if $P_V = -P_I$. In these cases the tip of the electric field vector describes a circle. On the other hand, linear polarization is obtained by setting $P_V = 0$. Now the tip of the electric vector oscillates along a segment whose inclination with respect to the x-axis is determined by the values of P_Q and P_U. In general, when none of the three quantities P_Q, P_U and P_V is zero, the tip of the electric vector describes an ellipse.

The description now given in terms of the polarization ellipse is however valid only for a plane, monochromatic wave which goes on indefinitely from $t = -\infty$ to $t = +\infty$. This is obviously a mathematical abstraction which, in general, has little to do with the physical world. A much more realistic description of a beam of radiation can be given only in terms of a statistical superposition of many wave-packets each having a limited extension in space and time. The beam thus loses its property of being monochromatic, becoming a quasi-monochromatic wave. Moreover, if the individual wave-packets do not share the same polarization properties, the polarization ellipse varies, statistically, in time. For such a beam of radiation it is then quite natural to generalize the previous definitions in the following form

$$P_I = \langle E_1^2 + E_2^2 \rangle = \langle \mathcal{E}_1^* \mathcal{E}_1 \rangle + \langle \mathcal{E}_2^* \mathcal{E}_2 \rangle \quad ,$$

$$P_Q = \langle E_1^2 - E_2^2 \rangle = \langle \mathcal{E}_1^* \mathcal{E}_1 \rangle - \langle \mathcal{E}_2^* \mathcal{E}_2 \rangle \quad ,$$

$$P_U = \langle 2E_1 E_2 \cos(\phi_1 - \phi_2) \rangle = \langle \mathcal{E}_1^* \mathcal{E}_2 \rangle + \langle \mathcal{E}_2^* \mathcal{E}_1 \rangle \quad ,$$

$$P_V = \langle 2E_1 E_2 \sin(\phi_1 - \phi_2) \rangle = i \left(\langle \mathcal{E}_1^* \mathcal{E}_2 \rangle - \langle \mathcal{E}_2^* \mathcal{E}_1 \rangle \right) \quad , \tag{2.1}$$

where the symbol $\langle ... \rangle$ means an average over the statistical distribution of the wave-packets.

Through the new definitions one can indeed describe a much larger set of physical situations. In particular, being now

$$P_I^2 \geq P_Q^2 + P_U^2 + P_V^2 \quad ,$$

it is possible for a particular beam of radiation to have $P_Q = P_U = P_V = 0$. As it can be easily derived from the equations, this implies

$$\langle \mathcal{E}_1^* \mathcal{E}_1 \rangle = \langle \mathcal{E}_2^* \mathcal{E}_2 \rangle \,, \qquad \langle \mathcal{E}_1^* \mathcal{E}_2 \rangle = 0 \quad ,$$

which means that the electric field components along the x and y-axis are, in average, equal and uncorrelated. Such a beam is a beam of "natural" radiation and its description has been made possible by the "averaging" operation over the different wave packets. It is just this operation that has been introduced by Stokes in the description of polarized radiation and the quantities defined in Eqs.(2.1) are, apart from a dimensional factor needed to transform the square of an electric field into a specific intensity, just the Stokes parameters. The older descriptions of polarization, like the one used by Fresnel, did not take into account this averaging process and were then suitable to treat only totally polarized beams of radiation.

The description of polarization presented above involves suitable averages of the electric vibrations along two orthogonal axes, x and y, perpendicular to the direction of propagation. In practice, with the remarkable exception of radio-polarimetry, the electric field of a radiation beam cannot be measured directly, and it is then necessary to introduce some operational definitions in order to relate the polarization properties of a beam to actual measurements that can be performed on the beam itself. To reach this aim, it is convenient to refer to the concept of ideal polarizing filters, such as the *ideal polarizer* and the *ideal retarder*. These ideal devices are defined by specifying their action on the electric field components along two orthogonal axes perpendicular to the direction of propagation. For the ideal polarizer one has

$$\begin{pmatrix} \mathcal{E}_a' \\ \mathcal{E}_b' \end{pmatrix} = e^{i\psi} \begin{pmatrix} 1 & 0 \\ 0 & 0 \end{pmatrix} \begin{pmatrix} \mathcal{E}_a \\ \mathcal{E}_b \end{pmatrix} = e^{i\psi} \begin{pmatrix} \mathcal{E}_a \\ 0 \end{pmatrix} \quad ,$$

where $\mathcal{E}_a$ and $\mathcal{E}_b$ are the components, at the entrance of the polarizer, of the electric field vector along the transmission axis and along the perpendicular axis, whereas $\mathcal{E}_a'$ and $\mathcal{E}_b'$ are the same components at the exit of the polarizer. As this equation shows, the electric field along the transmission axis is totally transmitted, whereas the transverse component is totally absorbed. The polarizer also manifests itself through a phase-factor, ψ, which is however completely inessential because it affects both components in the same way. For the ideal retarder, on the contrary, one has

$$\begin{pmatrix} \mathcal{E}_f' \\ \mathcal{E}_s' \end{pmatrix} = e^{i\psi} \begin{pmatrix} 1 & 0 \\ 0 & e^{i\delta} \end{pmatrix} \begin{pmatrix} \mathcal{E}_f \\ \mathcal{E}_s \end{pmatrix} = e^{i\psi} \begin{pmatrix} \mathcal{E}_f \\ e^{i\delta} \mathcal{E}_s \end{pmatrix} \quad ,$$

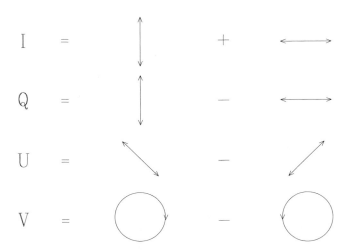

FIGURE 1. Pictorial representation of the Stokes parameters. The observer is supposed to face the radiation source.

where the notations are similar to those employed in the former equation and where the indices "f" and "s" stand, respectively, for the fast-axis and the slow-axis. The ideal retarder acts by introducing a supplementary phase factor, called *retardance* in the electric field component along the slow axis. If $\delta = \pi/2$, the retarder is also called a quarter-wave plate, if $\delta = \pi$, it is called a half-wave plate, and so on. It can be easily shown that the combination of a quarter-wave plate and a polarizer whose transmission axis is set at $+45°$ ($-45°$) from the fast axis of the plate acts as a filter for positive (negative) circular polarization.

Through the ideal polarizing filters it is possible to give a simple, operational definition of the Stokes parameters of a beam of radiation. Consider a beam and a reference direction in the plane perpendicular to the beam. One starts by setting an ideal polarizer with its transmission axis along the reference direction and measures the intensity of the beam at the exit of the polarizer, thus obtaining the value $I_{0°}$. The same operation is repeated three times after rotating the polarizer (in the counterclockwise direction facing the source) of the angles $45°$, $90°$, and $135°$, respectively, thus obtaining the values $I_{45°}$, $I_{90°}$, and $I_{135°}$. The ideal polarizer is then substituted by an ideal filter for positive circular polarization, the measured intensity at the exit of the filter being I_+, and by an ideal filter for negative circular polarization, the measured intensity being I_-. The operational definition of the four Stokes parameters, pictorially summarized in Fig. 1, is the following

$$I = I_{0°} + I_{90°} = I_{45°} + I_{135°} = I_+ + I_- \ ,$$

$$Q = I_{0°} - I_{90°} \ , \qquad U = I_{45°} - I_{135°} \ , \qquad V = I_+ - I_- \ .$$

By means of the properties of the ideal filters given previously, it is possible to relate the Stokes parameters with the quantities P_I, P_Q, P_U and P_V defined in Eqs.(2.1). When the reference direction introduced for the operational definition of the Stokes parameters coincides with the x-axis of the system introduced for the definition of the electric field components, one simply has

$$I = kP_I , \qquad Q = kP_Q , \qquad U = kP_U , \qquad V = kP_V ,$$

where k is a dimensional constant whose precise value is often irrelevant because only the ratios Q/I, U/I and V/I are generally measured in practice.

3. Polarization and Optical Devices: Jones Calculus and Mueller Matrices

The ideal polarizer and the ideal retarder that we have considered above, are just two examples of optical devices for which a linear relationship of the form

$$\begin{pmatrix} \mathcal{E}_1' \\ \mathcal{E}_2' \end{pmatrix} = \begin{pmatrix} a & b \\ c & d \end{pmatrix} \begin{pmatrix} \mathcal{E}_1 \\ \mathcal{E}_2 \end{pmatrix} \tag{3.2}$$

can be established. In this equation, the unprimed components of the electric field refer to the beam at the entrance of the optical device, whereas the primed components refer to the exit beam. Moreover, a, b, c, and d are four complex quantities that define the physical characteristics of the optical device. This equation is the basis of the so-called *Jones calculus*, a particular formalism for treating polarization phenomena systematically introduced in the scientific literature by Jones in the early 1940s. The two-component vectors containing the electric field (in complex notations) are called Jones vectors, whereas the 2×2 matrix containing the properties of the optical device is called the Jones matrix. Obviously, for a train of N optical devices one can simply build up the Jones matrix of the train by considering the product of N individual 2×2 matrices:

$$\begin{pmatrix} a & b \\ c & d \end{pmatrix} = \begin{pmatrix} a_N & b_N \\ c_N & d_N \end{pmatrix} \cdots \begin{pmatrix} a_2 & b_2 \\ c_2 & d_2 \end{pmatrix} \begin{pmatrix} a_1 & b_1 \\ c_1 & d_1 \end{pmatrix} ,$$

where the first optical device encountered by the beam is characterized by the index 1, the second by the index 2, and so on (in other words, the ordering of the matrices in the r.h.s. is opposite to the ordering in which the optical devices are inserted in the beam).

The relationship between the electric field components of the entrance and exit beams given by Eq.(3.2) can be easily translated into a relationship between the Stokes parameters. Using the definition of the Stokes parameters, one obtains, after some algebra, an equation of the form

$$S' = \mathbf{M} \, S , \tag{3.3}$$

where S is a 4-component vector constructed with the Stokes parameters of the entrance beam ($S^{\mathrm{T}} = (I, Q, U, V)$), S' has a similar meaning for the exit beam, and $\mathbf{M}$ is a 4×4 matrix given by

$$\frac{1}{2} \begin{pmatrix} a^*a + b^*b + c^*c + d^*d & a^*a - b^*b + c^*c - d^*d & 2\mathrm{Re}(a^*b + c^*d) & 2\mathrm{Im}(a^*b + c^*d) \\ a^*a + b^*b - c^*c - d^*d & a^*a - b^*b - c^*c + d^*d & 2\mathrm{Re}(a^*b - c^*d) & 2\mathrm{Im}(a^*b - c^*d) \\ 2\mathrm{Re}(a^*c + b^*d) & 2\mathrm{Re}(a^*c - b^*d) & 2\mathrm{Re}(a^*d + b^*c) & 2\mathrm{Im}(a^*d - b^*c) \\ -2\mathrm{Im}(a^*c + b^*d) & -2\mathrm{Im}(a^*c - b^*d) & -2\mathrm{Im}(a^*d + b^*c) & 2\mathrm{Re}(a^*d - b^*c) \end{pmatrix}$$

A 4×4 matrix as the one here introduced is usually referred to as a Mueller matrix. Such a matrix is made, in general, of 16 independent elements and the expression that we have derived above (which depends indeed on only 7 quantities –the real and imaginary parts of the 4 elements a, b, c, and d of the Jones matrix, minus an irrelevant phase that can be factorized in the same matrix) is a particular case of a Mueller matrix. In the following, we will refer to this particular case as the Jones-Mueller matrix.

The peculiarity of a Jones-Mueller matrix is contained in a subtle mathematical property which we state here without proof. If the determinant of the Jones matrix is non-zero, that is if

$$D = ad - bc \neq 0 \quad ,$$

then it follows that

$$|D|^2 \mathbf{M}^{-1} = \mathbf{X}\mathbf{M}^{\mathrm{T}}\mathbf{X} \quad ,$$

where $\mathbf{X}$ is the diagonal matrix defined by

$$\mathbf{X} = \mathbf{X}^{-1} = \begin{pmatrix} 1 & 0 & 0 & 0 \\ 0 & -1 & 0 & 0 \\ 0 & 0 & -1 & 0 \\ 0 & 0 & 0 & -1 \end{pmatrix} \quad . \tag{3.4}$$

Through this mathematical property it is possible to show an interesting result for the polarization properties of the entrance and exit beams connected by a Jones-Mueller matrix. Defining

$$\mathcal{P} = I^2 - Q^2 - U^2 - V^2 = S^{\mathrm{T}}\mathbf{X}S \; ; \qquad \mathcal{P}' = I'^2 - Q'^2 - U'^2 - V'^2 = S'^{\mathrm{T}}\mathbf{X}S' \quad , \tag{3.5}$$

one has, with easy transformations

$$\mathcal{P}' = S'^{\mathrm{T}}\mathbf{X}S' = S^{\mathrm{T}}\mathbf{M}^{\mathrm{T}}\mathbf{X}\mathbf{M}S = |D|^2 S^{\mathrm{T}}\mathbf{X}\mathbf{M}^{-1}\mathbf{M}S = |D|^2 \mathcal{P} \quad .$$

The equation connecting the first and last terms of this chain of equalities, which can be proved to be valid also in the case where $|D|^2 = 0$, shows that: a) if $\mathcal{P} \geq 0$, also $\mathcal{P}' \geq 0$; b) if $\mathcal{P} = 0$, then $\mathcal{P}' = 0$. Property a) means that a Jones-Mueller matrix is always a physical (or bona-fide) Mueller matrix, in the sense that it transforms physical polarization states ($\mathcal{P} \geq 0$) in physical polarization states ($\mathcal{P}' \geq 0$). Property b) shows that a totally polarized beam is always transformed by a Jones-Mueller matrix into another totally polarized beam. In other words a Jones-Mueller matrix is unable of describing depolarizing mechanisms and this clearly shows the limitations of the Jones calculus for handling a large variety of polarization phenomena. As an example, consider the case of an ideal depolarizer. The corresponding Mueller matrix is obviously given by an expression of the form

$$\mathbf{M}_{\text{ideal depolarizer}} = \begin{pmatrix} 1 & 0 & 0 & 0 \\ 0 & 0 & 0 & 0 \\ 0 & 0 & 0 & 0 \\ 0 & 0 & 0 & 0 \end{pmatrix} \quad ,$$

It can be easily proved that it is impossible to find a set of values for the quantities a, b, c, and d, such that, when substituted in the expression for the Jones-Mueller matrix, are capable of reproducing the Mueller matrix of the ideal polarizer.

Mueller matrices have a large variety of applications in physics and, more particularly, in astronomy. In many cases, one can even define the Mueller matrix of a telescope by analyzing the properties of each of its optical devices and then deducing the resulting matrix as the product of the matrices of each device. The "train property" outlined for the Jones matrices is obviously valid for the Mueller matrices too, so that one has, with evident notations

$$\mathbf{M} = \mathbf{M}_N...\mathbf{M_2}\mathbf{M_1} \quad .$$

An important problem about Mueller matrices, that often arises when one is trying to deduce the Mueller matrix of an optical device (or a combination of several optical devices) by means of experiments, is the following: given a 4×4 real matrix whose 16 elements are to be considered as quantities affected by experimental errors, is it a physical (or bona-fide) Mueller matrix, or not? This problem has been solved quite recently by means of a mathematical algorithm directly implemented in a code (Landi

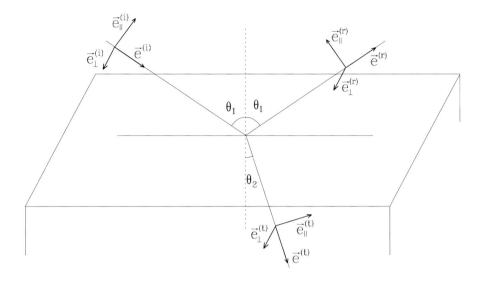

FIGURE 2. Reflection and refraction at the separation of two dielectrics.

Degl'Innocenti & del Toro Iniesta, 1998). As a curiosity we can mention the fact that the number of physical Mueller matrices is a very tiny fraction of all the 4×4 matrices that can be constructed. More precisely, calculations performed via a Montecarlo technique show that out of 10^6 matrices generated by setting the element M_{11} to 1 and all the other elements to random numbers bound in the interval $(-1, 1)$, only approximately two matrices turn out to be bona-fide Mueller matrices.

4. The Fresnel Equations

The simplest and commonest physical phenomenon where polarization processes enter into play is the ordinary reflection of a pencil of radiation on the surface of a dielectric medium. This phenomenon, which is generally accompanied by the related phenomenon of refraction, is described by the so-called Fresnel equations that can be derived as a direct consequence of the Maxwell equations. Referring to Fig. 2, we denote by n_1 and n_2 the index of refraction of the two media by θ_1 the angle of incidence (which is equal to the angle of reflection) and by θ_2 the angle of refraction. Considering, for the time being, the simplest case where both media (1 and 2) are dielectrics (which implies that n_1 and n_2 are real), and supposing $n_1 \leq n_2$, the angles θ_1 and θ_2 are connected by the usual Snell's law

$$n_1 \sin \theta_1 = n_2 \sin \theta_2 \quad . \tag{4.6}$$

The incident, the reflected, and the refracted ray all lie in the same plane which also contain the normal to the surface of separation between the two media (the so-called *incidence plane*). For each ray, a right-handed reference frame is introduced, with the third axis directed along the ray, the first axis lying in the plane of incidence, and the second axis being directed perpendicularly to the plane of incidence (and being then parallel to the surface of separation of the two media). The unit vectors are denoted respectively as $(\vec{e}_\parallel^{\,(i)}, \vec{e}_\perp^{\,(i)}, \vec{e}^{\,(i)})$ for the incident beam, $(\vec{e}_\parallel^{\,(r)}, \vec{e}_\perp^{\,(r)}, \vec{e}^{\,(r)})$ for the reflected beam, and $(\vec{e}_\parallel^{\,(t)}, \vec{e}_\perp^{\,(t)}, \vec{e}^{\,(t)})$ for the refracted (or transmitted) beam. Using similar notations for denoting the electric field components along the different axes, the laws of Fresnel are condensed by the following equations

$$\begin{pmatrix} \mathcal{E}_\parallel^{(r)} \\ \mathcal{E}_\perp^{(r)} \end{pmatrix} = \begin{pmatrix} r_\parallel & 0 \\ 0 & r_\perp \end{pmatrix} \begin{pmatrix} \mathcal{E}_\parallel^{(i)} \\ \mathcal{E}_\perp^{(i)} \end{pmatrix} \quad , \qquad \begin{pmatrix} \mathcal{E}_\parallel^{(t)} \\ \mathcal{E}_\perp^{(t)} \end{pmatrix} = \begin{pmatrix} t_\parallel & 0 \\ 0 & t_\perp \end{pmatrix} \begin{pmatrix} \mathcal{E}_\parallel^{(i)} \\ \mathcal{E}_\perp^{(i)} \end{pmatrix} \quad ,$$

where

$$r_\parallel = \frac{n_2 \cos\theta_1 - n_1 \cos\theta_2}{n_2 \cos\theta_1 + n_1 \cos\theta_2} \quad , \qquad r_\perp = \frac{n_1 \cos\theta_1 - n_2 \cos\theta_2}{n_1 \cos\theta_1 + n_2 \cos\theta_2} \quad ,$$

$$t_\parallel = \frac{2 n_1 \cos\theta_1}{n_2 \cos\theta_1 + n_1 \cos\theta_2} \quad , \qquad t_\perp = \frac{2 n_1 \cos\theta_1}{n_1 \cos\theta_1 + n_2 \cos\theta_2} \quad . \tag{4.7}$$

Since the equations now derived are in the form of "Jones equations" (cfr. Eq.(3.2)), it is easy to find the Mueller matrices corresponding to reflection and to refraction (or transmission). Taking into account Eq.(3.3) and choosing for each of the three rays the reference direction along $\vec{e}_\parallel$, we find for reflection

$$\mathbf{M}_{\text{reflection}} = \frac{1}{2} \begin{pmatrix} |r_\parallel|^2 + |r_\perp|^2 & |r_\parallel|^2 - |r_\perp|^2 & 0 & 0 \\ |r_\parallel|^2 - |r_\perp|^2 & |r_\parallel|^2 + |r_\perp|^2 & 0 & 0 \\ 0 & 0 & 2\mathrm{Re}(r_\parallel^* r_\perp) & 2\mathrm{Im}(r_\parallel^* r_\perp) \\ 0 & 0 & -2\mathrm{Im}(r_\parallel^* r_\perp) & 2\mathrm{Re}(r_\parallel^* r_\perp) \end{pmatrix} \quad ,$$

and, for transmission,

$$\mathbf{M}_{\text{transmission}} = \frac{n_2 \cos\theta_2}{n_1 \cos\theta_1} \frac{1}{2} \begin{pmatrix} |t_\parallel|^2 + |t_\perp|^2 & |t_\parallel|^2 - |t_\perp|^2 & 0 & 0 \\ |t_\parallel|^2 - |t_\perp|^2 & |t_\parallel|^2 + |t_\perp|^2 & 0 & 0 \\ 0 & 0 & 2\mathrm{Re}(t_\parallel^* t_\perp) & 2\mathrm{Im}(t_\parallel^* t_\perp) \\ 0 & 0 & -2\mathrm{Im}(t_\parallel^* t_\perp) & 2\mathrm{Re}(t_\parallel^* t_\perp) \end{pmatrix} \quad .$$

In this last equation, a supplementary factor $n_2 \cos\theta_2/(n_1 \cos\theta_1)$ has been introduced in front of the matrix to account for the fact that the energy that is contained, in the incident beam, within the infinitesimal angle $\mathrm{d}\theta_1$, is contained, after refraction, within the different infinitesimal angle $\mathrm{d}\theta_2$. On the other hand, from Snell's law (Eq.(4.6)), one has

$$\frac{\mathrm{d}\theta_1}{\mathrm{d}\theta_2} = \frac{n_2 \cos\theta_2}{n_1 \cos\theta_1} \quad .$$

An important property of the Fresnel equations is the fact that they are capable of describing, besides the phenomenon of reflection and refraction at the surface of two dielectrics, with the radiation propagating from the less refracting to the more refracting medium, also the inverse phenomenon where a pencil of radiation is propagating from a more refractive medium to a less refracting medium, and also the phenomenon of reflection on the surface of a metal. For treating these two supplementary cases, which require some further conventions, it is convenient to rewrite Eqs.(4.7) in the equivalent form

$$r_\parallel = \frac{n_2^2 u_1 - n_1^2 u_2}{n_2^2 u_1 + n_1^2 u_2} \quad , \qquad r_\perp = \frac{u_1 - u_2}{u_1 + u_2} \quad ,$$

$$t_\parallel = \frac{2 n_1 n_2 u_1}{n_2^2 u_1 + n_1^2 u_2} \quad , \qquad t_\perp = \frac{2 u_1}{u_1 + u_2} \quad , \tag{4.8}$$

where

$$u_1 = n_1 \cos\theta_1 \quad , \qquad u_2 = n_2 \cos\theta_2 \quad .$$

Consider first the case of two dielectrics with $n_1 > n_2$. A direct application of Snell's law (Eq.(4.6)) shows that, for values of θ_1 larger than the *critical angle*, θ_{cri}, defined by

$$\sin\theta_{\text{cri}} = \frac{n_2}{n_1} \quad,$$

Snell's law loses its meaning because one would find $\sin\theta_2 > 0$. In this case one is faced with the phenomenon of total reflection. It can be shown that, whereas the expressions for $t_\|$ and $t_\perp$ also lose their meaning, the quantities $r_\|$ and $r_\perp$ are still given by Eqs.(4.7), provided the following convention is adopted for u_2: applying formally Snell's law, one finds for u_2 the expression

$$u_2 = \sqrt{n_2^2 - n_1^2 \sin^2\theta_1} \quad.$$

Since the argument of the square root is negative, u_2 is pure imaginary and one has to impose the further convention that $\text{Im}(u_2) \geq 0$, so that

$$u_2 = +\mathrm{i}\sqrt{n_1^2 \sin^2\theta_1 - n_2^2} \quad.$$

Similarly, the phenomenon of reflection over a metallic surface can also be handled by the same equations. In this case the index of refraction, n_2, of the metal is a complex number (the imaginary part being connected with the exponential attenuation of the electric field during its propagation inside the metal). By convention, one has to impose $\text{Im}(n_2) \geq 0$, and $\text{Im}(u_2) \geq 0$. It has also to be remarked that the formulae for transmission lose their meaning also in this case.

As an application, let us consider the Mueller matrix for of a pencil of radiation that is propagating in air and is reflected over the surface of a dielectric. Indicating with n the index of refraction of the dielectric with respect to air ($n = n_2/n_1$), and taking into account Snell's law, one gets, from Eqs.(4.7) (or from Eqs.(4.8), which are totally equivalent for the case of a dielectric)

$$r_\| = \frac{n^2\cos\gamma - \sqrt{n^2 - \sin^2\gamma}}{n^2\cos\gamma + \sqrt{n^2 - \sin^2\gamma}} \quad, \qquad r_\perp = \frac{\cos\gamma - \sqrt{n^2 - \sin^2\gamma}}{\cos\gamma + \sqrt{n^2 - \sin^2\gamma}} \quad, \qquad (4.9)$$

where γ is the angle of incidence. Since, in this case, both $r_\|$ and $r_\perp$ are real, the Mueller matrix acquires the simpler form

$$\mathbf{M}_{\text{reflection}} = \frac{1}{2}\begin{pmatrix} r_\|^2 + r_\perp^2 & r_\|^2 - r_\perp^2 & 0 & 0 \\ r_\|^2 - r_\perp^2 & r_\|^2 + r_\perp^2 & 0 & 0 \\ 0 & 0 & 2r_\|r_\perp & 0 \\ 0 & 0 & 0 & 2r_\|r_\perp \end{pmatrix} \quad.$$

For an unpolarized incident beam, the reflected radiation turns out to be linearly polarized, being described by the Stokes parameters

$$\frac{Q}{I} = \frac{r_\|^2 - r_\perp^2}{r_\|^2 + r_\perp^2} \quad, \qquad \frac{U}{I} = 0 \quad, \qquad \frac{V}{I} = 0 \quad.$$

A simple analysis shows that, provided $n \geq 1$, one has, for any γ,

$$r_\|^2 \leq r_\perp^2 \quad,$$

so that the reflected radiation (having $Q/I \leq 0$) is linearly polarized perpendicularly to the incidence plane. In particular, when $r_\| = 0$, the reflected radiation is totally polarized. This happens for a particular value of the incidence angle that is called the *Brewster angle*. Looking for $r_\| = 0$, one finds the equation

$$n^2\cos\gamma_B = \sqrt{n^2 - \sin^2\gamma_B} \quad,$$

which is solved by

$$\gamma_B = \arctan(n) \quad .$$

For water ($n = 1.33$), $\gamma_B = 53°.1$, while for ordinary glass ($n = 1.51$), $\gamma_B = 56°.5$.

Similarly, for the refracted ray one has

$$t_{\|} = \frac{2\,n\cos\gamma}{n^2\cos\gamma + \sqrt{n^2 - \sin^2\gamma}} \quad , \qquad t_{\perp} = \frac{2\cos\gamma}{\cos\gamma + \sqrt{n^2 - \sin^2\gamma}} \quad ,$$

and the Mueller matrix is given by

$$\mathbf{M}_{\text{transmission}} = \frac{1}{2}\frac{\sqrt{n^2 - \sin^2\gamma}}{\cos\gamma}\begin{pmatrix} t_{\|}^2 + t_{\perp}^2 & t_{\|}^2 - t_{\perp}^2 & 0 & 0 \\ t_{\|}^2 - t_{\perp}^2 & t_{\|}^2 + t_{\perp}^2 & 0 & 0 \\ 0 & 0 & 2t_{\|}t_{\perp} & 0 \\ 0 & 0 & 0 & 2t_{\|}t_{\perp} \end{pmatrix} \quad .$$

For an unpolarized incident beam, the refracted radiation turns out to be linearly polarized, being described by the Stokes parameters

$$\frac{Q}{I} = \frac{t_{\|}^2 - t_{\perp}^2}{t_{\|}^2 + t_{\perp}^2} \quad , \qquad \frac{U}{I} = 0 \quad , \qquad \frac{V}{I} = 0 \quad .$$

A simple analysis shows that, provided $n \geq 1$, it is always verified, for any γ,

$$t_{\|}^2 \geq t_{\perp}^2 \quad ,$$

so that the reflected radiation (having $Q/I \geq 0$) is linearly polarized in the incidence plane. The fractional polarization increases monotonically with γ and reaches its maximum value for $\gamma = 90°$, where

$$\left(\frac{Q}{I}\right)_{\text{max}} = \frac{n^2 - 1}{n^2 + 1} \quad . \tag{4.10}$$

This limiting value is 0.28 for water and 0.39 for glass.

Similar considerations can be repeated for the opposite case where radiation is propagating from inside the dielectric medium to air. Again, for an unpolarized ray, incident at an angle $\gamma < \gamma_{\text{cri}}$, the ray reflected inside the dielectric is linearly polarized perpendicularly to the plane of incidence and it is totally polarized for an angle of incidence γ_B' given by

$$\gamma_B' = \arctan\left(\frac{1}{n}\right) \quad ,$$

whereas the ray transmitted outside the dielectric is linearly polarized in the plane of incidence and its polarization increases monotonically for increasing γ reaching its maximum value at $\gamma = \gamma_{\text{cri}}$. This maximum value is still given by Eq.(4.10).

When the incidence angle γ goes beyond the value of γ_{cri}, the radiation beam undergoes the phenomenon of total reflection. According to our previous discussion and to the conventions introduced, we have, from Eqs.(4.8)

$$r_{\|} = \frac{\cos\gamma - in\sqrt{n^2\sin^2\gamma - 1}}{\cos\gamma + in\sqrt{n^2\sin^2\gamma - 1}} \quad , \qquad r_{\perp} = \frac{n\cos\gamma - i\sqrt{n^2\sin^2\gamma - 1}}{n\cos\gamma + i\sqrt{n^2\sin^2\gamma - 1}} \quad ,$$

or, observing that both $r_{\|}$ and $r_{\perp}$ are complex numbers of the form $(a - ib)/(a + ib)$,

$$r_{\|} = e^{-2i\phi_1} \quad , \qquad r_{\perp} = e^{-2i\phi_2} \quad ,$$

where

$$\tan\phi_1 = \frac{n\sqrt{n^2\sin^2\gamma - 1}}{\cos\gamma} \quad , \qquad \tan\phi_2 = \frac{\sqrt{n^2\sin^2\gamma - 1}}{n\cos\gamma} \quad .$$

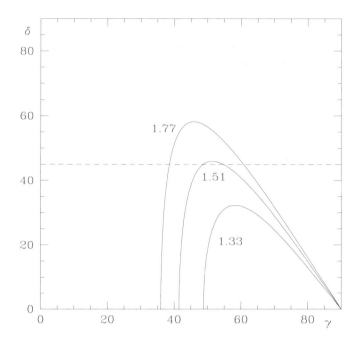

FIGURE 3. The retardance angle, δ, induced by total reflection inside a dielectric medium is plotted as a function of the incidence angle γ. The three curves are labeled by the value of the index of refraction. Note that for a glass ($n = 1.51$) the maximum value of δ is slightly larger than 45°. This allows the construction of a quarter-wave plate by means of two total reflections.

The Mueller matrix for total reflection is thus given by

$$\mathbf{M}_{\text{total reflection}} = \begin{pmatrix} 1 & 0 & 0 & 0 \\ 0 & 1 & 0 & 0 \\ 0 & 0 & \cos[2(\phi_1 - \phi_2)] & \sin[2(\phi_1 - \phi_2)] \\ 0 & 0 & -\sin[2(\phi_1 - \phi_2)] & \cos[2(\phi_1 - \phi_2)] \end{pmatrix} \quad .$$

This means that, in the phenomenon of total reflection, the dielectric behaves as a retarder having retardance $\delta = 2(\phi_1 - \phi_2)$, which through some algebra, can be expressed in the form

$$\delta = 2 \arctan\left(\frac{\cos\gamma \sqrt{n^2 \sin^2\gamma - 1}}{n \sin^2\gamma} \right) \quad .$$

A study of δ as function of γ (see Fig. 3) shows that δ is 0 either at $\gamma = \gamma_{\text{cri}}$ and at $\gamma = 90°$, and that it goes through a maximum at $\gamma_{\text{max}} = \arcsin[\sqrt{2/(1 + n^2)}]$, where it gets the value

$$\delta_{\text{max}} = 2 \arctan\left(\frac{n^2 - 1}{2n} \right) \quad .$$

Unfortunately, ordinary glasses have an index of refraction too small to get the possibility of constructing a quarter-wave plate ($\delta = 90°$) with a single (total) reflection. However, such a device can be constructed with two reflections, each introducing a retardance

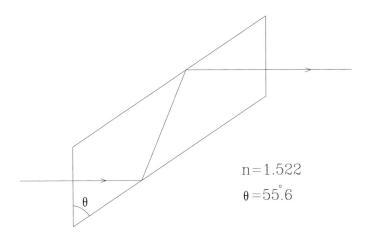

n=1.522

θ=55°.6

FIGURE 4. The Fresnel rhomb.

$\delta = 45°$. The incidence angle γ for such a device can be found by solving the equation

$$\frac{\cos\gamma\sqrt{n^2 \sin^2\gamma - 1}}{n \sin^2\gamma} = \tan(22°.5) = \sqrt{2} - 1 \quad,$$

which gives the two values

$$\gamma_\pm = \arcsin\sqrt{\frac{n^2 + 1 \pm \sqrt{n^4 - 2(7 - 4\sqrt{2})n^2 + 1}}{4(2 - \sqrt{2})n^2}} \quad.$$

For instance, for the glass BSC Crown whose refraction index is $n = 1.522$ at the wavelength of 5000 Å, one finds the two values $\gamma_- = 47°.37$ and $\gamma_+ = 55°.60$, the second one being preferred because the retardance has a smaller variation with the index of refraction. Cutting a prism of glass at the angle γ_+, Fresnel succeeded in obtaining, for the first time, an optical device capable of transforming linear polarization into circular polarization. Such a device, still in use today, is called a Fresnel rhomb (see Fig. 4).

The retardance is a slowly varying function of the index of refraction and –as a consequence– of wavelength. For the same BSC Crown glass, whose index of refraction in the visible region of the electromagnetic spectrum differs at maximum of $\Delta n = \pm 0.009$ with respect to its value at 5000 Å, the retardance of the Fresnel rhomb turns out to be contained between 89°.2 and 90°.8. This makes of the Fresnel rhomb an almost achromatic quarter-wave plate for all the visible range. Its main inconvenience stems from the fact that the retardance changes considerably as soon as the entrance beam is tilted, even of a small angle, with respect to its nominal direction. Therefore the Fresnel rhomb is not very suitable to operate on beams having a relatively wide angular aperture.

As a final application of the Fresnel equations, let us consider the case of the reflection on a metallic surface. If we set for simplicity $n_{\text{air}} = 1$, and we denote by n the (complex) index of refraction of the metal, we obtain for $r_\parallel$ and $r_\perp$ the same expressions as in the case of the reflection on a dielectric, with the difference that, in this case, both quantities are complex and one is obliged to be careful about conventions. For n we have to write

$$n = a + ib \quad, \tag{4.11}$$

with $b \geq 0$, and, for the square root, we have to take the determination which gives a positive imaginary part. Writing,

$$r_\| = \rho_\| e^{i\phi_\|} , \qquad r_\perp = \rho_\perp e^{i\phi_\perp} ,$$

the Mueller matrix for reflection on a metal is given by

$$\mathbf{M}_{\text{reflection}} = \frac{1}{2} \begin{pmatrix} \rho_\|^2 + \rho_\perp^2 & \rho_\|^2 - \rho_\perp^2 & 0 & 0 \\ \rho_\|^2 - \rho_\perp^2 & \rho_\|^2 + \rho_\perp^2 & 0 & 0 \\ 0 & 0 & 2\rho_\|\rho_\perp \cos(\phi_\| - \phi_\perp) & -2\rho_\|\rho_\perp \sin(\phi_\| - \phi_\perp) \\ 0 & 0 & 2\rho_\|\rho_\perp \sin(\phi_\| - \phi_\perp) & 2\rho_\|\rho_\perp \cos(\phi_\| - \phi_\perp) \end{pmatrix} .$$

Consider first the reflection for normal incidence ($\gamma = 0$). From Eqs.(4.9) one simply gets

$$r_\| = -r_\perp = \frac{n-1}{n+1} ,$$

and the reflection coefficient R is given by

$$R = \frac{1}{2} \left(|r_\||^2 + |r_\perp|^2 \right) = \left| \frac{n-1}{n+1} \right|^2 .$$

Substituting the expression for the index of refraction (Eq.(4.11)), one gets

$$R = \frac{(a-1)^2 + b^2}{(a+1)^2 + b^2} .$$

For silver, for instance, one has at the wavelength of 4959 Å (corresponding to photons of energy 2.5 eV) $a = 0.13$, $b = 2.88$, and one gets $R_{\text{silver}} = 0.95$. Similarly, for aluminum, being $a = 0.78$, $b = 5.84$, it follows $R_{\text{aluminum}} = 0.92$. This explains the high reflectivity of these two metals.

An important property of the reflection on metals is the fact that, differently from the case of the reflection on a dielectric, the elements of the Mueller matrix connecting the Stokes parameters U and V are different from zero. Moreover, for most metals and for most angles of incidence, the quantity $\phi_\| - \phi_\perp$ turns out to be negative. This shows that the reflection on a metal is a simple and efficient way of obtaining circular polarization of a given sign from linear polarization. Positive circular polarization can be obtained by reflecting on a metal a beam of radiation with positive U, and vice versa.

A further property of the reflection on metals is the fact that, for a suitable angle of incidence, the quantity $\phi_\| - \phi_\perp$ turns out to be $-90°$. In this case, the metal behaves, under reflection, as a kind of quarter-wave plate in the sense that it transforms U in V and V in $-U$. Such an incidence angle is called the "principal angle". A rather long algebraic analysis shows that the value of the principal angle, γ_{p}, is given by

$$\gamma_{\text{p}} = \arctan \sqrt{x_0} ,$$

where x_0 is the only positive solution of the third degree equation

$$x^3 - x^2 - \left[(a^2 + b^2)^2 - 2(a^2 - b^2) \right] x - (a^2 + b^2)^2 = 0 .$$

For photons of 2.5 eV (4959 Å), the principal angle for silver is $72°.6$ and for aluminum is $80°.6$.

5. Dichroism and Anomalous Dispersion

Though the etymology of the word is rather misleading, dichroism is a typical phenomenon of anisotropic media which consists in the fact that the absorption properties of the electric field which is propagating inside such media depends on the direction of the field, or, in other words, on the polarization state of the radiation. Anomalous dispersion is a related phenomenon which is due to the dephasing of the two components of the electric field vector in the propagation inside an anisotropic medium. The two phenomena can be unified by thinking about the two complex amplitudes $\mathcal{E}_a$ and $\mathcal{E}_b$ of the electric field along two orthogonal states of polarization. Dichroism is connected with the differential attenuation (or enhancement) of the modulus of $\mathcal{E}_a$ and $\mathcal{E}_b$ during the propagation, whereas anomalous dispersion is connected with the dephasing of the same quantities.

A description of these phenomena can be given by introducing the principal axes of the medium, characterized by the unit vectors $\vec{u}_\alpha$ ($\alpha = 1, 2, 3$), and the corresponding indexes of refraction n_α, which are, in general, complex numbers. A wave of angular frequency ω, and polarized along the direction $\vec{u}_\alpha$, propagates within the medium according to the equation

$$\vec{E}(\vec{x}, t) = E_\alpha \vec{u}_\alpha e^{i\frac{\omega}{c}(n_\alpha \vec{x}\cdot\vec{\Omega}-ct)} \quad .$$

Consider for simplicity the case of a medium where the principal axes $\vec{u}_1$ and $\vec{u}_2$ are two real, orthogonal unit vectors, and consider a wave propagating in the medium along the direction $\vec{\Omega}$ perpendicular to the plane of the two unit vectors. Denoting by s the coordinate measured along such a direction, and by $\mathcal{E}_1$ and $\mathcal{E}_2$ the components of the electric field along the unit vectors, one has

$$\frac{d\mathcal{E}_1}{ds} = i\frac{\omega}{c}n_1\mathcal{E}_1 \ , \qquad \frac{d\mathcal{E}_2}{ds} = i\frac{\omega}{c}n_2\mathcal{E}_2 \quad .$$

From these equations one obtains the transfer equation for the components of the polarization tensor

$$\frac{d(\mathcal{E}_i^* \mathcal{E}_j)}{ds} = \left(\frac{d\mathcal{E}_i^*}{ds}\right)\mathcal{E}_j + \mathcal{E}_i^*\left(\frac{d\mathcal{E}_j}{ds}\right) = i\frac{\omega}{c}(n_j - n_i^*)\mathcal{E}_i^* \mathcal{E}_j \quad ,$$

and next the transfer equations for the Stokes parameters. After some algebra, one obtains

$$\frac{d}{ds}\begin{pmatrix} I \\ Q \\ U \\ V \end{pmatrix} = -\begin{pmatrix} \eta_I & \eta_Q & 0 & 0 \\ \eta_Q & \eta_I & 0 & 0 \\ 0 & 0 & \eta_I & \rho_Q \\ 0 & 0 & -\rho_Q & \eta_I \end{pmatrix}\begin{pmatrix} I \\ Q \\ U \\ V \end{pmatrix} \quad ,$$

where

$$\eta_I = \frac{\omega}{c}\mathrm{Im}(n_1 + n_2) \ , \qquad \eta_Q = \frac{\omega}{c}\mathrm{Im}(n_1 - n_2) \ , \qquad \rho_Q = \frac{\omega}{c}\mathrm{Re}(n_1 - n_2) \quad .$$

These last equations show that, when the indexes of refraction n_1 and n_2 are different, the quantities η_Q and ρ_Q are non-zero. In particular, η_Q, which is proportional to the difference between the imaginary parts of the indexes of refraction, describes the dichroic properties of the medium, whereas ρ_Q, which is proportional to the difference between the real parts of the indexes of refraction, describes its anomalous dispersion properties. Acting alone, ρ_Q would induce a continuous oscillation between the Stokes parameters U and V, and it is usually referred to as the "pulsation" term.

The transfer equation that we have just derived is a particular case of a more general one that we just write here without proof. For an arbitrary anisotropic medium and an arbitrary direction of propagation, the transfer equation for the Stokes parameters acquires the form

$$\frac{d}{ds}\begin{pmatrix} I \\ Q \\ U \\ V \end{pmatrix} = -\begin{pmatrix} \eta_I & \eta_Q & \eta_U & \eta_V \\ \eta_Q & \eta_I & \rho_V & -\rho_U \\ \eta_U & -\rho_V & \eta_I & \rho_Q \\ \eta_V & \rho_U & \rho_Q & \eta_I \end{pmatrix}\begin{pmatrix} I \\ Q \\ U \\ V \end{pmatrix} \quad , \tag{5.12}$$

where

$$\eta_I = \text{Re}(G_{11}+G_{22}), \quad \eta_Q = \text{Re}(G_{11}-G_{22}), \quad \eta_U = \text{Re}(G_{12}+G_{21}), \quad \eta_V = \text{Im}(G_{12}-G_{21}),$$

$$\rho_Q = -\text{Im}(G_{11}-G_{22}), \quad \rho_U = -\text{Im}(G_{12}+G_{21}), \quad \rho_V = \text{Re}(G_{12}-G_{21}) \quad .$$

The quantity G_{ij} $(i,j = 1,2)$ appearing in these equations is connected to the indices of refraction n_α $(\alpha = 1,2,3)$ of the principal axes through the equation

$$G_{ij} = -\text{i}\frac{\omega}{c}\sum_\alpha n_\alpha(\vec{u}_\alpha \cdot \vec{e}_i)(\vec{u}_\alpha^* \cdot \vec{e}_j) \quad ,$$

where $\vec{e}_1$ and $\vec{e}_2$ are the unit vectors defining the Stokes parameters.

The transfer equation (Eq.(5.12)) describes in full generality the phenomena of dichroism (through the quantities η_Q, η_U, and η_V) and of anomalous dispersion (through the quantities ρ_Q, ρ_U, and ρ_V). In particular, the term ρ_V, acting alone, would induce, during the propagation, a continuous transformation between the Stokes parameters Q and U, or, in other words a rotation of the direction of linear polarization. This phenomenon, which is called *optical activity*, was discovered by François Arago in the light propagating along the optical axis of a quartz crystal, and, successively, by Jean-Baptiste Biot, who discovered such activity in many solutions of organic and inorganic substances (like, for instance, in sugary solutions). According to the direction of rotation of the plane of polarization, substances were classified into dextrogyrous and levogyrous, and it was found that all organic substances belong to the same class. Finally, the same phenomenon was discovered by Michael Faraday in the light travelling through a medium subjected to a strong magnetic field (aligned with the direction of propagation), and, since then, it is also known as the *Faraday effect*, or as *Faraday rotation*.

The matrix appearing in the transfer equation satisfies an important symmetry property. Indeed, writing Eq.(5.12) in the form

$$\frac{d}{ds}S = -\mathbf{K}S \quad ,$$

where S is the 4-component vector constructed with the Stokes parameters, it can be easily shown that the matrix $\mathbf{K}$ satisfies the symmetry property

$$\mathbf{X}\mathbf{K}^\text{T}\mathbf{X} + \mathbf{K} = 2\eta_I\mathbf{1} \quad ,$$

where $\mathbf{X}$ is the matrix defined in Eq.(3.4), and $\mathbf{1}$ is the 4×4 identity matrix. Through this property, one obtains for the quantity $\mathcal{P}$ defined in Eq.(3.5)

$$\frac{d}{ds}\mathcal{P} = \frac{d}{ds}S^\text{T}\mathbf{X}S = -S^\text{T}\left(\mathbf{K}^\text{T}\mathbf{X} + \mathbf{X}\mathbf{K}\right)S = -2\eta_I\mathcal{P} \quad .$$

This equation assures that, during the transfer, a physical Stokes vector (for which $\mathcal{P} \geq 0$) always remains a physical vector. Moreover, if the radiation beam is totally polarized ($\mathcal{P} = 0$), this property is conserved during the propagation.

6. Polarization in everyday life

Polarization is a rather obscure concept for the man on the street. This is not because polarization phenomena are not present in the world around us, but just because the human eye (differently from the eye of other living beings) is practically insensitive to the polarization of light, though some minor effects, due to the presence of a blue, dichroicly absorbing pigment in the *macula lutea* can indeed be observed under particular conditions (Heidinger's brush).

We have already seen that the light reflected from a dielectric surface is linearly polarized, the direction of polarization being the perpendicular to the plane of incidence. It then follows that sunlight reflected over the surface of the sea (or of a lake) is also linearly polarized and that such reflections can be extinguished, to a large extent, by means of a polarizing filter whose transmission axis is set along the vertical. This can be accomplished by wearing a particular type of sunglasses that can be commonly found in commerce and are usually referred to as Polaroids (Polaroid is a registered mark). The "lenses" of these sunglasses are nothing but polarizing filters suitably oriented by the manufacturer in such a way that the transmission axis coincides with the vertical axis.

With this simple device one can easily observe that the blue sky is strongly polarized, a phenomenon that is known since a long time and that is due to the scattering of sunlight by the air molecules. As we will see in Sect. 10, scattering by air molecules obeys the Rayleigh scattering law, which implies that, for a $90°$ scattering, the polarization of the scattered radiation is 100% linearly polarized, and that the direction of polarization is perpendicular to the scattering plane. To be clear on an example, let us suppose that the sun is setting exactly on the West. Looking towards the North or the South, one should then observe that the horizon sky is 100% linearly polarized along the vertical direction, and, looking towards the zenith, one should again observe a 100% polarized sky with the polarization direction along the circle joining South with North.

This would be exactly true if the radiation from the blue sky were only due to single scattering processes. In reality, there are secondary scattering processes which give rise to a further polarization signal, usually directed vertically (due to the fact that the photons' path between first and second scattering is mainly lying in the horizontal plane). This secondary effect can either add or subtract from the first one and the result is that, in general, the polarization of the blue sky never reaches the 100% value expected from single scattering processes.

A second phenomenon that contributes to lowering the polarization of the blue sky is the presence in the atmosphere of aerosols and other pollutant agents. Since these substances are made of particles having dimensions larger than the wavelength of light, the scattering obeys the law of Mie (instead of the law of Rayleigh) which results in much lower polarization efficiencies (see Sect. 11).

The polarization of the blue sky is an important physical phenomenon which is indeed used by several living beings (some species of insects, in particular) as a practical mean of orientation. Obviously, these insects are provided of a particular kind of eyes which allow them to observe the direction of polarization of the sky and to recover the sun's position. It has also been suggested that the navigators of Viking ships used a piece of Iceland spar to help them in finding the sun's direction in the heavily cloudy, northern skies.

Polarization also show up in other meteorological phenomena, like in the rainbow and in halos. The rainbow is produced by the refraction of the solar radiation by droplets of water (see Fig. 5). The primary arch is due to a process of refraction (the solar rays enter the droplet), an internal reflection (which takes place at an incidence angle less than the critical angle), and a final refraction (by which the solar ray exits the droplet). An analysis based on the Fresnel equations allows to deduce that the rainbow is linearly polarized, the polarization direction being parallel to the bow, and the fractional

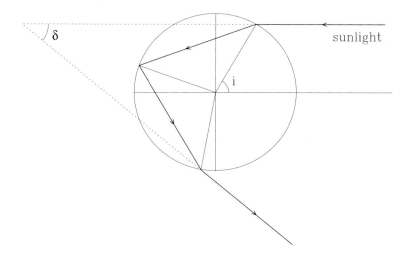

FIGURE 5. The rainbow phenomenon is due to the property that the deflection angle δ due to a droplet of water is stationary, with respect to small variations of the incidence angle i, for $i \simeq 59°$. The corresponding value of δ is $\simeq 42°$.

polarization being given by

$$P_{\text{rainbow}} = \frac{(2+n^2)^6 - 729\,n^4(2-n^2)^2}{(2+n^2)^6 + 729\,n^4(2-n^2)^2} \quad ,$$

where n is the index of refraction of water. Substituting the value of n, it is found that the polarization varies from 92% in the red ($n = 1.331$) to 94% in the violet ($n = 1.344$).

The halo, a somewhat less known phenomenon which shows up when ice crystals are present in the upper atmosphere, is due to the refraction of sunlight inside the crystals themselves. It shows up as a circular luminosity around the sun located at a distance of about 22°. The ice crystals have the shape of long prisms having an hexagonal cross section. A simple application of the Fresnel equations shows that the halo is linearly polarized, the polarization being directed perpendicularly to the halo, and the fractional polarization being given by

$$P_{\text{halo}} = \frac{\left(n\sqrt{3} + \sqrt{4-n^2}\right)^4 - \left(\sqrt{3} + n\sqrt{4-n^2}\right)^4}{\left(n\sqrt{3} + \sqrt{4-n^2}\right)^4 + \left(\sqrt{3} + n\sqrt{4-n^2}\right)^4} \quad ,$$

where n is the index of refraction of ice. Substituting $n = 1.31$, one finds that the halo polarization is of the order of 4%, a much less remarkable value than for the rainbow.

Finally, it is important to remark that polarimetry has a very large number of technological applications in many practical aspects of life. An example is the use of polarimetric techniques to measure the quality of sugar. The sugar is diluted in a solution at a known concentration and the rotatory power of the solution is then measured through an instrument that is called a *saccharimeter*. Another example concerns security systems that turn on when a beam of light is interrupted. By encoding the light beam and the receiver according to a particular state of polarization, it becomes almost impossible to substitute the original beam with a second one having the same polarimetric characteristics.

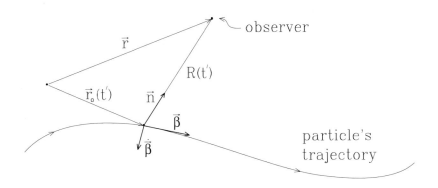

FIGURE 6. Motion of the radiating particle.

7. Polarization due to radiating charges

Consider a particle of charge e, moving in space according to the action of an ensemble of forces, and suppose that the position of the particle is specified, in an assigned reference system, by the function $\vec{r}_0(t)$. As a consequence of its motion, the particle generates in space a variable electromagnetic field which can be derived by solving the Maxwell equations. The result is contained in the well known equations that are usually referred to as the Liénard-Wiechert equations

$$\vec{E}(\vec{r}, t) = \frac{e}{\kappa^3(t')R^2(t')}[1 - \beta^2(t')][\vec{n}(t') - \vec{\beta}(t')]$$

$$+ \frac{e}{c\kappa^3(t')R(t')} \, \vec{n}(t') \times \left\{ [\vec{n}(t') - \vec{\beta}(t')] \times \dot{\vec{\beta}}(t') \right\} \quad,$$

$$\vec{B}(\vec{r}, t) = \vec{n}(t') \times \vec{E}(\vec{r}, t) \quad. \tag{7.13}$$

The meaning of the various symbols contained in these equations is the following (see Fig. 6): t' is the retarded time, or the time at which a signal propagating at the speed of light has to leave from the point $\vec{r}_0(t')$ to reach the point $\vec{r}$ at time t. It is implicitly defined by the equation

$$t = t' + \frac{R(t')}{c} \quad,$$

where

$$R(t') = |\vec{r} - \vec{r}_0(t')| \quad.$$

The other symbols are defined by

$$\beta(t') = \frac{\vec{v}(t')}{c} = \frac{\mathrm{d}\vec{r}_0(t')}{\mathrm{d}t'} \, , \qquad \dot{\vec{\beta}}(t') = \frac{\vec{a}(t')}{c} = \frac{\mathrm{d}\vec{\beta}(t')}{\mathrm{d}t'} = \frac{\mathrm{d}^2\vec{r}_0(t')}{\mathrm{d}t'^2} \quad,$$

$$\vec{n}(t') = \frac{\vec{r} - \vec{r}_0(t')}{R(t')} \, , \qquad \kappa(t') = 1 - \vec{n}(t') \cdot \vec{\beta}(t') \quad.$$

In other words, $\vec{\beta}$ is the velocity divided by c, the speed of light, $\dot{\vec{\beta}}$ is the acceleration, again divided by the speed of light, $\vec{n}$ is the unit vector pointing from $\vec{r}_0(t')$ to $\vec{r}$, and κ is a factor that becomes very important when the velocity of the particle is close to c.

The electric field given by Eq.(7.13) is composed of two terms. The first term goes as R^{-2} and is nothing but the generalization to relativity of the *Coulomb field*. It can be noticed that, for a static charge, one simply has $\vec{E}(\vec{r}) = (e/R^2)\vec{n}$ and $\vec{B}(\vec{r}) = 0$. The second term goes as R^{-1} and describes the the so-called *radiation field*. This field is proportional to the acceleration of the charge and is perpendicular to the unit vector $\vec{n}$. It follows that the associated value of the magnetic induction vector is also perpendicular to $\vec{n}$ and is equal, in modulus, to the electric field.

An order of magnitude estimate of the ratio between the Coulomb field and the radiation field shows that, if τ is a typical time interval for the variation of the velocity, such ratio is of the order of $\tau c/R$. On the other hand, $\tau c \simeq \lambda$, where λ is the wavelength of the radiation emitted by the charge, so that, for $R \gg \lambda$, or, in other words, in the so-called *radiation zone*, one can simply neglect the first term in the r.h.s of Eq.(7.13). Moreover, when the typical size, L, of the region where the particle is moving is much smaller than the distance R (a typical circumstance for astronomical observations), the dependence of $\vec{n}(t')$ on t' can be safely neglected. For $R \gg \lambda$ and $R \gg L$, one is thus left with the equation

$$\vec{E}(\vec{r}, t) = \frac{e}{c\kappa^3(t')R(t')}\, \vec{n} \times \left\{ [\vec{n} - \vec{\beta}(t')] \times \dot{\vec{\beta}}(t') \right\} \quad,$$

$$\vec{B}(\vec{r}, t) = \vec{n} \times \vec{E}(\vec{r}, t) \quad. \tag{7.14}$$

We now consider some applications of these equations to the non-relativistic regime. In this case, being $\beta \ll 1$, the equations for the radiating field simplify even further because the distinction between time t and retarded time t' becomes inessential, the two quantities being simply related by an equation of the form $t' = t - t_0$, t_0 being a constant (the transit time between the source and the observer). Introducing the symbol $\vec{a}$ for the acceleration, the equation for the radiation field is simply given by

$$\vec{E}(\vec{r}, t) = \frac{e}{c^2 R}\, \vec{n} \times (\vec{n} \times \vec{a}(t')) \quad,$$

$$\vec{B}(\vec{r}, t) = \vec{n} \times \vec{E}(\vec{r}, t) \quad.$$

To find the polarization of the radiation emitted by the accelerated charge we have, as usual, to introduce two unit vectors $\vec{e}_1(\vec{n})$ and $\vec{e}_2(\vec{n})$ in the plane perpendicular to the direction $\vec{n}$, in such a way that the the triplet $(\vec{e}_1(\vec{n})$, $\vec{e}_2(\vec{n})$, $\vec{n})$ form a right handed coordinate system. The electric field can thus be decomposed over the two unit vectors to give the two components $E_1(\vec{x}, t)$ and $E_2(\vec{x}, t)$. Taking into account that

$$\vec{n} \times (\vec{n} \times \vec{a}) = (\vec{n} \cdot \vec{a})\, \vec{n} - \vec{a} \quad,$$

and taking into account that both $\vec{e}_1$ and $\vec{e}_2$ are perpendicular to $\vec{n}$, one obtains

$$E_1(\vec{r}, t) = -\frac{e}{c^2 R}\vec{e}_1(\vec{n}) \cdot \vec{a}(t') \,, \qquad E_2(\vec{r}, t) = -\frac{e}{c^2 R}\vec{e}_2(\vec{n}) \cdot \vec{a}(t') \quad. \tag{7.15}$$

Independently of the regime considered (relativistic or non-relativistic), the "recipe" to find the Stokes parameters of the radiation emitted by the moving charge proceeds, in general, through the evaluation of the Fourier components of the quantities $E_1(\vec{r}, t)$ and $E_2(\vec{r}, t)$. Defining

$$\mathcal{E}_1(\omega) = \frac{1}{2\pi} \int_{-\infty}^{\infty} E_1(\vec{r}, t) e^{i\omega t}\, dt \quad,$$

$$\mathcal{E}_2(\omega) = \frac{1}{2\pi} \int_{-\infty}^{\infty} E_2(\vec{r}, t) e^{i\omega t}\, dt \quad, \tag{7.16}$$

with the inverse equations

$$E_1(\vec{r}, t) = \int_{-\infty}^{\infty} \mathcal{E}_1(\omega) e^{-i\omega t} \, d\omega \quad ,$$

$$E_2(\vec{r}, t) = \int_{-\infty}^{\infty} \mathcal{E}_2(\omega) e^{-i\omega t} \, d\omega \quad ,$$

the Stokes parameters at (angular) frequency ω (defined –in as far as the intensity is concerned– as the energy that flows, per unit time and per unit frequency interval, through the unit surface directed perpendicularly to the beam) are given by the equations

$$I_\omega = \mathcal{K}\Big(\langle \mathcal{E}_1^*(\omega)\mathcal{E}_1(\omega) \rangle + \langle \mathcal{E}_2^*(\omega)\mathcal{E}_2(\omega) \rangle \Big) \quad ,$$

$$Q_\omega = \mathcal{K}\Big(\langle \mathcal{E}_1^*(\omega)\mathcal{E}_1(\omega) \rangle - \langle \mathcal{E}_2^*(\omega)\mathcal{E}_2(\omega) \rangle \Big) \quad ,$$

$$U_\omega = \mathcal{K}\Big(\langle \mathcal{E}_1^*(\omega)\mathcal{E}_2(\omega) \rangle + \langle \mathcal{E}_2^*(\omega)\mathcal{E}_1(\omega) \rangle \Big) \quad ,$$

$$V_\omega = i\,\mathcal{K}\Big(\langle \mathcal{E}_1^*(\omega)\mathcal{E}_2(\omega) \rangle - \langle \mathcal{E}_2^*(\omega)\mathcal{E}_1(\omega) \rangle \Big) \quad , \tag{7.17}$$

where the dimensional constant $\mathcal{K}$ depends on the particular radiating process considered and can be deduced through the following considerations: the flux of radiation is given by the Poynting vector

$$\vec{S}(\vec{r}, t) = \frac{c}{4\pi} \vec{E}(\vec{r}, t) \times \vec{B}(\vec{r}, t) \quad ,$$

which can also be written in the form

$$\vec{S}(\vec{r}, t) = \frac{c}{4\pi} [E_1^2(\vec{r}, t) + E_2^2(\vec{r}, t)]\, \vec{n} \quad .$$

On the other hand, taking into account the Parseval theorem on the Fourier transforms, one has

$$\int_{-\infty}^{\infty} E_1^2(\vec{r}, t) dt = 2\pi \int_{-\infty}^{\infty} \mathcal{E}_1^*(\omega)\mathcal{E}_1(\omega) d\omega = 4\pi \int_{0}^{\infty} \mathcal{E}_1^*(\omega)\mathcal{E}_1(\omega) d\omega \quad ,$$

with a similar equation for the component $E_2(\vec{r}, t)$. These equations allow to express the total energy of the radiation, flowing across the unit surface in the time interval $(-\infty, \infty)$ and contained in the range of angular frequency $(\omega, \omega + d\omega)$ as

$$c\big(\langle \mathcal{E}_1^*(\omega)\mathcal{E}_1(\omega) \rangle + \langle \mathcal{E}_2^*(\omega)\mathcal{E}_2(\omega) \rangle \big) \quad .$$

Assuming that the fields $E_1(\vec{r}, t)$ and $E_2(\vec{r}, t)$ are due to a single pulse of radiation and that the source emits N_p such pulses for unit time, one finally gets that the constant $\mathcal{K}$ appearing in Eq.(7.17) is given by

$$\mathcal{K} = cN_p \quad . \tag{7.18}$$

A different expression holds when the radiation field is due to a charge which is oscillating with a periodic motion of frequency ω_0. In this case, instead of the Fourier transforms, it is better to use the Fourier components defined by

$$\tilde{\mathcal{E}}_1^{(n)} = \frac{1}{T} \int_{-T/2}^{T/2} E_1(\vec{r}, t)\, e^{in\omega_0 t} \, dt \quad ,$$

$$\tilde{\mathcal{E}}_2^{(n)} = \frac{1}{T} \int_{-T/2}^{T/2} E_2(\vec{r}, t)\, e^{in\omega_0 t} \, dt \quad .$$

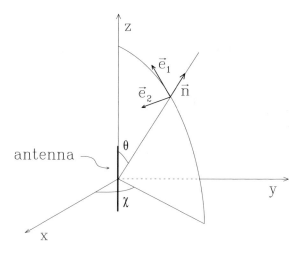

FIGURE 7. Geometry of the radiating antenna.

The Stokes parameters of the radiation at frequency $n\omega_0$ are then given by

$$I_\omega = \mathcal{K}' \left(\langle \tilde{\mathcal{E}}_1^{(n)*} \tilde{\mathcal{E}}_1^{(n)} \rangle + \langle \tilde{\mathcal{E}}_2^{(n)*} \tilde{\mathcal{E}}_2^{(n)} \rangle \right) \delta(\omega - n\omega_0) \quad ,$$

$$Q_\omega = \mathcal{K}' \left(\langle \tilde{\mathcal{E}}_1^{(n)*} \tilde{\mathcal{E}}_1^{(n)} \rangle - \langle \tilde{\mathcal{E}}_2^{(n)*} \tilde{\mathcal{E}}_2^{(n)} \rangle \right) \delta(\omega - n\omega_0) \quad ,$$

$$U_\omega = \mathcal{K}' \left(\langle \tilde{\mathcal{E}}_1^{(n)*} \tilde{\mathcal{E}}_2^{(n)} \rangle + \langle \tilde{\mathcal{E}}_2^{(n)*} \tilde{\mathcal{E}}_1^{(n)} \rangle \right) \delta(\omega - n\omega_0) \quad ,$$

$$V_\omega = i \mathcal{K}' \left(\langle \tilde{\mathcal{E}}_1^{(n)*} \tilde{\mathcal{E}}_2^{(n)} \rangle - \langle \tilde{\mathcal{E}}_2^{(n)*} \tilde{\mathcal{E}}_1^{(n)} \rangle \right) \delta(\omega - n\omega_0) \quad , \tag{7.19}$$

and the constant $\mathcal{K}'$ can be deduced through considerations similar to those developed above, that give

$$\mathcal{K}' = \frac{c}{2\pi} \quad . \tag{7.20}$$

Equations (7.17) and (7.19) are very general, and can be applied to a large variety of physical processes. In the following, they will be used to derive the polarization properties and the radiation diagram for a linear antenna, for Thomson and Rayleigh scattering, for bremsstrahlung radiation and for cyclotron and synchrotron radiation.

8. The Linear Antenna

Referring to Fig. 7, we consider a particle of charge e oscillating along the z-axis according to the law

$$z = Z \cos(\omega_0 t) \quad ,$$

where Z is the amplitude of the oscillation. From this equation one easily obtains

$$\vec{a} = -A \cos(\omega_0 t) \vec{k} \quad ,$$

where

$$A = \omega_0^2 Z \quad .$$

The components of the electric field emitted in the direction $\vec{n}$ follow from Eqs.(7.15) and from the geometry of Fig. 7:

$$E_1(\vec{r}, t) = \frac{eA}{c^2 R} \sin\theta \cos(\omega_0 t + \phi) \quad ,$$

$$E_2(\vec{r}, t) = 0 \quad ,$$

where ϕ is an inessential phase factor which depends on the distance from the radiating antenna. Taking into account that the Fourier component of $E_1(\vec{r}, t)$ corresponding to the first harmonic is $eA \sin\theta/(2c^2 R)$, and that all the other Fourier components are zero, and considering that the averaging process is here inessential, one obtains, substituting in Eqs.(7.19), and taking into account Eq.(7.20)

$$I_\omega = Q_\omega = \frac{e^2 A^2}{8\pi c^3 R^2} \sin^2\theta \, \delta(\omega - \omega_0) \quad ,$$

$$U_\omega = V_\omega = 0 \quad .$$

These equations imply that the radiation of emitted by a linear antenna is 100% linearly polarized, the polarization direction being contained in the plane defined by the direction of propagation of the radiation and the antenna. The same equations also give the radiation diagram, showing that the intensity of the radiation is proportional to $\sin^2\theta$. In particular, for the total emitted power one gets (the average of $\sin^2\theta$ over the sphere is $2/3$)

$$\mathcal{W} = \int_0^\infty d\omega \oint d\Omega R^2 I_\omega = \frac{e^2 A^2}{3c^3} = \frac{e^2 \omega_0^2 Z^2}{3c^3} \quad .$$

This expression coincides with the one that can be obtained by a direct application of the Larmor formula

$$\mathcal{W} = \frac{2e^2 a^2}{3c^3} \quad ,$$

by taking into account that the average of the square of the acceleration is by $A^2/2$.

9. Thomson scattering

Consider a free electron of charge $e = -e_0$, with $e_0 = 4.8 \times 10^{-10}$ u.e.s., and suppose that the electron is subjected to the action of a polarized electromagnetic wave of frequency ω propagating along the direction $\vec{n}'$ and characterized by the electric field components $\mathcal{E}_1'$ and $\mathcal{E}_2'$. Such components are defined on the couple of unit vectors $\vec{e}_1'$ and $\vec{e}_2'$ such that the triplet $(\vec{e}_1', \vec{e}_2', \vec{n}')$ is a right handed coordinate system (see Fig. 8).

The motion of the electron is described by the equation

$$\vec{a}(t) = -\frac{e_0}{m} \vec{E}'(t) \quad ,$$

where m is the electron mass and where

$$\vec{E}'(t) = \text{Re}(\vec{\mathcal{E}}' e^{-i\omega t}) \quad .$$

Using complex notations, one finds for the components of the (complex) acceleration (implicitly defined by $\vec{a}(t) = \text{Re}(\vec{\mathcal{A}} e^{-i\omega t})$) on the unit vectors $\vec{e}_1$ and $\vec{e}_2$

$$\mathcal{A}_1 = -\frac{e_0}{m} \vec{e}_1 \cdot (\mathcal{E}_1' \vec{e}_1' + \mathcal{E}_2' \vec{e}_2') \quad ,$$

$$\mathcal{A}_2 = -\frac{e_0}{m} \vec{e}_2 \cdot (\mathcal{E}_1' \vec{e}_1' + \mathcal{E}_2' \vec{e}_2') \quad , \tag{9.21}$$

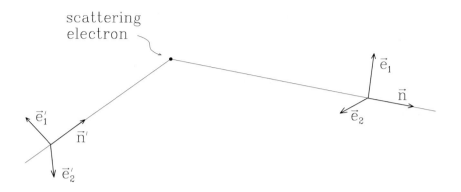

FIGURE 8. General geometry for Thomson scattering.

and taking into account that the components of the scattered field are

$$\mathcal{E}_i = \frac{e_0}{c^2 R}\mathcal{A}_i \quad ,$$

one obtains, in matrix form

$$\begin{pmatrix} \mathcal{E}_1 \\ \mathcal{E}_2 \end{pmatrix} = -\frac{r_c}{R} \begin{pmatrix} \vec{e}_1 \cdot \vec{e}_1' & \vec{e}_1 \cdot \vec{e}_2' \\ \vec{e}_2 \cdot \vec{e}_1' & \vec{e}_2 \cdot \vec{e}_2' \end{pmatrix} \begin{pmatrix} \mathcal{E}'_1 \\ \mathcal{E}'_2 \end{pmatrix} \quad ,$$

where r_c, the *classical radius of the electron*, is defined by

$$r_c = \frac{e_0^2}{mc^2} \quad .$$

The expressions now derived is nothing but the Thomson scattering law expressed in the Jones formalism. To pass to the Stokes parameters formalism, we have just to take into account Eqs.(3.2) and (3.3). After some algebra, one finds that the Stokes vector of the scattered radiation, $S(\vec{n})$, is connected to the Stokes vector of the incident radiation, $S'(\vec{n}')$, by the equation

$$S(\vec{n}) = \frac{2r_c^2}{3R^2} \mathbf{R}(\vec{n}, \vec{n}')S'(\vec{n}') \quad , \tag{9.22}$$

where the 4×4 matrix $\mathbf{R}$, usually referred to as the *Rayleigh scattering phase matrix*, is given by

$$\mathbf{R}(\vec{n}, \vec{n}') = \frac{3}{4} \begin{pmatrix} a^2 + b^2 + c^2 + d^2 & a^2 - b^2 + c^2 - d^2 & 2(ab + cd) & 0 \\ a^2 + b^2 - c^2 - d^2 & a^2 - b^2 - c^2 + d^2 & 2(ab - cd) & 0 \\ 2(ac + bd) & 2(ac - bd) & 2(ad + bc) & 0 \\ 0 & 0 & 0 & 2(ad - bc) \end{pmatrix} \quad ,$$

with

$$a = \vec{e}_1 \cdot \vec{e}_1' \,, \qquad b = \vec{e}_1 \cdot \vec{e}_2' \,, \qquad c = \vec{e}_2 \cdot \vec{e}_1' \,, \qquad d = \vec{e}_2 \cdot \vec{e}_2' \,.$$

This expression can be simplified by suitably choosing the polarization unit vectors. The simplest choice is to define the unit vectors $\vec{e}_1$ and $\vec{e}_1'$ as being perpendicular to the scattering plane, and the unit vectors $\vec{e}_2$ and $\vec{e}_2'$ as lying in the scattering plane (see Fig. 9). With this choice, one obtains for a scattering at the angle Θ

$$a = 1 \,, \qquad b = 0 \,, \qquad c = 0 \,, \qquad d = \cos\Theta \quad ,$$

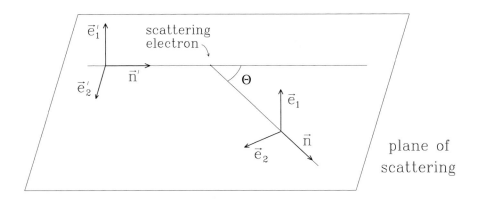

FIGURE 9. Particular geometry for Thomson scattering. In this geometry, the expression of the scattering phase matrix simplifies considerably.

and one gets

$$\mathbf{R}(\Theta) = \frac{3}{4} \begin{pmatrix} 1 + \cos^2 \Theta & \sin^2 \Theta & 0 & 0 \\ \sin^2 \Theta & 1 + \cos^2 \Theta & 0 & 0 \\ 0 & 0 & 2\cos \Theta & 0 \\ 0 & 0 & 0 & 2\cos \Theta \end{pmatrix} .$$

In particular, for the scattering at an angle Θ of an unpolarized radiation beam of intensity I', one has

$$I(\Theta) = \frac{r_c^2}{2R^2}(1 + \cos^2 \Theta)I' , \qquad Q(\Theta) = \frac{r_c^2}{2R^2}\sin^2 \Theta I' ,$$

$$U(\Theta) = 0 , \qquad V(\Theta) = 0 .$$

It follows that the fractional linear polarization of the scattered radiation is given, as a function of Θ, by

$$\frac{Q(\Theta)}{I(\Theta)} = \frac{\sin^2 \Theta}{1 + \cos^2 \Theta} ,$$

which implies that the linear polarization is always positive (polarization perpendicular to the scattering plane), and that it is as high as 100% for $\Theta = 90°$. Moreover, integrating over a sphere of radius R, one finds that the total energy radiated by the electron per unit time, $\mathcal{W}$, is given by

$$\mathcal{W} = \int I(\Theta) \, d\Omega = \sigma_T I' ,$$

where σ_T, the so-called *Thomson cross section*, is defined by

$$\sigma_T = \frac{8\pi}{3} r_c^2 = \frac{8\pi e_0^4}{3m^2 c^4} = 6.65 \times 10^{-25} \text{cm}^2 .$$

10. Rayleigh scattering

Rayleigh scattering is quite similar to Thomson scattering, with the only difference that electron is not free but bound in an atom or a molecule. From the point of view of classical physics the bound electron can be described through a simple model, due to Lorentz, in which it is assumed that the action of the cloud of charge present in the atom (or the molecule) on the electron can be schematized as a restoring force of the form $\vec{F} = -k\vec{x}$, where k is a constant and $\vec{x}$ is the position of the electron with respect to the center of gravity of the charges. The law of motion of the bound electron under the action of an electric field of frequency ω is then

$$\frac{\mathrm{d}^2 \vec{x}}{\mathrm{d}t^2} = -\omega_0^2 \vec{x} - \frac{e_0}{m}\vec{E}'(t) \quad ,$$

where $\omega_0 = \sqrt{k/m}$, and where

$$\vec{E}'(t) = \mathrm{Re}(\vec{\mathcal{E}}' e^{-i\omega t}) \quad .$$

The stationary solution of this equation is easily found. From the expression of $\vec{x}(t)$, one then finds the acceleration by taking the second derivative with respect to time. The result for the components of the (complex) acceleration along the unit vectors $\vec{e}_1$ and $\vec{e}_2$ is the following

$$\mathcal{A}_1 = -\frac{e_0}{m}\frac{\omega^2}{\omega^2 - \omega_0^2}\vec{e}_1 \cdot (\mathcal{E}_1'\vec{e}_1' + \mathcal{E}_2'\vec{e}_2') \quad ,$$

$$\mathcal{A}_2 = -\frac{e_0}{m}\frac{\omega^2}{\omega^2 - \omega_0^2}\vec{e}_2 \cdot (\mathcal{E}_1'\vec{e}_1' + \mathcal{E}_2'\vec{e}_2') \quad .$$

This expression is very similar to the one that we have previously obtained for Thomson scattering (cfr. Eq.(9.21)), and indeed it reduces to to the former when $\omega_0 = 0$ (case of the free electron). Repeating the same arguments as in the previous section, one finally arrives to the following expression for the Stokes parameters of the scattered radiation (cfr. Eq.(9.22))

$$S(\vec{n}) = \frac{2r_c^2}{3R^2}\frac{\omega^4}{(\omega^2 - \omega_0^2)^2} \mathbf{R}(\vec{n}, \vec{n}')S'(\vec{n}') \quad ,$$

where $\mathbf{R}(\vec{n}, \vec{n}')$ is the same as in the case of Thomson scattering.

For Rayleigh scattering we have thus obtained the same result as for Thomson scattering, with the only difference that the Thomson cross section has to be substituted by the Rayleigh cross section given by

$$\sigma_\mathrm{R} = \frac{\omega^4}{(\omega^2 - \omega_0^2)^2}\sigma_\mathrm{T} = \frac{\omega^4}{(\omega^2 - \omega_0^2)^2}\frac{8\pi}{3}r_c^2 \quad .$$

An important aspect of Rayleigh scattering is the fact that, for $\omega_0 \gg \omega$, the cross section results in being proportional to ω^4. Since this is a good approximation for nitrogen and oxygen molecules (the major constituents of the earth atmosphere), and since skylight is nothing by sunlight scattered by such molecules, it follows that the sky is blue, and, for the same reason, the sun is red at sunrise and at sunset.

11. A Digression on Mie Scattering

In the former Sections we have considered the polarization properties of the radiation scattered by a single electron (either free or bound). Here, we want to discuss, mainly in a qualitative way, the phenomenon of scattering on a macroscopic particle having typical dimensions comparable or larger than the wavelength of the radiation. This is a rather important phenomenon because such particles are the major constituent of various media of astrophysical interest, such as planetary atmospheres, atmospheres of fully evolved red giants, planetary nebulae, infrared objects, and the interstellar medium.

The problem of finding the polarization characteristics of the radiation diffused by a macroscopic particle is very complicated. One starts by describing the physical properties of the material of the particle by assigning its index of refraction, which is, in general, a complex number (possibly also depending on the polarization of the radiation), and by assigning the geometrical shape and dimensions of the particle. Only in the simplest cases, like for instance when the index of refraction is real and independent of direction, and the shape of the particle is spherical, the problem can be solved analytically, but the solution remains very involved and requires the introduction of several special functions. This particular case is often referred to as *Mie scattering* from the name of the physicist who developed this research field in the early years of 1900.

To understand the basic physical aspects of Mie scattering, one can think about two different processes that contribute to modify the results obtained in the former section. Suppose first that the electric field of the incoming radiation is not modified by the material composing the particle (this will be indeed true only when the index of refraction of the material differs very little from unity, namely when $|n - 1| \ll 1$). Any volume element of the particle, dV, will thus harbour an elementary electric dipole given by

$$\mathrm{d}\vec{P} = \frac{1}{4\pi}(n^2 - 1)\vec{E}' \, \mathrm{d}V \simeq \frac{1}{2\pi}(n - 1)\vec{E}' \, \mathrm{d}V \quad .$$

This elementary dipole oscillates with the frequency ω of the incoming radiation beam and can be considered by all means as an accelerated charge. The electric field radiated by the elementary dipole can thus be obtained directly through Eq.(7.15), provided the formal substitution $e\vec{a} \rightarrow -\omega^2 \, \mathrm{d}\vec{P}$ is applied and provided an integral is performed over the volume of the particle. Obviously, this implies that also the scattered beam is not modified by the material composing the particle during the internal propagation. The important physical fact is now that all the oscillating dipoles are not in phase, because, if the dimensions of the particle are comparable or larger than the wavelength, at different points within the particle the incoming radiation field has different phases. Moreover, one has also to take into account that the radiation field scattered along the direction $\vec{n}$ by the different elementary dipoles suffers different phase lags (see Fig. 10).

Decomposing the electric fields of the incoming and scattered beams along the same unit vectors as in the previous sections, one finally obtains

$$\begin{pmatrix} \mathcal{E}_1 \\ \mathcal{E}_2 \end{pmatrix} = \frac{(n - 1)\omega^2 V}{c^2 R}\Phi(\vec{n}, \vec{n}') \begin{pmatrix} \vec{e}_1 \cdot \vec{e}_1' & \vec{e}_1 \cdot \vec{e}_2' \\ \vec{e}_2 \cdot \vec{e}_1' & \vec{e}_2 \cdot \vec{e}_2' \end{pmatrix} \begin{pmatrix} \mathcal{E}'_1 \\ \mathcal{E}'_2 \end{pmatrix} \quad ,$$

where V is the volume of the particle, and where the phase factor Φ is given by

$$\Phi(\vec{n}, \vec{n}') = \frac{1}{V} \int e^{\mathrm{i}\phi(\vec{r})} \, \mathrm{d}V \quad ,$$

with

$$\phi(\vec{r}) = \frac{\omega}{c}\vec{r} \cdot (\vec{n}' - \vec{n}) \quad .$$

Performing the same transformations as in the previous sections, the Stokes parameters of the scattered radiation can be expressed in the form

$$S(\vec{n}) = \frac{2}{3}\frac{(n - 1)^2\omega^4 V^2}{c^4 R^2}|\Phi(\vec{n}, \vec{n}')|^2 \, \mathbf{R}(\vec{n}, \vec{n}')S'(\vec{n}') \quad ,$$

where $\mathbf{R}(\vec{n}, \vec{n}')$ is the same as in the case of Thomson and Rayleigh scattering.

The scattering process that has been considered here is generally referred to as the Rayleigh-Gans scattering. It has to be remarked that it brings to the same values for the fractional polarization of the scattered radiation as in the cases of Thomson and of Rayleigh scattering. The main difference with respect to the two formers cases is due to the presence of the phase factor which brings to an overall modulation of the scattered

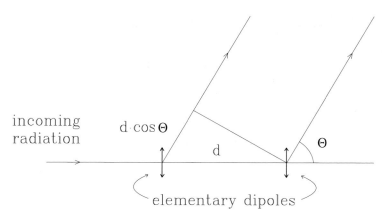

scattered radiation

incoming radiation

d·cosϴ

d

ϴ

elementary dipoles

FIGURE 10. The radiation beams scattered by the two dipoles have a well defined phase difference. Such dephasing is one of the basic physical phenomena by which Mie scattering differs from Thomson and Rayleigh scattering.

radiation. For forward scattering, in particular, the phase factor is equal to unity, irrespectively of the geometrical shape of the particle. This property is generally lost for any other direction and the phase factor turns out to be (in modulus) less than unity. This implies that forward scattering is always preferred in Rayleigh-Gans scattering.

The phase factor can be computed analytically only for few geometrical shapes of the particles. For a sphere of radius a one obtains

$$\Phi_{\text{sphere}}(\vec{n}, \vec{n}') = \frac{3}{y^3}(\sin y - y \cos y) \quad ,$$

where

$$y = 2\frac{\omega a}{c} \sin \frac{\Theta}{2} = \frac{4\pi a}{\lambda} \sin \frac{\Theta}{2} \quad .$$

A plot of the function Φ_{sphere} is shown in Fig. 11.

Apart from the physical phenomenon illustrated above and which is connected to the phase differences among the elementary dipoles, there is another phenomenon that enters into play when the condition $|n - 1| \ll 1$ is not verified. This is the fact that the electric field is now deeply modified inside the particle, either in its propagation direction, in its polarization properties and in its phase, which now also depends on the path that has been travelled inside the particle. Consider, as an extreme example, the case of a droplet of water having a radius much larger than the wavelength. In this case one can apply the laws of geometrical optics and one finds for polarization the result of the rainbow. As we have stated earlier, these cases can be considered only through more complicated approaches. In general, it turns out that the law of scattering is no longer described by the Rayleigh scattering phase matrix and that negative linear polarization can also be found for particular combinations of the parameters.

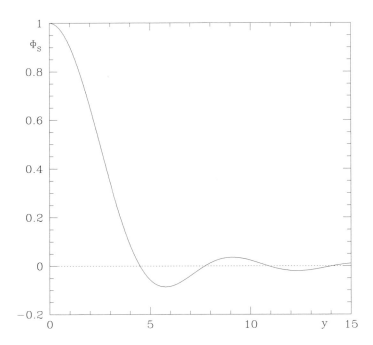

FIGURE 11. Plot of Φ_{sphere} as a function of y. The first zero corresponds to $y = 4.49$.

12. Bremsstrahlung Radiation

The radiation emitted in the process of collision between a fast electron and a heavy nucleus is usually refereed to under the name of bremsstrahlung radiation. In this section we are going to determine the polarization of the bremsstrahlung radiation emitted by a monochromatic and unidirectional beam of non-relativistic electrons. Referring to Fig. 12, consider an electron of velocity v which is passing by a nucleus of charge Ze_0 with impact parameter b, and let us assume that the electron is moving sufficiently fast that the deviation of its trajectory from a straight line can be neglected. This is verified if the inequality

$$\frac{1}{2}mv^2 \gg \frac{Ze_0^2}{b} \quad,$$

is satisfied, which implies

$$b \gg b_{\text{min}} \quad,$$

where

$$b_{\text{min}} = \frac{2Ze_0^2}{mv^2} \quad. \tag{12.23}$$

Under the straight trajectory approximation, the position of the electron, in a reference frame x, y, z centered on the nucleus, is given by

$$\vec{x}(t) = b\cos\varphi\,\vec{i} + b\sin\varphi\,\vec{j} + vt\,\vec{k} \quad,$$

where $\vec{i}$, $\vec{j}$, and $\vec{k}$ are three unit vectors pointing along the axes x, y, and z, respectively, φ is the azimuth angle specifying the geometry of the collision, and t is time measured from the instant when the electron is crossing the x-y plane. The acceleration of the

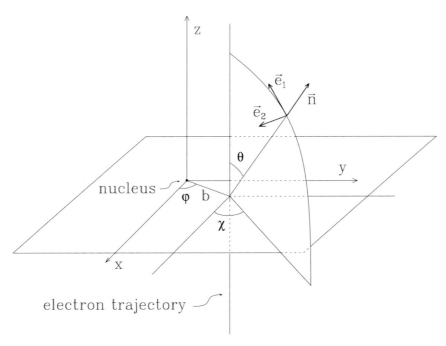

FIGURE 12. Geometry of bremsstrahlung radiation.

electron can be easily derived from the Coulomb force. One has

$$\vec{a}(t) = \frac{Ze_0^2}{m} \frac{1}{(b^2 + v^2 t^2)^{3/2}} (-b \cos\varphi \, \vec{i} - b \sin\varphi \, \vec{j} - vt \, \vec{k}) \quad .$$

We now consider the radiation propagating along the direction $\vec{n}$ and we introduce the unit vectors $\vec{e}_1$ and $\vec{e}_2$ according to the usual conventions (the triplet $(\vec{e}_1, \vec{e}_2, \vec{n})$ has to form a right-handed reference frame). If θ and χ are the polar and azimuth angles relative to the direction $\vec{n}$, one has

$$\vec{n} = \sin\theta \cos\chi \, \vec{i} + \sin\theta \sin\chi \, \vec{j} + \cos\theta \, \vec{k} \quad ,$$

$$\vec{e}_1 = -\cos\theta \cos\chi \, \vec{i} - \cos\theta \sin\chi \, \vec{j} + \sin\theta \, \vec{k} \, , \qquad \vec{e}_2 = \sin\chi \, \vec{i} - \cos\chi \, \vec{j} \quad .$$

Decomposing the acceleration along the unit vectors $\vec{e}_1$ and $\vec{e}_2$, and applying Eq.(7.15) one finds the components of the radiation field

$$E_1(\vec{r}, t) = \frac{Ze_0^3}{mc^3 R} \frac{1}{(b^2 + v^2 t^2)^{3/2}} \, [b \cos\theta \cos(\varphi - \chi) - vt \sin\theta] \quad ,$$

$$E_2(\vec{r}, t) = \frac{Ze_0^3}{mc^3 R} \frac{1}{(b^2 + v^2 t^2)^{3/2}} \, b \sin(\varphi - \chi) \quad .$$

Differently from the cases that we have considered so far, the electric field doesn't show an oscillatory, sinusoidal behaviour. To obtain the polarization properties and, at the same time, the spectral properties of the radiation, it is necessary to pass through the Fourier transforms of the components of the electric field defined in Eq.(7.16). Introducing the functions

$$F(z) = \int_0^{\pi/2} \cos(z \tan x) \cos x \, dx \, , \qquad G(z) = \int_0^{\pi/2} \sin(z \tan x) \sin x \, dx \quad ,$$

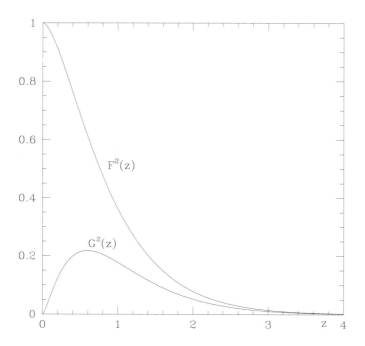

FIGURE 13. Plot of $F^2(z)$ and $G^2(z)$ as a function of z.

one gets

$$\mathcal{E}_1(\omega) = \frac{Ze_0^3}{\pi mc^2 bvR} \left[\cos\theta \cos(\varphi - \chi)\, F(z) - \mathrm{i} \sin\theta\, G(z) \right] \quad ,$$

$$\mathcal{E}_2(\omega) = \frac{Ze_0^3}{\pi mc^2 bvR}\, \sin(\varphi - \chi)\, F(z) \quad ,$$

where

$$z = \frac{\omega b}{v} \quad .$$

The functions $F(z)$ and $G(z)$ can be computed numerically. The behavior of their squared values as a function of z is shown in Fig. 13.

The Stokes parameters of the radiation emitted in the direction $\vec{n}$ can be obtained through Eqs.(7.17) and (7.18). In this case, the constant $\mathcal{K}$ is given by $\mathcal{K} = c N_c$, where N_c is the number of collisions per unit time. After some algebra one gets

$$I_\omega(\vec{n}) = \mathcal{C} \left\{ \left[\cos^2\theta \cos^2(\varphi - \chi) + \sin^2(\varphi - \chi) \right] F^2(z) + \sin^2\theta G^2(z) \right\} \quad ,$$

$$Q_\omega(\vec{n}) = \mathcal{C} \left\{ \left[\cos^2\theta \cos^2(\varphi - \chi) - \sin^2(\varphi - \chi) \right] F^2(z) + \sin^2\theta G^2(z) \right\} \quad ,$$

$$U_\omega(\vec{n}) = \mathcal{C}\, 2 \cos\theta \cos(\varphi - \chi) \sin(\varphi - \chi) F^2(z) \quad ,$$

$$V_\omega(\vec{n}) = -\mathcal{C}\, 2 \sin\theta \sin(\varphi - \chi) F(z) G(z) \quad ,$$

where

$$\mathcal{C} = \frac{Z^2 e_0^6 N_c}{\pi^2 m^2 c^3 b^2 v^2 R^2} \quad .$$

The expressions now derived refer to collisions having well defined values for the impact geometrical parameters φ and b. When averaging over φ, the Stokes parameters U_ω and V_ω vanish and one is left with the simpler expressions

$$I_\omega(\vec{n}) = \frac{\mathcal{C}}{2}\left[(1+\cos^2\theta)F^2(z) + \sin^2\theta\, G^2(z)\right] , \qquad Q_\omega(\vec{n}) = -\frac{\mathcal{C}}{2}\sin^2\theta\left[F^2(z) - G^2(z)\right] .$$

A final sum over all the impact parameters b has to be performed. This is done by integrating the expressions given above in db, and taking into account that for a beam of electrons having number density n_e and velocity v, one has

$$N_c = n_e v\, 2\pi b\, db .$$

The integrals in db of I_ω and Q_ω, if performed between 0 and ∞, are diverging integrals because for $b \to 0$ the integrands behave as b^{-1}. This inconsistency is due to the approximation of the straight line trajectories that we have introduced at the beginning of our calculations. Since this approximation is not justified for $b < b_{min}$, where b_{min} is the quantity defined in Eq.(12.23), one can simply avoid the divergence by extending the integrals in db between b_{min} and ∞. By so doing, one obtains an approximated expression that can be improved only by means of more involved calculations. Defining

$$\mathcal{F} = \int_{b_{min}}^{\infty} \frac{1}{b}F^2\left(\frac{\omega b}{v}\right)\, db , \qquad \mathcal{G} = \int_{b_{min}}^{\infty} \frac{1}{b}G^2\left(\frac{\omega b}{v}\right)\, db , \tag{12.24}$$

one gets

$$I_\omega(\vec{n}) = \mathcal{C}'\left[(1+\cos^2\theta)\mathcal{F} + \sin^2\theta\mathcal{G}\right] , \qquad Q_\omega(\vec{n}) = -\mathcal{C}'\sin^2\theta(\mathcal{F} - \mathcal{G}) ,$$

where

$$\mathcal{C}' = \frac{Z^2 e_0^6 n_e}{\pi m^2 c^3 v R^2} .$$

As can be argued by an inspection to Fig. 13, and as confirmed by direct numerical calculations, for any angular frequency ω that is substantially contributing to the emissivity, the inequality $\mathcal{G} \ll \mathcal{F}$ is always verified. This implies that, to a good approximation, the fractional polarization of the radiation emitted in direction $\vec{n}$ is independent of frequency and is given by

$$\frac{Q_\omega(\vec{n})}{I_\omega(\vec{n})} = -\frac{\sin^2\theta}{1 + \cos^2\theta} .$$

This equation implies that the radiation is totally linearly polarized when emitted in the plane perpendicular to the velocity of the colliding beam, the direction of polarization being contained in the same plane. Vice versa, the radiation emitted along the direction of the colliding beam is not polarized. Concerning the radiation diagram, we have a θ-dependency of the form $(1 + \cos^2\theta)$ showing that the radiation emitted towards the "poles" is twice as much the radiation emitted in the "equatorial plane".

Finally, it has to be remarked that the equations now derived can also be used to give an order of magnitude for the total power emitted by bremsstrahlung radiation. This can be achieved by taking a drastic approximation on the behavior of the function $F^2(z)$ appearing in the integral of Eq.(12.24). Supposing that the function is unity from 0 up to a maximum b-value, denoted by b_{max}, and such that

$$b_{max} = \frac{v}{\omega} ,$$

one simply obtains

$$\mathcal{F} = \ln\left(\frac{b_{max}}{b_{min}}\right) . \tag{12.25}$$

Supposing, moreover, that $\mathcal{G} = 0$, integrating the emitted intensity over a sphere of radius

R, and substituting for $b_{\min}$ and $b_{\max}$ the values given by Eqs.(12.23) and (12.25), one obtains the power for the emission of bremsstrahlung radiation in the form

$$W_\omega = \frac{16}{3} \frac{Z^2 e_0^6 n_e}{m^2 c^3 v} \ln \left(\frac{mv^3}{2 Z e_0^2 \omega} \right) \quad .$$

A deeper analysis due to Landau & Lifchitz (1966) shows that this formula is correct in the limit of low frequencies, provided the factor $1/2$ in the argument of the logarithm is substituted by the factor $2/C$, where $C = 1.781..$ is the exponential of the Euler constant. In the opposite limit of high frequencies, the formula has to be modified by multiplying the r.h.s. for the factor $\pi/\sqrt{3}$ and by dropping the logarithm, thus to obtain the frequency independent expression.

$$W_\omega = \frac{16\pi}{3\sqrt{3}} \frac{Z^2 e_0^6 n_e}{m^2 c^3 v} \quad .$$

An important consequence of the frequency-independence of this expression for large values of ω is the appearance, in the classical theory of bremsstrahlung, of a kind of *ultraviolet catastrophe* similar to the one that is met in the black-body theory. Indeed, defining the total power of the radiation emitted through the equation

$$\mathcal{W} = \int_0^\infty W_\omega \, d\omega \quad ,$$

one finds that the integral diverges. This is because quantum effects have not been taken into account. The most important consequence of quantum effects is the appearance of a threshold value, $\omega_{\max}$, for the frequency of the emitted photons, which is given by

$$\hbar \omega_{\max} = \frac{1}{2} m v^2 \quad .$$

As on order of magnitude we thus have

$$\mathcal{W} \simeq \frac{Z^2 e_0^6 n_e}{m^2 c^3 v} \omega_{\max} = \frac{Z^2 e_0^6 n_e}{m^2 c^3 v} \frac{m v^2}{2\hbar} \quad .$$

This equation allows to define a cross-section for bremsstrahlung emission by dividing $\mathcal{W}$ by the flux of energy of the colliding electrons. Since this flux is given by

$$F_{\text{electrons}} = n_e v \frac{1}{2} m v^2 \quad ,$$

one finds

$$\sigma_{\text{B}} = \frac{Z^2 \alpha}{\beta^2} r_{\text{c}}^2 \quad ,$$

where α is the fine structure constant, $\alpha = e_0^2/(\hbar c) \simeq 1/137.06...$, $\beta = v/c$, and r_{c} is the classical radius of the electron.

13. Cyclotron Radiation

A non relativistic electron, moving in a constant magnetic field $\vec{B}$, is subjected to the Lorentz force. Its motion is described by the equation

$$\frac{d^2 \vec{x}}{dt^2} = -\frac{e_0}{mc} \frac{d\vec{x}}{dt} \times \vec{B} \quad .$$

In a right-handed reference system (x, y, z), whose z-axis is pointing along the direction of the magnetic field, the most general solution of this equation is the following

$$x = x_0 + A \cos(\omega_{\text{C}} t + \phi) \, , \qquad y = y_0 + A \sin(\omega_{\text{C}} t + \phi) \, , \qquad z = z_0 + v_\parallel t \quad ,$$

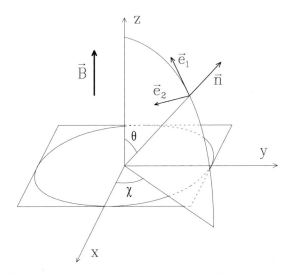

FIGURE 14. Geometry for cyclotron radiation.

where

$$\omega_C = \frac{e_0 B}{mc}$$

is the so-called *cyclotron frequency*, and where x_0, y_0, z_0, A, ϕ, and $v_\parallel$ are 6 constants of integration. The electron describes a circular trajectory in the x-y plane with a superimposed uniform translation along the z-axis (see Fig. 14). The circular motion takes place at a velocity $v_\perp = A\omega_C$ and the direction of motion is counterclockwise for an observer seeing the motion from the positive z-axis.

The components of the acceleration of the electron on the unit vectors $\vec{e}_1$ and $\vec{e}_2$ (also defined in Fig. 14) can be easily computed. This gives,

$$a_1 = v_\perp \omega_C \cos\theta \cos(\omega_C t + \phi - \chi), \qquad a_2 = v_\perp \omega_C \sin(\omega_C t + \phi - \chi).$$

The components of the electric field vector of the emitted radiation are given by Eq.(7.15), and the expression of the Stokes parameters can be deduced from Eqs.(7.19) and (7.20) taking into account that $E_1(\vec{r}, t)$ and $E_2(\vec{r}, t)$ are purely sinusoidal functions of frequency ω_C. One gets

$$\begin{pmatrix} I_\omega \\ Q_\omega \\ U_\omega \\ V_\omega \end{pmatrix} = \frac{1}{8\pi} \frac{e_0^2 v_\perp^2 \omega_C^2}{c^3 R^2} \delta(\omega - \omega_C) \begin{pmatrix} 1 + \cos^2\theta \\ -\sin^2\theta \\ 0 \\ -2\cos\theta \end{pmatrix}.$$

This equation shows that cyclotron radiation is elliptically polarized. In particular, it is circularly polarized when emitted along the direction of the magnetic field vector, and linearly polarized (the direction of polarization being perpendicular to the magnetic field) when emitted in the plane perpendicular to the field. The total power of the emitted radiation is easily found by integrating the expression of I_ω over frequencies and over a sphere of radius R. The result is

$$\mathcal{W} = \frac{2}{3} \frac{e_0^2 v_\perp^2 \omega_C^2}{c^3},$$

a formula that can also be expressed in the alternatives forms

$$W = \frac{2}{3} \frac{e_0^4 v_\perp^2 B^2}{m^2 c^5} = \frac{2}{3} \beta_\perp r_c^2 v_\perp B^2 \quad ,$$

where $\beta_\perp = v_\perp/c$ and where r_c is the classical radius of the electron. The last equation also allows the introduction of a suitable cross-section. Recalling that the magnetic energy density is $B^2/(8\pi)$, the energy flux "swept" by the electron in its accelerated motion is $v_\perp B^2/(8\pi)$. The magnetic energy is thus transformed into electromagnetic energy with a cross section given by

$$\sigma_C = \frac{16\pi}{3} \beta_\perp r_c^2 \quad .$$

14. Synchrotron Radiation

As we have seen in the former section, a non-relativistic electron (or any other charged particle) moving in a constant magnetic field describes a helical trajectory characterized by the frequency ω_C. At he same time, the electron radiates a monochromatic wave of the same frequency and with the polarization characteristics described above.

For relativistic particles, the physical characteristics of the motion of the electron remain the same, except that the cyclotron frequency is changed into a lower frequency (now depending on the velocity of the particle), called the synchrotron frequency and defined by

$$\omega_S = \frac{e_0 B}{\gamma m c} \quad ,$$

where γ, the Lorentz factor, is given by

$$\gamma = \frac{1}{\sqrt{1 - \beta^2}} \quad ,$$

with $\beta = v/c$.

On the contrary, the spectral characteristics of the radiation, and its polarization properties as well, result in being deeply modified by the relativistic motion, especially for values of β approaching unity (ultrarelativistic case). This is mainly due to the so-called *beaming effect*, which is nothing but the high-velocity limit of another phenomenon, aberration, well known in astronomy. Consider an accelerated charge that is radiating in space and suppose, for the sake of simplicity, that the radiation diagram is isotropic in the rest frame of the particle. If $\vec{v}$ is the velocity of the particle, a ray that, in the rest frame, is radiated along the direction $\vec{n}$, will result, in the laboratory frame, to propagate along the direction $\vec{n}\,'$. The components of $\vec{n}$ and $\vec{n}\,'$ along the directions parallel and perpendicular to $\vec{v}$ are connected by the Lorentz transformations

$$n'_\| = \frac{n_\| + \beta}{1 + \beta n_\|} \quad , \qquad n'_\perp = \frac{n_\perp}{\gamma(1 + \beta n_\|)} \quad .$$

Denoting by θ and θ' the angles that the unit vectors $\vec{n}$ and $\vec{n}\,'$ form with the direction of $\vec{v}$, one has

$$\tan \theta' = \frac{n'_\perp}{n'_\|} = \frac{\sin \theta}{\gamma(\cos \theta + \beta)} \quad .$$

In the limit of $\beta \ll 1$, these formulae just describe the phenomenon of aberration. On the contrary, if we assume $\beta \simeq 1$, the factor γ in the denominator makes the angle θ' to be very small for almost any value of θ. In particular, assuming $\theta = \pi/2$, one gets

$$\theta' = \arctan \frac{1}{\gamma\beta} \simeq \frac{1}{\gamma} = \sqrt{1 - \beta^2} \simeq \sqrt{2(1 - \beta)} \quad .$$

This means that the radiation that, in the rest frame, is emitted in a full hemisphere is now concentrated in a small cone having aperture $1/\gamma$, with a reduction in solid angle from 2π to π/γ^2. This is the beaming effect which has very important consequences on synchrotron radiation.

The analysis of the polarization characteristics of synchrotron radiation follows from the relativistic equations given in Sect.7. Without going into a full analysis, we just give here a simple illustration by assuming that the electron is moving into a circular orbit (we neglect the motion along the z-axis). In the geometry of Fig. 14, already used for treating cyclotron radiation, we can suppose, without losing in generality, that the position at time t' of the radiating electron is given by

$$\vec{x}(t') = A \cos\omega_S t' \, \vec{i} + A \sin\omega_S t' \, \vec{j} \quad ,$$

where $A = c\beta/\omega_S$. From these expressions we can compute the electric field components along the unit vectors $\vec{e}_1$ and $\vec{e}_2$ relative to a direction $\vec{n}$ contained in the x-z plane ($\vec{n} = \sin\theta\,\vec{i} + \cos\theta\,\vec{k}$). Using Eq.(7.14), we obtain after some algebra

$$E_1(\vec{r}, t) = \frac{e_0}{cR} \frac{1}{\kappa^3(t')} \beta\omega_S \cos\theta \cos\omega_S t' \quad ,$$

$$E_2(\vec{r}, t) = \frac{e_0}{cR} \frac{1}{\kappa^3(t')} \beta\omega_S (\sin\omega_S t' - \beta \sin\theta \cos 2\omega_S t') \quad ,$$

where t is connected to t' by the equation (apart from an inessential time t_0)

$$t = t' + \frac{\beta}{\omega_S} \sin\theta(1 - \cos\omega_S t') \quad ,$$

and where

$$\kappa(t') = 1 + \beta \sin\theta \sin\omega_S t' \quad .$$

Fig. 15 shows a plot of the electric field components E_1 and E_2 as a function of time for various values of θ. The figure, obtained for $\beta = 0.8$, shows that, for $\theta = 90°$, the radiation is concentrated in very narrow pulses having a typical width of the order of one tenth of the period, and that the component E_1 of the electric field is identically zero. This means that the spectrum of the radiation is very broad, extending to frequencies much higher than the synchrotron frequency ω_S (up to $\simeq 10\omega_S$ in this case), and that the radiation is totally linearly polarized the direction of polarization being perpendicular to the magnetic field. Going to smaller values of the angle θ, the amplitude of the electric field substantially decreases, and the same happens for the polarization because the other polarization component appears. Finally, for for $\theta = 0°$, there are no complications due to retarded times and the equations directly show that the electric field is purely sinusoidal (at frequency ω_S) and circularly polarized (as in the case of cyclotron radiation).

The full analysis of the polarization of synchrotron radiation is rather complex and will not be extended here any further. As a hint for a deepening of the subject, we can just mention that the analysis would proceed by finding the Fourier components of the electric field amplitudes E_1 and E_2, and then finding the Stokes parameters as a function of frequency by means of Eqs.(7.19). When the calculations are extended to comply also for the case where the pitch-angle of the particle is different from $90°$ (the pitch-angle is the angle between the trajectory of the particle and the direction of the magnetic field), and after the results have been averaged over all the possible values of the pitch-angle,

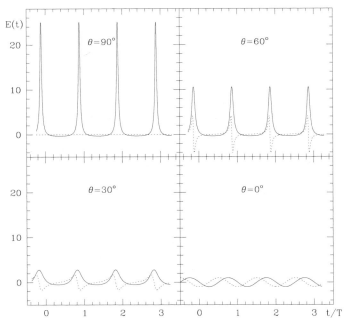

FIGURE 15. The electric field components of synchrotron radiation along the unit vector $\vec{e}_1$ (solid line) and $\vec{e}_2$ (dotted line) are plotted as a function of time for various values of θ. The unit vectors and the angle θ are defined as in Fig. 14. The relativistic electron has velocity $\beta = 0.8$. The electric field is given in arbitrary units. Time is given in units of the period T of the circular motion.

it is found that synchrotron radiation is strongly linearly polarized (typical values of the fractional polarization being of the order of 75%), the direction of polarization being perpendicular to the magnetic field.

15. Polarization in spectral lines

In the spectropolarimetric analysis of several astrophysical objects, it is commonly observed that spectral lines are polarized, the polarization signature being, in general, different from line to line and variable along the line profile. Line polarization may be accompanied, or not, by continuum polarization. As typical examples, we can just mention that this phenomenon is present in the spectra of the following objects: sunspots and solar active regions, higher layers of the solar atmosphere (including the chromosphere and the corona), magnetic stars, magnetic white dwarfs, stars with extended envelops, broad line regions of active galactic nuclei, and astrophysical masers.

Though the physical mechanisms involved are different from object to object, yet it is possible, at least in principle, to treat the general phenomenona underlying the processes of generation and transfer of polarized radiation in spectral lines within a unique theoretical framework which is based on the description of the radiation field through the four Stokes parameters, considered as functions of point P, direction $\vec{\Omega}$ and time t,

$$I(\mathrm{P}, \vec{\Omega}, t),\ Q(\mathrm{P}, \vec{\Omega}, t),\ U(\mathrm{P}, \vec{\Omega}, t),\ V(\mathrm{P}, \vec{\Omega}, t)\ ,$$

and on the description of the atomic system, interacting with the radiation field, through the density-matrix elements, again considered as functions of point P and time t,

$$\rho_{nm}(\mathrm{P}, t) = \langle n|\rho(\mathrm{P}, t)|m\rangle \quad ,$$

where $|n\rangle$ and $|m\rangle$ are eigenvectors of the atomic Hamiltonian, $\mathcal{H}_{\mathrm{A}}$, describing the atomic system

$$\mathcal{H}_{\mathrm{A}}|n\rangle = \epsilon_n|n\rangle \quad .$$

The symbol n appearing as the argument of the "ket" $|n\rangle$ represents, in general, the set of quantum numbers which are necessary to fully characterize the eigenvectors of $\mathcal{H}_{\mathrm{A}}$. For the hydrogen atom, we have, for instance, $|n\rangle \to |n, l, j, m\rangle$, where n, l, j, and m represent, respectively, the principal quantum number, the azimuthal quantum number, the total angular momentum quantum number, and the magnetic quantum number. Similarly, with self-evident notations, for an arbitrary atom in L-S coupling, for an atom with hyperfine structure, and for a molecule, we have that the "ket" $|n\rangle$ has to be replaced, respectively, by the "kets" $|\alpha, L, S, J, M\rangle$, $|\alpha, L, S, J, I, F, f\rangle$, and $|\mathrm{X}, v, \Lambda, S, N, J, M\rangle$.

The theoretical framework as of above, which is based on a perturbative development of non-relativistic quantum electrodynamics, has the advantage of providing, by means of a unique approach, both the radiative transfer equations for polarized radiation and the statistical equilibrium equations for the density-matrix elements of the material system interacting with the radiation field. The influence of collisions can also be added to the relevant equations. The theory, basically developed in the late 1970's and in the early 1980's, is still scattered in the literature, though a monograph by Landi Degl'Innocenti & Landolfi is now in preparation and will be soon made available to the scientific community. For the time being, the interested reader is referred to the papers by Landi Degl'Innocenti (1983,1984,1985) for the outline of the unifying approach, to Landi Degl'Innocenti & Landi Degl'Innocenti (1972, 1975) for early derivations of the transfer equations for polarized radiation from quantum electrodynamics, to the work by Bommier (1977, 1980) for the derivation of the statistical equilibrium equations for the density-matrix elements in optically thin plasmas, to Landi Degl'Innocenti, Bommier and Sahal-Bréchot (1990, 1991) for coupling the two sets of equations in optically thick plasmas, and to Bommier & Sahal-Bréchot (1991) and Bommier (1991) for a reconsideration of the unified approach by means of an alternative formalism based on the S-matrix theory.

Unfortunately, the density-matrix operator, introduced in the unified theory to describe the physical state of the atomic system interacting with the radiation field, is a tool rather unfamiliar to astrophysicists. This explains why the theory of polarization in spectral lines, though developed more than 20 years ago, still stands as a research tool which is used in practice only by a very restricted number of devoted scientists. In the following we will try to clarify on some simple examples the physical meaning of the density-matrix operator.

16. Density Matrix and Atomic Polarization

In the standard theory which is commonly used to address the problem of line formation in astrophysical plasmas (either optically thin or optically thick), the physical state of the atom (or molecule) interacting with the radiation field is generally described by assigning the populations n_i of its different energy levels, where i is an index running from 1 to N_{lev}, the total number of energy levels considered in the model atom. However, when polarization phenomena have to be accounted for, this simple description turns out, in most cases, to be insufficient and one has to introduce a deeper description which has to account for the spatial degeneracy of the energy levels. Referring for instance to the simplest case of a multi-level atom and neglecting the possible presence of hyperfine structure and of magnetic fields, each of its energy levels has an intrinsic degeneracy with respect to the magnetic quantum number M, where M is an integer (or half-integer) such that $-J \leq M \leq J$, J being the angular momentum quantum number of the level. It has to be remarked that M is the eigenvalue of the projection of the total angular momentum of the atom, $\vec{J}$, along an arbitrarily chosen quantization axis (the z-axis). In this deeper description, one has to specify the single populations of each of the M-sublevels of the N_{lev} energy levels of the atomic system. These quantities are nothing but the diagonal elements of the density-matrix operator. For each of the energy levels, one has then to introduce the $(2J + 1)$ quantities

$$\rho_J(M, M) = \langle JM | \rho | JM \rangle \quad .$$

Though, in many cases, this is sufficient to fully specify the physical state of the atom, there are in general more complicated situations where also the non-diagonal matrix elements of the density operator play an essential role and have to be specified as well. These quantities, that are defined by

$$\rho_J(M, M') = \langle JM | \rho | JM' \rangle \quad ,$$

are the so-called *coherences* or *phase relationships* and bring the total number of quantities necessary to describe the physical state of a single atomic level to $(2J + 1)^2$. Even for a particularly low value of J such as $J = 1$, the number of quantities necessary to fully specify the physical situation of an atom increases by one order of magnitude with respect to the "non polarized case".

The full description of an atomic system, in the general case where polarization phenomena are accounted for, thus requires, for each level of the model-atom, the specification of a matrix. When such a matrix is not proportional to the identity matrix, the atom is said to be *polarized* (or to show *atomic polarization*) in the specific level considered.

Atomic polarization can be introduced in an atomic system (an atom or a molecule) either by collisions with a collimated beam of fast particles or by anisotropic illumination by an external source. In both cases the radiation re-emitted by the atomic system is polarized. In the first case one speaks about *impact polarization* a phenomenon which will not be considered any longer in this course. The second phenomenon, which is more widespread in the astrophysical context, will be illustrated in the following on some simple examples.

We consider first the case of an atomic transition between two atomic levels, the lower level having angular momentum $J_\ell = 0$ and the upper level having angular momentum $J_u = 1$. We also suppose the atom to be irradiated by a unidirectional, unpolarized radiation beam characterized by a given intensity I_0 at the transition frequency ν_0. Assuming for simplicity the quantization axis to be directed along the direction of the beam, it follows, due to the transversality character of the electric field, that the only transitions that can be induced by the incident radiation are those satisfying the selection rule $\Delta M = \pm 1$. The atom thus results in being pumped only in the two sublevels $M_u = 1$ and $M_u = -1$, which entails the presence of atomic polarization in the upper level. The situation is schematically illustrated in Fig. 16a, the number of dots drawn on each sublevel being proportional to its population. The figure shows that in this case the atom is strongly polarized.

When a polarized atom de-excites, the radiation emitted is, in general, also polarized. To find its polarization properties, it is enough to take into account that a quantum

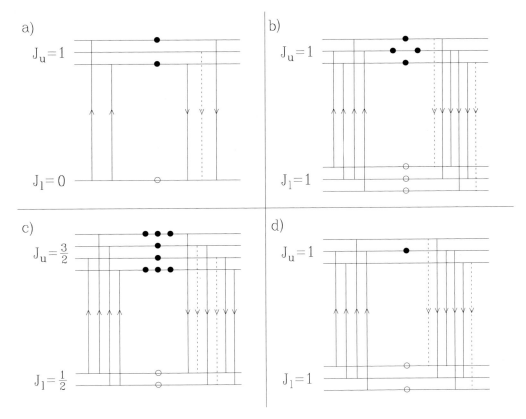

FIGURE 16. Pictorial illustration of atomic polarization in four different atomic models. The number of circles drawn on each M-sublevel is proportional to the relative population of the sublevel itself. Full vertical lines identify transitions with $\Delta M = \pm 1$, dotted lines transitions with $\Delta M = 0$.

transition where the magnetic quantum number undergoes a variation $\Delta M = 0$ behaves, in as far as the radiation diagram and the polarization properties are concerned, as a linear antenna, whereas a transition $\Delta M = \pm 1$ behaves as a circular antenna, the direction of rotation of the electric current depending on the sign of ΔM. By means of this analogy, and taking into account the results obtained in Sects. 8 and 13 for the linear antenna and for cyclotron radiation, respectively, it is possible to establish a simple formula for the polarization of the radiation emitted along the direction $\vec{n}$ forming with the quantization axis an arbitrary angle θ. Assuming as the reference direction for positive Q the unit vector $\vec{e}_1$ perpendicular to $\vec{n}$ and lying in the plane formed by $\vec{n}$ and the quantization direction (coincident with the direction of the pumping beam of radiation), the ratio Q/I of the radiation emitted by the atom can be expressed, for an arbitrary transition $J_u \to J_\ell$, in the form

$$\frac{Q}{I} = \frac{\mathcal{A}\sin^2\theta - \mathcal{B}\sin^2\theta}{\mathcal{A}\sin^2\theta + \mathcal{B}(1+\cos^2\theta)} \quad , \tag{16.26}$$

where

$$\mathcal{A} = \sum_{\Delta M=0} n(M_u)\mathcal{S}(M_u, M_\ell) \,, \qquad \mathcal{B} = \frac{1}{2}\sum_{\Delta M=\pm 1} n(M_u)\mathcal{S}(M_u, M_\ell) \quad .$$

In these equations, $n(M_u)$ is the number density of atoms in the magnetic sublevel M_u, $S(M_u, M_\ell)$ is the relative strength of the transition between the sublevel M_u and the sublevel M_ℓ, which is given by

$$S(M_u, M_\ell) = 3 \begin{pmatrix} J_u & J_l & 1 \\ -M_u & M_\ell & \Delta M \end{pmatrix}^2 \quad,$$

and the sums have to be extended to all the possible transitions between magnetic sub-levels. For the case of the transition considered above, a direct application of Eq.(16.26) brings to the result

$$\frac{Q}{I} = -\frac{\sin^2 \theta}{1 + \cos^2 \theta} \quad,$$

which means that in a 90° scattering event the radiation is totally polarized, the direction of polarization being perpendicular to the scattering plane.

Similar considerations can be repeated for the other transitions that are illustrated in panels (b) and (c) of Fig. 16. For the transition $J_\ell = 1$, $J_u = 1$ of panel (b), all transitions have the same strength $S = 1/2$ (except for the transition $M_u = 0 \rightarrow M_\ell = 0$ which is forbidden), and it follows that the sublevel $M_u = 0$ has a population which is twice the populations of the sublevels $M_u = \pm 1$. The polarization of the radiation emitted in the de-excitation process follows again from Eq.(16.26) and is given by

$$\frac{Q}{I} = -\frac{\sin^2 \theta}{5 + \cos^2 \theta} \quad.$$

For the transition $J_\ell = 1/2$, $J_u = 3/2$ of panel (c), on the other hand, one has to take into account that the strengths of the different transitions are no longer equal, being given by

$$S(3/2, 1/2) = 3/4 \,, \quad S(1/2, 1/2) = 1/2 \,, \quad S(1/2, -1/2) = 1/4 \,,$$

$$S(-1/2, 1/2) = 1/4 \,, \quad S(-1/2, -1/2) = 1/2 \,, \quad S(-1/2, -3/2) = 3/4 \quad.$$

As a consequence, the population of the sublevels $M_u = \pm 3/2$ turns out to be three times the population of the sublevels $M_u = \pm 1/2$, and the polarization of the emitted radiation, still given by Eq.(16.26), results

$$\frac{Q}{I} = -\frac{3 \sin^2 \theta}{7 + 3 \cos^2 \theta} \quad.$$

In the absence of relaxation mechanisms, the same pumping process which generates atomic polarization in the upper level is capable of producing similar effects in the lower level too. The results previously derived for the transition $J_\ell = 1$, $J_u = 1$ are based on the assumption that the sublevels M_ℓ of the lower level are, *a priori*, equally populated. When this hypothesis is released, the situation changes drastically and is illustrated in panel (d) of Fig. 16. Atomic polarization is now present both in the upper level *and in the lower level*, and the polarization of the emitted radiation is given by

$$\frac{Q}{I} = -\frac{\sin^2 \theta}{1 + \cos^2 \theta} \quad.$$

The simple examples considered here illustrate how pumping mechanisms due to the illumination of an atomic system by an anisotropic radiation field are capable of intro-ducing atomic polarization in the different levels. In particular, the last example shows the importance of lower-level polarization in determining the polarization characteris-tics of the radiation scattered by an atomic system. The degree of polarization of the radiation scattered at 90° by a two-level atomic system with $J_\ell = 1$ and $J_u = 1$ out of a unidirectional and unpolarized radiation beam changes from 20% to 100% when lower-level atomic polarization is either neglected or accounted for.

The diagonal elements of the density matrix have a simple physical interpretation in terms of populations of the magnetic sublevels. The physical interpretation of the non-diagonal elements (the so-called *coherences*) is more subtle. In general, it can be stated that a particular coherence, for instance the coherence $\rho_J(M, M')$ is different from zero when the atomic wave-function presents a well defined phase relationship between the pure quantum states $|JM\rangle$ and $|JM'\rangle$, and though this may seem a very particular circumstance, it can be easily shown that, on the contrary, coherences are be rather commonly met in practice. Probably, the simplest example that illustrates this statement is a pumping experiment in the transition $J_\ell = 0$, $J_u = 1$ by a directional beam of *polarized* radiation. Aligning the quantization axis with the direction of the pumping beam, the only sublevels that can be excited by the radiation are the sublevels $M_u = 1$ and $M_u = -1$. In particular, if the pumping beam is totally circularly polarized in one direction, only the sublevel $M_u = 1$ is excited; if it is totally circularly polarized in the opposite direction, only the sublevel $M_u = -1$ is excited, and if the beam is not polarized, both sublevels are excited, but without phase relationships between them because a natural beam of radiation can be considered as the *incoherent* superposition of positive and negative circular polarization. On the contrary, if the pumping beam is linearly polarized, it has to be considered as the *coherent* superposition of two beams of opposite circular polarizations, and it is just this coherence that is transferred to the density matrix of the upper level.

Another fact that shows the ineluctable presence of non-diagonal density-matrix elements in many cases of interest is the law of transformation of the density-matrix under a rotation of the reference system. If R is the rotation that carries the old reference system (x, y, z) in the new reference system $(x', y'z')$, the new eigenvectors $|JM\rangle_{\text{new}}$ are connected to the old ones $|JM\rangle_{\text{old}}$ by the equation

$$|JM\rangle_{\text{new}} = \sum_{M'} \mathcal{D}^J_{M'M}(R)|JM'\rangle_{\text{old}} \quad,$$

where $\mathcal{D}^J_{M'M}(R)$ is the ordinary rotation matrix. This transformation law implies a similar transformation law on the density-matrix elements. Directly from the definition we have

$$\left[\rho_J(M, M')\right]_{\text{new}} = \sum_{NN'} \mathcal{D}^J_{NM}(R)^* \mathcal{D}^J_{N'M'}(R) \left[\rho_J(N, N')\right]_{\text{old}} \quad. \tag{16.27}$$

This equation implies that even if the density matrix is diagonal in the old reference system, it turns out to be non-diagonal in the new one.

In many cases, it is convenient to introduce, instead of the "standard" density-matrix elements, $\rho_J(M, M')$, some suitable linear combinations, the so-called irreducible tensors of the density matrix, $\rho^K_Q(J)$. These quantities, that are often referred to as *statistical tensors*, or *multipole moments* (of the density matrix) are defined by

$$\rho^K_Q(J) = \sum_{MM'} (-1)^{J-M} \sqrt{2K+1} \begin{pmatrix} J & J & K \\ M & -M' & -Q \end{pmatrix} \rho_J(M, M') \quad,$$

where the integers K and Q satisfy the inequalities

$$0 \le K \le 2J \,, \qquad -K \le Q \le K \quad.$$

The statistical tensors obey a simpler transformation law under rotations of the reference system. Indeed, for the statistical tensors Eq.(16.27) is substituted by the following one

$$\left[\rho_Q^K(J)\right]_{\text{new}} = \sum_{Q'} \mathcal{D}_{Q'Q}(R)^* \left[\rho_{Q'}^K(J)\right]_{\text{old}} \quad .$$

Apart from this simple transformation law, the statistical tensors turn out to be very useful, especially when dealing with physical systems characterized by particular symmetries. Consider, for instance, an atom which is irradiated by a cylindrically symmetric, unpolarized radiation field, like the radiation field typical of the quiet solar atmosphere. In the absence of magnetic fields, taking the vertical direction as the quantization axis (the z-axis), it can easily be shown that all the irreducible tensors ρ_Q^K are zero, except for those having $Q = 0$ and K even. An atomic level with $J = 4$, for instance, needs, in general, 81 density-matrix components for its full description. In the case of a cylindrically symmetric radiation field, only the three irreducible components, ρ_0^0, ρ_0^2, and ρ_0^4 are enough for its description.

Among the different irreducible tensors, $\rho_0^0(J)$ plays a particular role because it is proportional to the overall population of the J-level. From its definition it follows that

$$\rho_0^0(J) = \frac{1}{\sqrt{2J+1}} \sum_M \rho_J(MM) = \frac{1}{\sqrt{2J+1}} \mathcal{N}_J \quad ,$$

where $\mathcal{N}_J$ is the total population of level J. All the other tensors $\rho_Q^K(J)$ describe atomic polarization. In particular, those with $K = 1$ describe the so-called *atomic orientation*, whereas those with $K = 2$ describe the so-called *atomic alignment*. Finally, any irreducible tensor with $Q = 0$ is the linear combination of populations of different magnetic sublevels, whereas any tensor with $Q \neq 0$ is the linear combinations of coherences between all the pairs of sublevels, $|JM\rangle$ and $|JM'\rangle$, such that $M - M' = Q$. For the particular case $J = 1$, the explicit expression of the irreducible tensors with $Q > 0$ is given below. The irreducible tensors with $Q < 0$ are connected to the former by the conjugation property $\rho_{-Q}^K = \rho_Q^{K\,*}$.

$$\rho_0^0 = \frac{1}{\sqrt{3}}\left[\rho(1,1) + \rho(0,0) + \rho(-1,-1)\right] , \qquad \rho_0^1 = \frac{1}{\sqrt{2}}\left[\rho(1,1) - \rho(-1,-1)\right] ,$$

$$\rho_0^2 = \frac{1}{\sqrt{6}}\left[\rho(1,1) - 2\rho(0,0) + \rho(-1,-1)\right] , \qquad \rho_1^1 = -\frac{1}{\sqrt{2}}\left[\rho(1,0) + \rho(0,-1)\right] ,$$

$$\rho_1^2 = -\frac{1}{\sqrt{2}}\left[\rho(1,0) - \rho(0,-1)\right] , \qquad \rho_2^2 = \rho(1,-1) \quad .$$

17. Radiative Transfer and Statistical Equilibrium Equations

For the standard case where polarization phenomena are neglected, the radiative transfer equation for the intensity, $I_\nu(\vec{\Omega})$, of the radiation at frequency ν propagating along the direction $\vec{\Omega}$ is generally written in the form

$$\frac{\mathrm{d}I_\nu(\vec{\Omega})}{\mathrm{d}s} = -\left(k^{(\mathrm{A})} - k^{(\mathrm{S})}\right)I_\nu(\vec{\Omega}) + \epsilon \quad , \tag{17.28}$$

where $k^{(\mathrm{A})}$ is the absorption coefficient, $k^{(\mathrm{S})}$ is the negative absorption coefficient due to stimulated emission, and ϵ is the emission coefficient. These three quantities are frequency-dependent. In the neighbouring of a spectral line, they result from the contribution of a continuous term (due to bound-free and free-free transitions) plus a term due to the bound-bound transition corresponding to the given line. For this last term, $k^{(\mathrm{A})}$ is proportional to n_ℓ, the number density of atoms in the lower level of the bound-bound

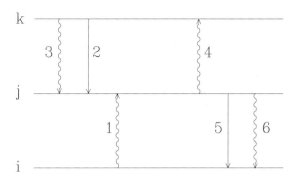

FIGURE 17. The population of level j is affected by six different radiative rates, numbered from 1 to 6. Straight lines refer to spontaneous emission, while wavy lines refer to absorption (arrows pointing upward) or stimulated emission (arrows pointing downward). Transfer rates are represented by transitions ending into level j, while relaxation rates are represented by transitions originating from level j.

transition, whereas $k^{(\mathrm{S})}$ and ϵ are proportional to n_u, the number density of atoms in the upper level of the same transition. In non-LTE problems, the transfer equation has to be supplemented by the statistical equilibrium equations for the level populations n_i, $(i = 1, 2, ..., N_{\mathrm{lev}})$ where N_{lev} is the number of levels considered in the model-atom. Considering only radiative rates, for any level there are three types of *transfer rates* and three types of *relaxation rates*, as shown in Fig. 17, and the statistical equilibrium equations are of the form

$$\frac{\mathrm{d}n_j}{\mathrm{d}t} = T_\mathrm{A}(i \to j)\, n_i + T_\mathrm{E}(k \to j)\, n_k + T_\mathrm{S}(k \to j)\, n_k$$

$$- R_\mathrm{A}(j \to k)\, n_j - R_\mathrm{E}(j \to i)\, n_j - R_\mathrm{S}(j \to i)\, n_j \quad , \qquad (17.29)$$

where T_A describes the processes of absorption from lower levels, T_E, and T_S the processes of spontaneous and stimulated emission from upper levels, R_A the processes of absorption towards upper levels, and R_E and R_S the processes of spontaneous and stimulated emission towards lower levels. These rates are numbered from 1 to 6 in Fig. 17. It has to be remarked that the rates T_A, T_S, R_A, and R_S are proportional to the radiation field intensity at the frequency of the transition. For instance,

$$T_\mathrm{A}(i \to j) = B_{ij}\, J(\nu_{ji}) \quad ,$$

where B_{ij} is the Einstein coefficient for the transition, and $J(\nu_{ji})$ is the solid angle average of the specific intensity at the transition frequency ν_{ji}.

The radiative transfer equation and the statistical equilibrium equations get generalized, in the "polarized case", by more complicated but substantially equivalent equations. From the theory outlined in Sect. 15, it follows that the radiative transfer equation (Eq.(17.28)) is generalized by the following equation

$$\begin{pmatrix} I_\omega(\vec{\Omega}) \\ Q_\omega(\vec{\Omega}) \\ U_\omega(\vec{\Omega}) \\ V_\omega(\vec{\Omega}) \end{pmatrix} = - \begin{pmatrix} \eta_I & \eta_Q & \eta_U & \eta_V \\ \eta_Q & \eta_I & \rho_V & -\rho_U \\ \eta_U & -\rho_V & \eta_I & \rho_Q \\ \eta_V & \rho_U & -\rho_Q & \eta_I \end{pmatrix} \begin{pmatrix} I_\omega(\vec{\Omega}) \\ Q_\omega(\vec{\Omega}) \\ U_\omega(\vec{\Omega}) \\ V_\omega(\vec{\Omega}) \end{pmatrix} + \begin{pmatrix} \epsilon_I \\ \epsilon_Q \\ \epsilon_U \\ \epsilon_V \end{pmatrix} \quad ,$$

where

$$
\begin{pmatrix}
\eta_I & \eta_Q & \eta_U & \eta_V \\
\eta_Q & \eta_I & \rho_V & -\rho_U \\
\eta_U & -\rho_V & \eta_I & \rho_Q \\
\eta_V & \rho_U & -\rho_Q & \eta_I
\end{pmatrix} =
$$

$$
= \begin{pmatrix}
\eta_I^{(A)} & \eta_Q^{(A)} & \eta_U^{(A)} & \eta_V^{(A)} \\
\eta_Q^{(A)} & \eta_I^{(A)} & \rho_V^{(A)} & -\rho_U^{(A)} \\
\eta_U^{(A)} & -\rho_V^{(A)} & \eta_I^{(A)} & \rho_Q^{(A)} \\
\eta_V^{(A)} & \rho_U^{(A)} & -\rho_Q^{(A)} & \eta_I^{(A)}
\end{pmatrix} -
\begin{pmatrix}
\eta_I^{(S)} & \eta_Q^{(S)} & \eta_U^{(S)} & \eta_V^{(S)} \\
\eta_Q^{(S)} & \eta_I^{(S)} & \rho_V^{(S)} & -\rho_U^{(S)} \\
\eta_U^{(S)} & -\rho_V^{(S)} & \eta_I^{(S)} & \rho_Q^{(S)} \\
\eta_V^{(S)} & \rho_U^{(S)} & -\rho_Q^{(S)} & \eta_I^{(S)}
\end{pmatrix} . \qquad (17.30)
$$

The first matrix in the r.h.s., containing the quantities with apex (A), is the analogue, in the polarized case, of the absorption coefficient $k^{(A)}$ of Eq.(17.28). Similarly, the second matrix, containing the quantities with apex (S), is the analogue of the negative absorption coefficient $k^{(S)}$. Finally, the vector $(\epsilon_I, \epsilon_Q, \epsilon_U, \epsilon_V)^{\mathrm{T}}$ is the analogue of the emission coefficient ϵ. In the neighbourhood of a spectral line, all these quantities contain two terms: a term due to processes in the continuum (bound-free and free-free transitions) plus a term due to the bound-bound transition corresponding to the given line. For this last contribution, the quantities $\eta_I^{(A)}$, $\eta_Q^{(A)}$, $\eta_U^{(A)}$, $\eta_V^{(A)}$, $\rho_Q^{(A)}$, $\rho_U^{(A)}$, and $\rho_V^{(A)}$ depend linearly on the statistical tensors $\rho_Q^K(J_\ell)$ of the lower level of the transition, whereas the quantities $\eta_I^{(S)}$, $\eta_Q^{(S)}$, $\eta_U^{(S)}$, $\eta_V^{(S)}$, $\rho_Q^{(S)}$, $\rho_U^{(S)}$, and $\rho_V^{(S)}$, and the quantities ϵ_I, ϵ_Q, ϵ_U, and ϵ_V depend linearly on the statistical tensors $\rho_Q^K(J_u)$ of the upper level.

For treating non-LTE problems, as in the "non-polarized case", the transfer equation has to be supplemented by the statistical equilibrium equation for the statistical tensors. In strict analogy with Eq.(17.29), the equation is of the form (see Fig. 17 for the meaning of the symbols i, j, and k)

$$
\frac{d\rho_Q^K(J_j)}{dt} = \sum_{K'Q'} T_{\mathrm{A}}(i; K'Q' \to j; KQ)\, \rho_{Q'}^{K'}(J_i) + \sum_{K'Q'} T_{\mathrm{E}}(k; K'Q' \to j; KQ)\, \rho_{Q'}^{K'}(J_k)
$$

$$
+ \sum_{K'Q'} T_{\mathrm{S}}(k; K'Q' \to j; KQ)\, \rho_{Q'}^{K'}(J_k) - \sum_{K'Q'} R_{\mathrm{A}}(j; KQ, K'Q' \to k)\, \rho_{Q'}^{K'}(J_j)
$$

$$
- \sum_{K'Q'} R_{\mathrm{E}}(j; KQ, K'Q' \to i)\, \rho_{Q'}^{K'}(J_j) - \sum_{K'Q'} R_{\mathrm{S}}(j; KQ, K'Q' \to i)\, \rho_{Q'}^{K'}(J_j) . \qquad (17.31)
$$

The full expressions for the different rates cannot be written down explicitly here. Just to give an example, the transfer rate for absorption to level j from lower level i is given by

$$
T_{\mathrm{A}}(i; K'Q' \to j; KQ) = (2J_i + 1)B(J_i \to J_j) \sum_{K_{\mathrm{r}} Q_{\mathrm{r}}} \sqrt{3(2K+1)(2K'+1)(2K_{\mathrm{r}}+1)}
$$

$$
\times (-1)^{K'+Q'}
\begin{pmatrix} J_j & J_i & 1 \\ J_j & J_i & 1 \\ K & K' & K_{\mathrm{r}} \end{pmatrix}
\begin{pmatrix} K & K' & K_{\mathrm{r}} \\ -Q & Q' & -Q_{\mathrm{r}} \end{pmatrix} J_{Q_{\mathrm{r}}}^{K_{\mathrm{r}}}(\nu_{ji}) ,
$$

where $J_{Q_{\mathrm{r}}}^{K_{\mathrm{r}}}(\nu_{ji})$, with $K_{\mathrm{r}} = 0, 1, 2$, and $Q_{\mathrm{r}} = -K_{\mathrm{r}}, ..., K_{\mathrm{r}}$, is the radiation field tensor at the frequency ν_{ji}. The nine independent components of this tensor are integrals over the solid angle of the Stokes parameters weighted by suitable angular functions. In the case of a cylindrically symmetric, unpolarized radiation field, only two components are different from zero, namely

$$J_0^0(\nu) = \oint I_\nu(\vec{\Omega}) \frac{d\Omega}{4\pi} , \qquad J_0^2(\nu) = \frac{1}{\sqrt{2}} \oint (3\cos^2\theta - 1) I_\nu(\vec{\Omega}) \frac{d\Omega}{4\pi} ,$$

where θ is the angle between $\vec{\Omega}$ and the symmetry axis of the radiation field.

The radiative transfer and the statistical equilibrium equations that have been sketched here stand at the basis of a large number of physical applications. They constitute, without any doubt, the most direct generalization to spectro-polarimetry of the "classical" equations that have been used for a long time for the interpretation of the spectra of stars and other astrophysical plasmas.

18. The Amplification Condition in Polarized Radiative Transfer

Stimulated emission provides the possibility of obtaining very intense and directive beams of electromagnetic radiation when some sort of pumping process is capable of determining an inversion of populations between two levels that are connected by an electric dipole transition. This phenomenon is at the basis of the artificial devices that are nowadays known as lasers and masers, these acronyms standing for "light (or microwave) amplification by stimulated emission of radiation", respectively.

Masers are also known to exist in several astrophysical environments and it has to be expected on general physical grounds that the phenomena of dichroism and anomalous dispersion may play an important role in the physics of these objects.

When polarization is neglected, radiative transfer is simply described by Eq.(17.28). In order to get amplification during the propagation, the coefficient of the intensity in the r.h.s. of the same equation has to be negative. This immediately yields the amplification condition (or masing condition) in the form

$$k = k^{(A)} - k^{(S)} < 0 .$$

When polarization phenomena are accounted for, the radiative transfer equation is more complicated, being given by Eq.(17.30). In this case, to determine the amplification properties of the medium, one has to evaluate the eigenvalues of the propagation matrix. A direct calculation shows that two out of the four eigenvalues are real, whereas the other two are complex. Their explicit expression is the following (Landi Degl'Innocenti & Landi Degl'Innocenti, 1985).

$$\lambda_1 = \eta_I - \Lambda_+ , \qquad \lambda_2 = \eta_I + \Lambda_+ ,$$

$$\lambda_3 = \eta_I - i\Lambda_- , \qquad \lambda_4 = \eta_I + i\Lambda_- ,$$

where

$$\Lambda_+ = \sqrt{\sqrt{\frac{(\eta^2 - \rho^2)^2}{4} + (\vec{\eta} \cdot \vec{\rho})^2} + \frac{\eta^2 - \rho^2}{2}} ,$$

$$\Lambda_- = \sqrt{\sqrt{\frac{(\eta^2 - \rho^2)^2}{4} + (\vec{\eta} \cdot \vec{\rho})^2} - \frac{\eta^2 - \rho^2}{2}} ,$$

and where the formal vectors $\vec{\eta}$ and $\vec{\rho}$ are defined by

$$\vec{\eta} = (\eta_Q, \eta_U, \eta_V) , \qquad \vec{\rho} = (\rho_Q, \rho_U, \rho_V) .$$

Each of the four eigenvalues corresponds to a particular eigenvector (a polarization mode of the propagation matrix), and it can be shown that the polarization modes corresponding to the eigenvalues λ_1 and λ_2 are totally polarized (in the usual sense where $I^2 - Q^2 - U^2 - V^2 = 0$).

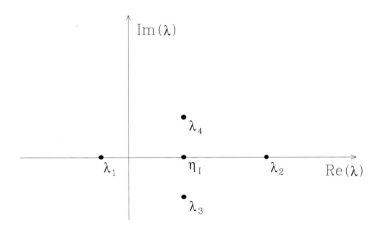

FIGURE 18. The four eigenvalues of the propagation matrix are found at the vertices of a diamond in the complex plane. The particular case drawn in the figure is labelled as case (c) in the text.

The masing condition has now to be imposed on the eigenvalues, or, more precisely, on their real parts. In the complex plane of the variable λ the four eigenvalues are located at the four vertices of a diamond whose center is on the real axis at $\lambda = \eta_I$, and whose horizontal and vertical diagonals are given, respectively, by $2\Lambda_+$ and $2\Lambda_-$ (see Fig. 18). Overlooking special cases, there are four different alternatives concerning the signs of the eigenvalues:

$$(a) \qquad \lambda_1 < \eta_I < \lambda_2 < 0 \quad,$$

$$(b) \qquad \lambda_1 < \eta_I < 0 < \lambda_2 \quad,$$

$$(c) \qquad \lambda_1 < 0 < \eta_I < \lambda_2 \quad,$$

$$(d) \qquad 0 < \lambda_1 < \eta_I < \lambda_2 \quad.$$

In case (a) all four eigenvalues are negative, so there is masing action in the four polarization modes. The number of negative eigenvalues decreases to 3 in case (b), and to 1 in case (c). Finally, in case (d) masing action is impossible because all four eigenvalues are positive. A necessary and sufficient condition for masing action in polarized radiative transfer is thus summarized by the inequality

$$\lambda_1 < 0 \quad.$$

In concluding this section, it is important to notice that the presence of dichroism (or of anomalous dispersion, or of both) has the universal tendency of lowering the condition for masing action, or, in other words, the production of maser radiation is made easier, in a given plasma, by the presence of dichroism. This can be understood by looking to Fig. 18. If one supposes, in a kind of *gedankenexperiment*, of artificially suppressing dichroism and anomalous dispersion (by equating to zero all the non-diagonal elements of the propagation matrix), the diamond degenerates into a single point located in the half-plane $\lambda > 0$, and the masing action thus disappears. It has then be stated on general physical grounds that nature has a tendency to prefer *dichroic masers* to ordinary masers.

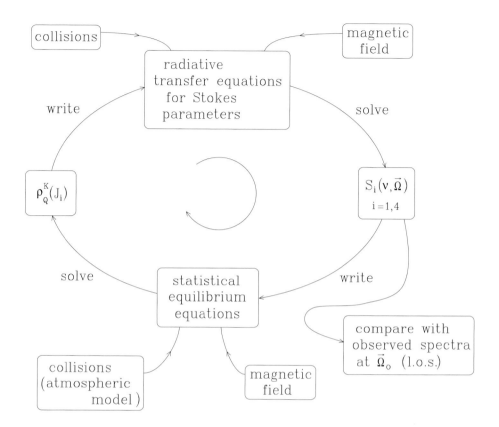

FIGURE 19. The self-consistent loop for the solution of the non-LTE problem of the 2nd kind. Note that collisions and the magnetic field play a double role, entering both into the radiative transfer equations (splitting and broadening the profiles) and into the statistical equilibrium equations (through the magnetic term and the collisional rates that have to be added to the radiative rates of Eq.(17.31)).

19. Coupling Radiative Transfer and Statistical Equilibrium Equations

It has been realized since a long time that the interpretation of the spectra of stars and of other astrophysical objects cannot be carried on, in many cases, within the framework of the LTE (Local Thermodynamical Equilibrium) hypothesis. On the contrary, non-LTE calculations, based on the self-consistent solution of the radiative transfer and of the statistical equilibrium equations, are often required.

Similarly, for the interpretation of spectro-polarimetric observations it is often necessary to find the self-consistent solution of the basic equations that we have discussed in Sect. 17, namely, the transfer equations for the Stokes parameters and the statistical equilibrium equations for the density-matrix components (or, alternatively, for the statistical tensors).

This procedure, that can be simply referred to as *Non-LTE of the 2nd kind* is schematically summarized in Fig. 18. Referring for simplicity to the standard case of a plane-parallel stellar atmosphere whose thermodynamical parameters are specified by a suitable

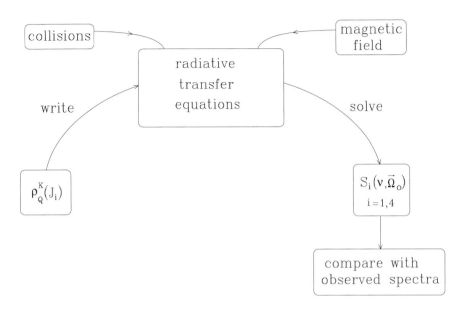

FIGURE 20. The self-consistency loop is avoided by the LTE assumption which sets the density-matrix elements to fixed values, independent of the radiation field.

model, one starts by assuming a zero-order solution for the local values of the density-matrix elements as a function of optical depth. This allows finding the expression of the various coefficients appearing in the radiative transfer equation for polarized radiation, namely the dichroism coefficients (η_I, η_Q, η_U, and η_V), the anomalous dispersion coefficients (ρ_Q, ρ_U, and ρ_V), and the emission coefficients in the four Stokes parameters (ϵ_I, ϵ_Q, ϵ_U, and ϵ_V). The radiative transfer equations can thus be solved for any direction, thus yielding the values of the Stokes parameters as a function of frequency, of optical depth, and of direction. From the Stokes parameters one can then find the expressions of the radiation field tensor at any optical depth, and, consequently, the expressions of the different rates, namely the transfer rates (T_A, T_E, and T_S), and the relaxation rates (R_A, R_E, and R_S), appearing in the statistical equilibrium equations for the density-matrix elements. Such equations, once implemented by adding the collisional rates, can then be solved to find the first-order solution of the density-matrix elements. The procedure can then be iterated until the self-consistent solution is reached.

Obviously, the search for the self-consistent solution is non-trivial, and only quite recently it has been possible to generalize to the problem of non-LTE of the 2nd kind highly-convergent, iterative methods that had been previously developed to treat the ordinary non-LTE multilevel problem (Trujillo Bueno & Landi Degl'Innocenti, 1997, Trujillo Bueno, 1999, Manso Sainz & Trujillo Bueno, 2001). These radiative transfer techniques look very promising for the interpretation of solar disk observations and, in particular, for the interpretation of the second solar spectrum in quiet and active regions (see the review by Trujillo Bueno, 2001).

Fortunately, the self-consistent solution of the non-LTE loop turns out to be unnecessary for the interpretation of different types of spectro-polarimetric observations. Indeed,

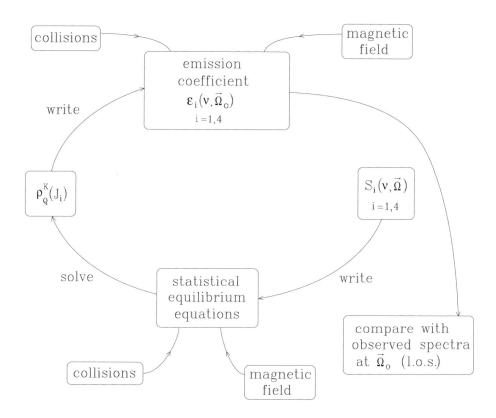

FIGURE 21. In an optically thin plasma, like for instance in a prominence, the radiation field which illuminates the atomic system is known a priori. Hence there is no need of solving the self-consistency loop of Fig. 19.

when the LTE hypothesis is satisfied, the density-matrix elements can be simply deduced at any optical depth from the atmospheric model. Given the temperature T, the statistical tensors are given by

$$\rho_Q^K(J_i) = \delta_{K,0}\delta_{Q,0}\sqrt{2J_i + 1}\ \exp\left[-\frac{E(J_i)}{k_B T}\right]\frac{1}{Z(T)}\quad,$$

where $E(J_i)$ is the energy of level J_i and where $Z(T)$ is a function which depends on the normalization condition imposed on the density-matrix. When the density-matrix is normalized to unity, one has

$$Z(T) = \sum_i \sqrt{2J_i + 1}\ \exp\left[-\frac{E(J_i)}{k_B T}\right]\quad.$$

As schematically shown in Fig. 20, one only needs to solve the radiative transfer equations to obtain the Stokes parameters of the radiation emerging from the stellar atmosphere.

This kind of procedure is at the basis of a whole branch of research in solar physics, namely *solar magnetometry*, or the "art" of measuring magnetic fields through spectro-polarimetric observations in photospheric lines. Indeed, the LTE hypothesis is generally satisfied to describe the process of line formation of the spectral lines that are commonly employed for the diagnostic of solar magnetic fields at the photospheric level.

Another case where the self-consistent solution of the non-LTE loop is unnecessary, is the case of an optically thin plasma, illuminated by a a radiation field whose spectral and directional properties are known a priori.

As shown in Fig. 21, one now only needs to solve the statistical equilibrium equations for the density-matrix elements. Once these quantities are known, a simple substitution yields the emission coefficient in the four Stokes parameters, hence the polarization profiles of the emerging radiation, apart from a factor proportional to the optical depth τ of the emitting plasma (with $\tau \ll 1$).

This procedure is used for the interpretation of resonance polarization observations and for the diagnostic of magnetic fields in prominences and in the corona.

Acknowledgments

For the redaction of this course I have drawn on several textbooks or monographs which I feel necessary to quote here also to give to the interested reader some bibliographical basis for a possible deepening of the subject. Apart from the volumes explicitly quoted in the text (Landau & Lifchitz, 1966, Swyndell, 1975), I wish to acknowledge: for Sects. 2-3, Shurcliff (1962), Clarke & Grainger (1971), and Collett (1993); for Sect. 4, Born & Wolf (1964), and Jenkins & White (1976); for Sect. 6, Minnaert (1959), and Leroy (1998); for Sects. 7 to 14, Rybicki & Lightman (1979); for Sect. 11, Van de Hulst (1981).

REFERENCES

BABCOCK, H.W. 1947 Zeeman effect in stellar spectra. *The Astrophys. Journ.* **105**, 105–119

BOMMIER, V. 1977 Étude théorique de l'effet Hanle; traitement du cas de la raie D$_3$ de l'Hélium en vue de la détermination du champ magnétique des protubérances solaires. Thèse de 3ème cycle, Université de Paris

BOMMIER, V. 1980 Quantum theory of the Hanle effect II: Effect of level-crossings and anti-level crossings on the polarization of the D$_3$ helium line of solar prominences. *Astron. Astrophys.* **87**, 109–120

BOMMIER, V. 1991 Derivation of the radiative transfer equation for line polarization studies in the presence of magnetic field in astrophysics. *Ann. Phys. Fr.* **16**, 599–622

BOMMIER, V. & SAHAL-BRÉCHOT, S. 1991 Derivation of the master equation for the atomic density matrix for line polarization studies in the presence of magnetic field and depolarizing collisions in astrophysics. *Ann. Phys. Fr.* **16**, 555-598

BORN, M. & WOLF, E. 1994 Principle of Optics. Pergamon Press, Oxford

CLARKE, D. & GRAINGER, J.F. 1971 Polarized Light and Optical Measurements. Pergamon Press, Oxford

COLLETT, E. Polarized Light. Marcel Dekker Inc., New York, Basel

HALE, G.E. 1908 On the probable existence of a magnetic field in sun-spots. *The Astrophys. Journ.* **28**, 315–343

JENKINS, F.A. & WHITE, H.E. 1976 Fundamentals of Optics. McGraw-Hill, New York, etc.

KEMP, J.C., SWEDLUND, J.B., LANDSTREET & J.D., ANGEL, J.R.P 1970 Discovery of circularly polarized light from a white dwarf. *The Astrophys. Journ. Lett.* **161**, L77–L79

LANDAU, L. & LIFCHITZ, E. 1966 Théorie du Champ. Éditions Mir, Moscou

LANDI DEGL'INNOCENTI, E. 1983 Polarization in spectral lines. I: A unifying theoretical approach. *Solar Phys.* **85**, 3–31

LANDI DEGL'INNOCENTI, E. 1984 Polarization in spectral lines. III: Resonance polarization in the non-magnetic, collisionless regime. *Solar Phys.* **91**, 1–26

LANDI DEGL'INNOCENTI, E. 1985 Polarization in spectral lines. IV: Resonance polarization in the Hanle effect, collisionless regime. *Solar Phys.* **102**, 1–20

LANDI DEGL'INNOCENTI, E. & LANDI DEGL'INNOCENTI, M. 1972 Quantum theory of line formation in a magnetic field. *Solar Phys.* **27**, 319–329

LANDI DEGL'INNOCENTI, E. & LANDI DEGL'INNOCENTI, M. 1975 Transfer equations for polarized light. *Il Nuovo Cimento* **27B**, 134–144

LANDI DEGL'INNOCENTI, E. & LANDI DEGL'INNOCENTI, M. 1985 On the solution of the radiative transfer equations for polarized radiation. *Solar Phys.* **97**, 239–250

LANDI DEGL'INNOCENTI, E. & DEL TORO INIESTA, J.C. 1998 Physical significance of experimental Mueller matrices. *Journ. Opt. Soc. Amer* **A 15**, 533–537

LANDI DEGL'INNOCENTI, E., BOMMIER & V., SAHAL BRÉCHOT, S. 1990 Resonance line polarization and the Hanle effect in optically thick media. I. Formulation for the two-level atom. *Astron. Astrophys.* **235**, 459–471

LANDI DEGL'INNOCENTI, E., BOMMIER, V. & SAHAL BRÉCHOT, S. 1991 Resonance line polarization for arbitrary magnetic fields in optically thick media. I. Basic formalism for a 3-dimensional medium. *Astron. Astrophys.* **244**, 391–400

LEROY, J.L. 1998 La Polarisation de la Lumière et l'Observation Astronomique. Gordon and Breach, Amsterdam

MANSO SAINZ, R. & TRUJILLO BUENO, J. 2001 Modelling the scattering line polarization in the CaII IR triplet. In *Advanced Solar Polarimetry*, (ed. M. Sigwarth). ASP Conference Series, in press.

MINNAERT, M. 1959 Light and Colour in the Open Air. Bell, London

RYBICKI, G.B. & LIGHTMAN, A.P. 1979 Radiative Processes in Astrophysics. John Wiley & Sons, New York, etc.

SHURCLIFF, W.A. 1962 Polarized Light. Harvard University Press, Harvard, Mass.

STOKES, G.G. 1852 On the composition and resolution of streams of polarized light from different sources. *Trans. Cambridge Phil. Soc.* **9**, 399-416

SWYNDELL, W. 1975 Polarized Light. Dowden, Hutchinson & Ross, Inc., Stroudsbourgh, Pennsylvania

TRUJILLO BUENO, J. 1999 Towards the modelling of the second solar spectrum. In *Solar Polarization*, (eds. K.N. Nagendra and J.O. Stenflo), pp. 73-96. Kluwer Academic Publishers.

TRUJILLO BUENO, J. 2001 Atomic polarization and the Hanle effect. In *Advanced Solar Polarimetry*, (ed. M. Sigwarth). ASP Conference Series, in press.

TRUJILLO BUENO, J. & LANDI DEGL'INNOCENTI, E. 1997 Linear polarization due to lower-level depopulation pumping in stellar atmospheres. *Astrophys. Journ. Lett.* **482**, L183–L186

VAN DE HULST, H.C. 1981 Light Scattering by Small Particles. Dover, New York

Polarized Radiation Diagnostics
of Solar Magnetic Fields

By **JAN OLOF STENFLO**

Institute of Astronomy, ETH Zentrum, CH-8092 Zurich, Switzerland

The Sun is unique as an astrophysics laboratory because we can spatially resolve its structures in great detail and apply sophisticated diagnostic techniques that require high spectral resolution. The magnetic flux in the solar atmosphere occurs in extremely fragmented, nearly fractal form, with a range of spatial scales that extend well beyond the angular resolution limit of current telescopes and into the optically thin regime. The magnetic field leaves various kinds of "fingerprints" in the polarized spectrum. In the past only the fingerprints of the Zeeman effect have been used, but more recently new, highly sensitive imaging polarimeters have given us access to other physical effects. In particular a wealth of previously unknown spectral structures due to coherent scattering processes have been uncovered. These phenomena show up in linear polarization as a new kind of spectrum (the so-called "second solar spectrum"), which bear little resemblance to the ordinary intensity spectrum. Magnetic fields modify the coherent scattering processes and produce polarized spectral signatures that greatly extend the diagnostic range of the Zeeman effect. This diagnostic window has just been opened, and we are only now beginning to develop the needed diagnostic tools and apply them to learn about previously "invisible" aspects of solar magnetic fields.

1. The Sun's magnetic field — An introductory overview

1.1. *Role of magnetic fields in astrophysics*

Most of the matter in the universe, like stars, nebulae, and interstellar matter, consists of *plasma*, partially ionized gas with high electrical conductivity. Due to the strong induction effects, the magnetic fields and the medium in which they are embedded evolve together as one single entity. Magnetic fields are amplified by the motions of the medium, and they back-react on the medium via the Lorentz force. The equations describing the dynamics are therefore coupled to the Maxwell equations. This complex interplay between the vector quantities of the magnetic and velocity fields and the scalar quantities of temperature and density is responsible for much of the structuring and variability that we see in the universe on short and intermediate time scales. The magnetic fields control star formation and the dynamics of the interstellar medium, govern stellar activity, and are responsible for particle acceleration, like cosmic rays. The Sun provides us with an astrophysics laboratory, in which these various physical processes can be explored in detail. For this reason the Sun has often been called the "Rosetta stone" of astrophysics.

Although the solar plasma occupies similar locations in a temperature-density diagram as laser plasmas, Tokamak plasmas, and gas discharge plasmas (Petrasso 1990), one cannot simulate solar conditions in the laboratory. The reason is that the physics depends on the spatial dimensions (cf. Alfvén & Fälthammar 1963). If the linear scale is changed by a factor γ, Maxwell's equations demand that the time scale is also changed by the same factor. Imposing the condition that the energies are left unchanged, the product between the electric field $\vec{E}$ and the length scale ℓ must be left invariant, which implies that $\vec{E}$ scales with γ^{-1}. Maxwell's equations then demand that the magnetic field $\vec{B}$ also scales with γ^{-1}. If we for instance would want to simulate in the laboratory the conditions in a small sunspot, with a size of 10,000 km, life time of 10^6 s (about two weeks), and a field strength of a few thousand G, and the laboratory plasma to be used

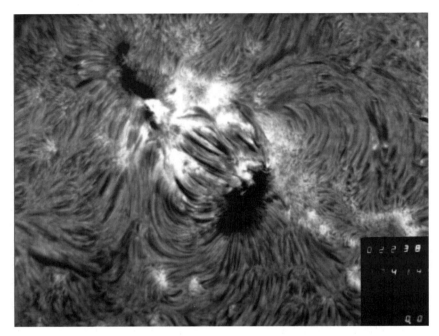

FIGURE 1. Image of the solar chromosphere taken in the hydrogen Hα line. Since the chromosphere is magnetically controlled, the emission features align themselves along the magnetic field lines connecting two sunspots in a newly emerged active region. Courtesy of H. Zirin, Big Bear Solar Observatory.

has a dimension of 1 m, then the scaling factor γ is 10^7, which means that the field strength should be a few times 10^{10} G over a life time of 0.1 s. Laboratory analogs of solar and other cosmic plasmas are therefore far out of reach of current technology. The Sun is a unique physical laboratory, in which a domain of physics can be explored that is not accessible by other experimental means.

1.2. *Role of magnetic fields for the structuring and thermodynamics of the solar atmosphere*

It is evident from Hα filtergrams with their striated fibril patterns (cf. Fig. 1), or from the appearance of the corona, with its complex streamer structure (cf. Fig. 2), that the outer solar atmosphere is magnetically controlled. In contrast, white-light images of the Sun show a granular pattern of the solar surface, which represents a convective pattern produced by non-magnetic hydrodynamic processes. Still this lower part of the solar atmosphere is magnetized, but here the magnetic fields play a more passive role.

The transition between a hydrodynamically controlled and a magnetically controlled atmosphere depends on the plasma β, which represents the ratio between the gas pressure and the magnetic pressure of the plasma. Since the gas pressure decreases with height almost exponentially, much faster than the magnetic field, the plasma β suddenly drops way below unity when we go from the photosphere to the chromosphere. In the photospheric layers, where the bulk of the visible spectrum is formed, the temperature decreases outwards, reaching a minimum, and then increases steeply in the chromosphere, until temperatures of 1–2 million degrees are reached in the corona. In the chromosphere and corona the magnetic forces dominate entirely over the gas dynamic forces and dictate how the plasma can move and how the heating processes are channeled.

FIGURE 2. Eclipse photographs reveal that the Sun is a magnetized sphere. The shape of the coronal structures are governed by magnetic fields that are anchored in the Sun. Courtesy of G. Newkirk, Jr., High Altitude Observatory, Boulder, Colorado.

Due to the extremely high electrical conductivity of the solar plasma the magnetic field lines can generally be considered to be "frozen" to the gas — the field and the plasma move together as one single medium. Inside the Sun the field lines are passively tangled up and amplified by the convective motions and differential rotation. Magnetic flux tubes are highly buoyant and try to overcome their own tension forces that hold them down, to float up to the surface. When they emerge into the photosphere they form bipolar magnetic regions that can be observed. The corresponding field lines are firmly anchored in the solar interior but arch into the corona to connect the opposite polarities in the photosphere. When the polarity separation is sufficiently large the field lines arch so high that they can be pulled out by the expanding corona. As the corona is maintained at temperatures in the million degree range by various heating processes, its thermal gas pressure is kept so high that it cannot be contained by gravitation. It therefore has to expand into a supersonic wind that blows through the solar system, much beyond the orbit of Pluto, before it is braked by the interstellar medium. In the outer corona the solar wind pressure can overcome the magnetic tension forces and open the magnetic arches to form coronal streamers.

In terms of their topological connections the coronal field can be classified in terms of "open" and "closed" magnetic structures, which determine the temperature-density structure of the corona as seen by X-ray telescopes (cf. Fig. 3). The closed loops form magnetic "bottles", in which the plasma is trapped and where heat accumulates. They appear as intensely bright structures in X-ray radiation. In the open, diverging magnetic field regions, on the other hand, the plasma can freely escape as the solar wind. These regions appear dark in X-rays and are the so-called "coronal holes".

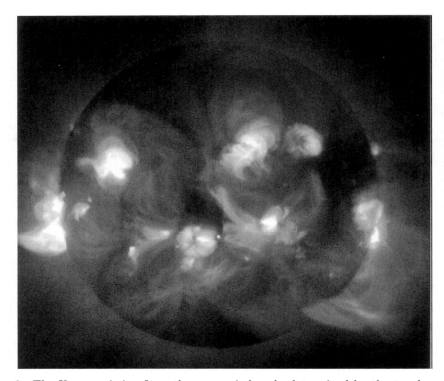

FIGURE 3. The X-ray emission from the corona is largely determined by the topology of the magnetic field. While the bright regions correspond to closed magnetic "bottles" or loops, the dark regions or coronal holes correspond to open magnetic fields, where the plasma can escape into interplanetary space. Courtesy of Y. Uchida, University of Tokyo.

1.3. *Dynamo generation of cosmic magnetic fields and the origin of the Sun's activity cycle*

Dynamo theory provides a general framework for understanding how macroscopic magnetic fields are produced in various cosmical objects, like in the Earth and planets, in the Sun and stars, or in the Milky Way and other galaxies. Magnetic fields may be amplified by induction effects in electrically conducting media (plasmas). Turbulent motions can amplify the magnetic field on small scales, but to build up large-scale magnetic fields one needs some large-scale ordering principle. This is provided by the Coriolis forces if the medium is rotating. They break the left-right symmetry of the turbulent convection and make it cyclonic, with opposite screw directions in the two hemispheres. It is thus the interaction between magnetic fields, turbulence, and rotation that forms the basis of dynamo operation. The planets, stars and galaxies are rotating and have turbulent and electrically conducting regions. For the galaxies it is the ionized gas in the interstellar medium that is turbulent and conductive, for the planets it is their semi-fluid iron cores that are slowly convective.

Depending on the details of its operation, the dynamo can produce either a steady state or an oscillating state. The Sun presents us with a unique astrophysics laboratory, in which we can explore the dynamo processes in detail. The solar activity cycle can be seen as the expression of a dynamo that oscillates with a period of 22 years, the magnetic "Hale cycle" (which is twice as long as the sunspot activity cycle, since the magnetic polarities reverse each 11 years). More specifically the Sun's global magnetic field oscillates between

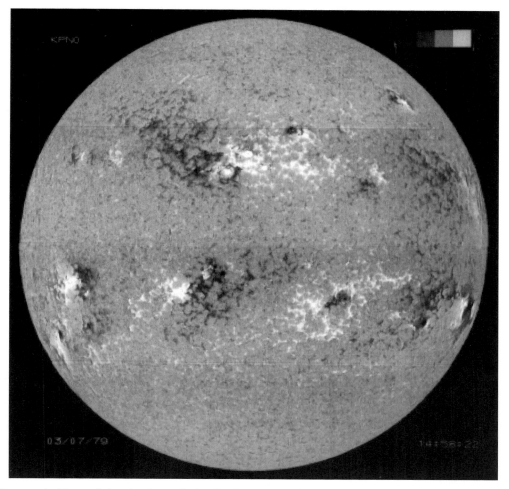

FIGURE 4. Maps of the circular polarization in a spectral line sensitive to the Zeeman effect are called magnetograms, since they show the spatial distribution of the line-of-sight component of the magnetic field. In the bright areas the field is directed towards the observer, in the dark areas it is directed away. Heliographic north is at top, east is at left. The global pattern of the magnetic fields evolve with the 22 yr Hale cycle. Courtesy of J.W. Harvey, National Solar Observatory, Tucson.

a poloidal and a toroidal state. Starting with a poloidal, dipole-like field (which only has components in meridional planes), the differential rotation of the Sun winds up the field lines and thereby builds up a toroidal (azimuthal) field. When toroidal flux ropes float up to the surface, they form bipolar sunspot groups. Cyclonic convection acting on the toroidal field in a statistical manner induces contributions to a new poloidal field that has opposite polarity with respect to the previous field. Through turbulent diffusion the new poloidal field lines spread by a 3-D random walk process to become a new global, dipole-type field that replaces the old one and has the opposite polarity. The resulting complexity of the pattern of the surface magnetic fields is illustrated in Figs. 4 and 5. The magnetic flux can be mapped through the use of the Zeeman effect, but it has a highly intermittent structure, with "building blocks" or flux elements that are of small scales

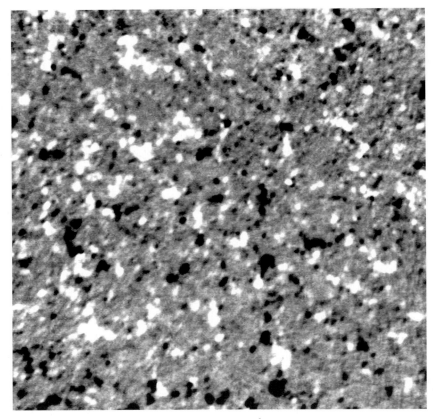

FIGURE 5. "Deep" magnetogram of a $280 \times 280\,\mathrm{arcsec}^2$ portion of the quiet Sun at the center of the solar disk, obtained on July 1, 1998, by Jongchul Chae at Big Bear Solar Observatory (BBSO). The opposite-polarity magnetic fluxes are represented by the brighter and darker patches against a neutral grey background. Courtesy of BBSO / New Jersey Institute of Technology.

beyond the spatial resolution limit of current telescopes. The main diagnostic problem is to determine the intrinsic properties of the spatially unresolved magnetic fields.

Since toroidal flux ropes are highly buoyant in the solar convection zone, it is now believed that the solar dynamo primarily operates in the stably stratified overshoot region just below the convection zone, where the dynamo-generated fields can be stored long enough to be sufficiently amplified by the dynamo processes. Inversions of helioseismic observations have shown that the Sun's angular rotation has a fairly abrupt transition near the bottom of the convection zone, from being differential above to rigid below. The large radial shear in the rotation rate is favorable for the generation of toroidal fields from poloidal ones. For a more comprehensive introduction to solar physics we refer to the book "The Sun" by Stix (1989).

2. Diagnostic techniques — An introductory overview

2.1. *Remote sensing of the Sun: Statement of the inversion problem*

The information about the physical conditions on the Sun and stars is encoded in the radiation that we receive from them. The signatures of the different physical parameters are imprinted in subtle and non-linear ways on the profiles of the many spectral lines that make stellar spectra so richly structured. The information is encoded not only in the intensity but also in the state of polarization of the radiation as a function of wavelength. The full state of polarization can be completely specified by the four Stokes parameters I, Q, U, and V, which will be defined below. Together they form a 4-vector. The task of the observations is to record this 4-vector with the highest possible precision, spectral, angular, and temporal resolution. From the observables measured in this way we want to extract the entangled information and deduce the physical state of the remote region, where the observed radiation originates. This is the inversion problem.

The direct problem of calculating synthetic Stokes spectra from a given model of a stellar atmosphere is of course much simpler. The problem of remote diagnostics is the inverse of this and is usually approached stepwise. First we use the observations to select between rivaling models. The remaining models may be described in terms of free parameters, which are constrained by the observations. With increasing quality of the observational constraints increasingly realistic and sophisticated models can be introduced, while keeping them sufficiently simple so that their free parameters can be fixed in a unique way by the observations.

Since stellar surfaces are in general not resolved and the photon flux is smaller, the models used to invert the observations must necessarily be crude in comparison with those of solar observations, for which both the solar disk and the solar spectrum can be resolved in great detail. The level of ambition for realism is therefore enormously much higher in the solar than in the stellar case (cf. the monograph by Stenflo 1994).

2.2. *Accessibility of different atmospheric layers*

The interior of stars are opaque to electromagnetic radiation, which diffuses outwards from the energy-producing core in a random-walk process of absorption and emission until an outer region is reached, where the mean free path of the photons increases to infinity and they can leave the star. The transition region where this occurs and which is the region accessible to direct observation is what we call a stellar atmosphere. The photons emitted from a certain depth in the atmosphere are attenuated by absorption by the factor $e^{-\tau_\lambda}$, where τ_λ is the wavelength-dependent optical depth. The probability is highest that the photons we see have come from optical depth unity. The corresponding geometrical depth is highly wavelength dependent. While the continuous spectrum comes from the deepest layers, the radiation in spectral lines emanates from different layers, depending on the line strength, ionization and excitation potentials, position within the line profile, temperatures, densities, etc. This has great diagnostic advantages. With combinations of lines that are formed at different heights in the atmosphere we may reconstruct the height variation of the physical parameters of the atmosphere. With combinations of lines that are formed at the same heights but have different non-linear responses to the various physical parameters we may remove ambiguities in the interpretation and untangle the various parameters. In particular we can extract information beyond the angular resolution limit in observations of solar Stokes spectra.

2.3. *Tools and concepts of the measurement problem*

2.3.1. *Operational definition of the Stokes parameters*

The Stokes parameters, which provide a complete description of a partially polarized light beam, can be defined in a variety of ways, which are all equivalent but differently suited for use in different contexts. Thus we can use definitions in terms of classical electromagnetic theory or in terms of quantum theory. Here we will limit ourselves to the conceptually simplest and most concrete definition, namely the operational one in

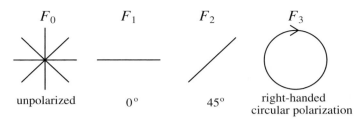

F_0 F_1 F_2 F_3

unpolarized $0°$ $45°$ right-handed circular polarization

FIGURE 6. Symbolic representation of the four idealized filters F_k used for the operational definition of the four Stokes parameters.

terms of four idealized filters in a measurement process. This definition most directly relates to the design and understanding of polarimetric observations.

The meaning of the four Stokes parameters I, Q, U, and V can be visualized by considering the effect of four different, idealized filters, F_k. Here index $k = 0, 1, 2, 3$ refers to each of I, Q, U, and V, respectively. The filter properties are illustrated in Fig. 6. While filter F_0 represents empty space without any absorbing effects, F_1 and F_2 transmit linear polarization with the electric vector at position angles 0 and $45°$, respectively, and F_3 transmits right-handed circular polarization. Since each of F_1, F_2, and F_3 block the orthogonal polarization state, they reduce the intensity of an unpolarized beam by a factor of two. The four intensity readings I_k transmitted by each filter uniquely determine the full polarization state of the incident beam. In terms of the Stokes parameters,

$$
\begin{aligned}
I_0 &= I\,, \\
I_1 &= \tfrac{1}{2}(I + Q)\,, \\
I_2 &= \tfrac{1}{2}(I + U)\,, \\
I_3 &= \tfrac{1}{2}(I + V)\,.
\end{aligned}
\tag{2.1}
$$

Let us for convenience denote the four Stokes parameters by S_k and form a Stokes vector

$$
\vec{I} = \begin{pmatrix} S_0 \\ S_1 \\ S_2 \\ S_3 \end{pmatrix} \equiv \begin{pmatrix} I \\ Q \\ U \\ V \end{pmatrix}\,.
\tag{2.2}
$$

Their relation to the measurements with the four filters can be expressed as $S_k = 2I_k - I_0$. If we for F_1 and F_2 had instead chosen filters that transmit linear polarization at $90°$ and $-45°$, and for F_3 left-handed circular polarization, then the signs of Q, U, and V would change. From this it follows that Stokes I represents the intensity, Stokes Q the intensity difference between horizontal and vertical linear polarization, Stokes U the intensity difference between linear polarization at $+$ and $-45°$, Stokes V the intensity difference between right- and left-handed circular polarization.

2.3.2. *Physical meaning of partial polarization*

Next let us introduce a set of orthogonal basis vectors $\vec{e}_1$ and $\vec{e}_2$ in a plane perpendicular to the light beam. At any point in space the electric vector $\vec{E}$ can be decomposed as

$$
\vec{E} = \mathrm{Re}\,(E_1 \vec{e}_1 + E_2 \vec{e}_2)\,,
\tag{2.3}
$$

where

$$
E_k = E_{0k} e^{-i\omega t}\,, \quad k = 1, 2,
\tag{2.4}
$$

and E_{0k} are complex amplitudes.

Now we define the *Jones vector* $\vec{J}$ as

$$\vec{J} = \begin{pmatrix} E_1 \\ E_2 \end{pmatrix}. \tag{2.5}$$

Since each of the complex numbers E_1 and E_2 has a real and an imaginary part, the Jones vector is, like the Stokes vector, characterized by a set of four parameters. There is a fundamental difference, however. While the Stokes vector can represent partially polarized light, the Jones vector can only represent 100 % elliptically polarized light. The reason is the following:

Each wave train (photon) is always 100 % elliptically polarized. In nature (except in lasers) we are however dealing with ensembles of mutually uncorrelated photons with different polarization states, since they have been created by stochastically independent atomic processes. A linear superposition of Jones vectors still represents 100 % elliptically polarized light, since the superposition is coherent or phase-preserving. To describe partially polarized light we need an *incoherent* superposition of the photons.

This is achieved by the superposition of bilinear products of Jones vectors for each individual wave train. In a bilinear product the factor $e^{-i\omega t}$ disappears. We can then define the 2×2 coherency matrix $\vec{D}$ of the radiation field by

$$\vec{D} = \langle \vec{J}\vec{J}^\dagger \rangle, \tag{2.6}$$

where $\vec{J}^\dagger$ denotes the adjoint of $\vec{J}$ (transposition and complex conjugation of $\vec{J}$), and the bracket represents averaging over a statistical ensemble of uncorrelated photons.

The relation between the coherency matrix and the Stokes parameters is given by

$$\vec{D} = \frac{1}{2} \begin{pmatrix} I + Q & U + iV \\ U - iV & I - Q \end{pmatrix}. \tag{2.7}$$

2.3.3. *Mueller calculus*

Let us now consider a wave train that enters a medium as Jones vector $\vec{J}$ and exits it as $\vec{J}'$. The relation between them can be described by the complex 2×2 matrix $\vec{w}$:

$$\vec{J}' = \vec{w}\vec{J}. \tag{2.8}$$

$\vec{w}$ is a property of the medium, and does not depend on whether we make a coherent or incoherent superposition of the many wave trains. Then the coherency matrix transforms as

$$\vec{D}' = \vec{w}\vec{D}\vec{w}^\dagger. \tag{2.9}$$

The transformation of the Stokes vector $\vec{I}$ by a medium can be described by the 4×4 *Mueller matrix* $\vec{M}$:

$$\vec{S}' = \vec{M}\vec{S}. \tag{2.10}$$

Using the relation between the Stokes parameters and the coherency matrix, we can express the Mueller matrix in terms of the Jones matrix $\vec{w}$. The resulting relation is

$$\vec{M} = \vec{T}\vec{W}\vec{T}^{-1}, \tag{2.11}$$

where the physical properties of the medium are contained in the matrix

$$\vec{W} = \vec{w} \otimes \vec{w}^* = \begin{pmatrix} w_{11}w_{11}^* & w_{11}w_{12}^* & w_{12}w_{11}^* & w_{12}w_{12}^* \\ w_{11}w_{21}^* & w_{11}w_{22}^* & w_{12}w_{21}^* & w_{12}w_{22}^* \\ w_{21}w_{11}^* & w_{21}w_{12}^* & w_{22}w_{11}^* & w_{22}w_{12}^* \\ w_{21}w_{21}^* & w_{21}w_{22}^* & w_{22}w_{21}^* & w_{22}w_{22}^* \end{pmatrix} \tag{2.12}$$

(the symbols $\otimes$ and $*$ denote tensor product and complex conjugation, respectively), and

$\vec{T}$ is a purely mathematical transformation matrix without physical contents, given by

$$\vec{T} = \begin{pmatrix} 1 & 0 & 0 & 1 \\ 1 & 0 & 0 & -1 \\ 0 & 1 & 1 & 0 \\ 0 & -i & i & 0 \end{pmatrix},$$

$$\vec{T}^{-1} = \frac{1}{2} \begin{pmatrix} 1 & 1 & 0 & 0 \\ 0 & 0 & 1 & i \\ 0 & 0 & 1 & -i \\ 1 & -1 & 0 & 0 \end{pmatrix}. \tag{2.13}$$

The Mueller matrix is a very flexible tool that describes a linear transformation of the Stokes vector in very general contexts. It is for instance used in the radiative transfer equation for the Stokes vector, both as the 4×4 absorption matrix that contains the Zeeman effect, and the scattering matrix that can include both coherent and incoherent scattering, partial frequency redistribution, and the Hanle effect. The Mueller matrix is also an important tool to characterize optical instruments. For an optical system, like a telescope, spectrograph, liquid crystal modulator, etc., one obtains the total polarization matrix $\vec{M}$ of the system through simple matrix multiplication of the Mueller matrices $\vec{M}_i$ of each individual optical component:

$$\vec{M} = \vec{M}_n \vec{M}_{n-1} \ldots \vec{M}_2 \vec{M}_1. \tag{2.14}$$

The whole optical system can then be treated as a "black box" characterized by a single 4×4 matrix, which may be calibrated for instance by inserting polarizing filters in front of the "black box", and measuring the Stokes vector at the exit of the system. For more details, see Stenflo (1994).

2.3.4. *Observational obstacles: Instrumental polarization, seeing, flat field, spatial, spectral, and temporal resolutions, photon statistics*

Using various types of polarization optics, including electrooptical polarization modulators, it is possible to measure the Stokes vector accurately. However, many spurious effects that originate outside the polarimeter system may degrade the measurements. Great care has to be taken to minimize these effects.

The aim is to have a polarimeter whose accuracy is only limited by photon statistics, which represents the fundamental limit for a given telescope system. The main spurious sources of noise come from seeing in the earth's atmosphere and from pixel-to-pixel sensitivity variations (gain table) in the detector. Since the seeing occurs at frequencies below a few hundred Hz, we can avoid spurious seeing-induced polarization effects by modulating the state of polarization in the kHz range. This demands that the detector system is equipped with fast buffers, between which the images in the different polarization states can be shifted around at kHz frequencies. Such systems now exist even for large-size CCD detectors (cf. Povel 1995). They allow the elimination of not only seeing but also gain table noise in the fractional polarization images Q/I, U/I, and V/I, since the identical pixels are used for the different polarization states. In this way it has been possible to routinely obtain polarization accuracies of better than 10^{-5} in the degree of polarization in combination with high spectral resolution (Stenflo & Keller 1996, 1997). At this level of accuracy entirely new physical effects on the Sun are brought out, which otherwise would be drowned by noise.

Spurious seeing-induced polarization signals can also be avoided without modulation by using a polarizing beam splitter that gives two orthogonally polarized images, which have identical seeing distortions since they are recorded simultaneously. Although the fractional polarization obtained as the ratio between the difference and the sum image would then be free from seeing-induced features, it is affected by the gain table, which cannot be determined by flat-field calibration to much better than 0.5 %. Since solar polarimetry almost always needs much higher precision, the measuring system should not be dependent on how accurately the flat field can be determined.

It is possible to eliminate the gain-table problem with a beam splitter system by making a second exposure in which the polarization states of the two images have been reversed by a wave plate, and combine the four images in a certain way. For this technique to work the two beams have to have identical distortions by the aberrations in the optical system. Also one beam splitter and two separate exposures only give us Stokes I and one of the other Q, U, or V parameters, but not all four of them. For Stokes vector polarimetry it is therefore inferior to fast modulation systems.

The Stokes vector that is measured is further corrupted by the polarizing properties of the telescope optics that precedes the optical package for polarization analysis. This instrumental polarization leads to cross talk between the Stokes parameters in a way that can be described by the Mueller matrix of the telescope. Unfortunately all of the world's major solar telescopes have obliquely reflecting elements and produce large and time-varying instrumental polarization. Real-time optical compensation has proven impractical for achieving the required precision, and telescope calibration with polarization optics in front of the telescope is also hardly feasible for large-aperture telescopes. Determination of the instrumental cross talk is therefore usually done with a combination of techniques, including calibration recordings in solar regions where the qualitative properties of the polarization signals (like their symmetries) can be assumed to be known and therefore may be used to determine certain cross talk terms and place constraints on the models of the telescope optical system.

Even if we overcome all these obstacles, we are fundamentally limited by the number of photons received. It is a common misunderstanding that the Sun always provides us with enough photons for our observations, in contrast to stellar observations, which are photon starved. However, the surface brightness of a stellar disk is independent of distance and depends only on the effective temperature of the surface. If we would barely resolve a solar-type star with one angular resolution element, then we would receive from that distant star the same number of photons as we receive from the nearby solar disk within the same resolution element. The difference is that we can fit many angular resolution elements on the solar disk, in contrast to stellar disks, for which very few have been resolved at all.

We may characterize an observational program in terms of a 4-D parameter space, spanned by the three resolutions (angular, spectral, and temporal), and the polarimetric (or photometric) accuracy. In the design of observing programs we always make major trade-offs between these four observing parameters, depending on the scientific objectives. Even for the largest solar telescopes that will be built in the foreseeable future (the next few decades) major trade-offs will still be necessary. Polarimetry with an accuracy of 10^{-5} requires according to Poisson statistics 10^{10} collected electrons, or, taking into account the typical optical transmission and detector efficiency, on the order of 10^{12} photons per spectral resolution element. This can only be achieved by making large trade-offs with the temporal and spatial resolutions, as illustrated in Fig. 7.

2.4. *Physical effects producing polarization signatures: Scattering polarization, Zeeman effect, Hanle effect*

Polarization is produced when the spatial symmetry is broken in the physical process that generates the radiation that we observe. The symmetry breaking can be caused by macroscopic magnetic and electric fields, by non-symmetrical components in a telescope

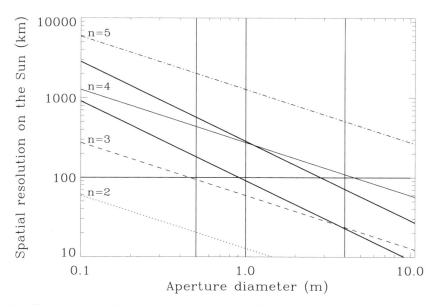

FIGURE 7. Illustration of the polarimetric accuracy that can be reached (limited by photon statistics), as a function of the diameter of the telescope aperture and the spatial resolution of the observations, assuming a spectral resolution of 300,000. The best spatial resolution that can be attained with a given aperture diameter is the diffraction limit given by the two thick, slanted lines for the wavelengths 1.56 μm (upper line) and 5000 Å (lower line). The thin, slanted lines connect points with equal polarimetric accuracy 10^{-n}, labeled by the parameter n. The horizontal line marks the 100 km scale, which is approximately the pressure scale height and the photon mean free path at the level of formation of the continuous spectrum. It is the next crucial length scale that we need to resolve in solar physics. From Stenflo (1999).

(e.g. oblique reflection), or by an anisotropic excitation process (radiative or collisional). Collisional excitation by directed particle beams would only occur in connection with solar flares and may lead to so-called impact polarization. Scattering at free electrons makes the white-light corona linearly polarized. Anisotropic radiative scattering also occurs down in the photosphere due to the limb darkening of the solar disk. For symmetry reasons such scattering polarization is zero at disk center and increases monotonically as we move towards the limb. The whole solar spectrum, both lines and continuum, is polarized by such scattering in the solar atmosphere, but only recently sufficiently sensitive instrumentation (ZIMPOL, **Z**urich **Im**aging **Pol**arimeter; cf. Povel 1995) has been developed that allows us to explore this scattering polarization. An example of what ZIMPOL reveals to us is seen in Fig. 8.

The appearance of the linearly polarized spectrum that is produced by coherent scattering processes is entirely different from the ordinary intensity spectrum. It is as if the Sun has presented us with an entirely new type of spectrum, and we have to start over again to identify the structures that we see. This new spectrum is therefore generally referred to as the "second solar spectrum", a name first introduced by Ivanov (1991).

An external magnetic field causes the atomic energy levels to split into different sublevels, and the emitted radiation gets polarized. This phenomenon is called the Zeeman effect (cf. Fig. 9). When atoms in a magnetic field scatter radiation via bound-bound transitions, the phase relations or quantum interferences between the Zeeman-split sublevels give rise to polarization phenomena that go under the name Hanle effect. Although

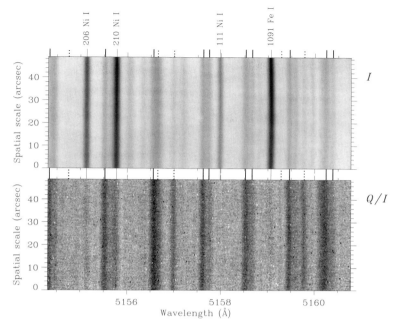

FIGURE 8. The top panel shows the intensity (Stokes I) spectrum, the bottom panel the simultaneously recorded degree of linear polarization (Stokes Q/I). The spectrograph slit has been placed parallel to and 5 arcsec inside the solar limb in a quiet region on the Sun. The linear polarization is here exclusively due to coherent scattering processes. In this portion of the spectrum the dominant contributors are molecules in the solar atmosphere, namely C_2 (marked by the thick, solid tick marks) and MgH (marked by the dashed tick marks). Due to its entirely different appearance, the linearly polarized spectrum that is produced by coherent scattering is called the "second solar spectrum".

Zeeman splitting of the atomic levels is a necessary requisite for the Hanle effect to occur, the usual terminology is to let "Zeeman effect" refer, besides the splitting of the energy levels, to the set of polarization phenomena that occur in the absence of coherence (or quantum interference) effects, while "Hanle effect" (Hanle 1924; Moruzzi & Strumia 1991) refers to the set of polarization phenomena that are produced by coherent scattering in a magnetic field.

The ordinary Zeeman effect, without atomic coherences produced by scattering, manifests itself in the circular polarization as the longitudinal Zeeman effect, which responds to the line-of-sight component of the magnetic field, in linear polarization as the transverse Zeeman effect, which carries information on the field component perpendicular to the line of sight. From measurements of the full Stokes vector it is thus possible in principle to derive both the strength and orientation of the field vector.

The Hanle effect occurs exclusively where there is scattering polarization. It represents the modification of the scattering polarization by magnetic fields. It responds to magnetic fields for which the ordinary Zeeman effect is insensitive or even "blind", like weak magnetic fields or fields of mixed polarities within the angular resolution element. The Hanle and Zeeman effects therefore complement each other in a rather ideal way.

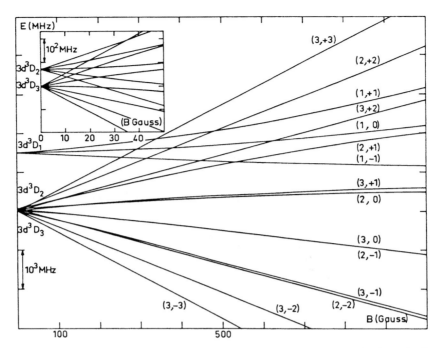

FIGURE 9. Energy level diagram for the excited magnetic sublevels involved in the formation of the He I D₃ 5876 Å line. Each line in the diagram is labeled by its (J, M) quantum numbers. The blown-up portion shows the region where level crossings occur. In this Paschen-Back regime the relation between splitting and field strength is no more linear. From Bommier (1980).

3. Zeeman-effect diagnostics

3.1. *Observational signatures of the Zeeman effect: Stokes spectra with FTS and imaging polarimeters*

With modern imaging polarimeters it is now possible to record images of the full Stokes vector as four images, one for each Stokes parameter. This can be done either via a narrow-band filter, which gives monochromatic images showing the spatial morphology at a selected wavelength, or via a spectrograph, showing the spectral dimension and one spatial dimension. The overall spectral signatures can be seen clearly in such Stokes images obtained in the spectral focus (cf. Fig. 10). The transverse Zeeman effect produces line profile signatures in Stokes Q and U that are nearly symmetric around the line centers. The σ components appear in the blue and red line wings with equal signs, the π component at line center with opposite sign. In contrast, the longitudinal Zeeman effect, which governs the appearance of the Stokes V images, has in first approximation anti-symmetric line profiles, since the σ components in the blue and red line wings have opposite signs.

The shapes of the polarized line profiles can be studied in greater detail in spectral recordings with a Fourier Transform Spectrometer (FTS), which provides spectrally fully resolved spectra with high signal-to-noise ratio but at the expense of much reduced temporal and spatial resolution (cf. Fig. 11). One striking feature of the Stokes V spectra is that the profiles are not entirely anti-symmetric, but that in most cases the blue wing lobe has a larger amplitude and area than the red wing lobe. This Stokes V asymmetry varies with excitation potential, line strength and center-to-limb distance

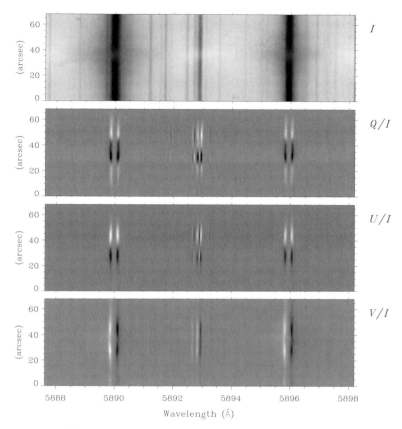

FIGURE 10. Example of Stokes vector spectro-polarimetry showing the Zeeman-effect signatures in the four Stokes parameters. The two strong spectral lines at 5890 and 5896 Å are the Na I D_2 and D_1 lines, which are formed in the lower chromosphere. The spectrograph slit is crossing a large sunspot. While the longitudinal Zeeman effect gives anti-symmetric profile signatures in Stokes V, the transverse Zeeman effect gives symmetric profile signatures in Stokes Q and U. From Stenflo et al. (2001).

in ways that can be explained and modelled in terms of correlated spatial gradients of the velocity field and the magnetic field. Observations with higher spatial and temporal resolutions show large fluctuations of the Stokes V asymmetries, which are averaged down when low resolution is used.

Another striking effect in the Stokes V spectra concerns the relative amplitudes of the polarization in the different lines. Since the polarization signals seen outside active regions are generally small, one might be led to believe that the magnetic fields there are weak. If each of the σ components is Zeeman-shifted by $\Delta\lambda_H$ to either side of the line center, and we denote the signals in right and left circular polarization by $I_\pm$, so that

$$I_\pm = \tfrac{1}{2}(I \pm V)\,,\tag{3.15}$$

then for weak fields

$$I_\pm(\Delta\lambda) \approx I_0(\Delta\lambda \mp \Delta\lambda_H \cos\gamma)\,,\tag{3.16}$$

where I_0 represents the intensity profile in the absence of magnetic fields, and γ is the angle between the field vector and the line of sight. Since for weak fields the line

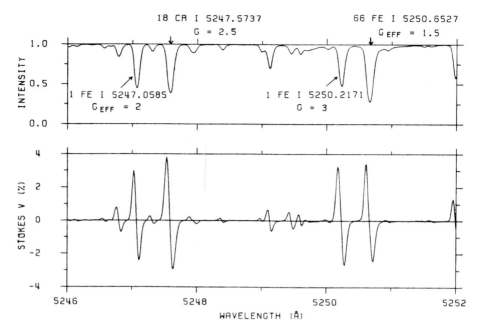

FIGURE 11. Stokes I and V spectra recorded in a strong plage near disk center with the FTS polarimeter of the National Solar Observatory. ¿From Stenflo et al. (1984).

broadening due to the Zeeman effect is insignificant, I_0 should be practically identical to the observed Stokes I profile. A Taylor expansion then gives

$$V \approx -\frac{\partial I}{\partial \lambda} \Delta \lambda_H \cos \gamma \,. \tag{3.17}$$

Since

$$\Delta \lambda_H = 4.67 \times 10^{-13} g_{\text{eff}} \lambda^2 B \,, \tag{3.18}$$

where the field strength B should be given in G and the wavelength in Å, we would expect that the observed Stokes V amplitudes scale with $g_{\text{eff}} \partial I/\partial \lambda$ in a solar region with a given line-of-sight component $B \cos \gamma$. The observations however show that this is generally not the case.

The two iron lines Fe I 5250.22 and and 5247.06 Å are an example of such apparently "anomalous" behavior. They have the same line strength and excitation potential, and they belong to the same atomic multiplet. The only real difference is in their Landé factors, 3.0 for the 5250 Å line, 2.0 for the 5247 Å line. One would then expect their V amplitudes to be in proportion 3:2, but a glance at the FTS spectrum shows that the observed ratio is much closer to 1:1. The reason for this discrepancy is the circumstance that the magnetic fields are not spatially resolved, and that much of the unresolved magnetic flux is not weak but clumped in a strong-field state. For weak fields the Stokes V signal is proportional to the field strength, but when the Zeeman splitting becomes comparable to the line width saturation sets in. The deviation from linearity sets in earlier for lines with larger Landé factors. Therefore the V amplitude ratio between the 5250 and 5247 Å lines is smaller than expected under the weak-field assumption. While the V amplitudes scale with the magnetic flux integrated over the angular resolution element, the amplitude ratio in this special case contains information on the intrinsic field strengths of the unresolved, clumped field elements. This field strength can be orders of magnitude larger than the average field strength (flux divided by the area of the resolution element).

Let us next compare the two iron lines at 5247.06 and 5250.65 Å. Although the latter line has a smaller Landé factor it has a larger Stokes V amplitude. This behavior cannot be explained in terms of Zeeman saturation, which is small for both lines, but is the result of different temperature sensitivities. The line weakening is more pronounced for the 5247.06 Å line because of its lower excitation potential (0.09 eV) as compared with that (2.20 eV) of the 5250.65 Å line.

Further differential effects can be found when comparing lines of different strengths, which are formed at different heights in the atmosphere. The height of formation also varies with wavelength within a single line profile.

All such differential effects in the spectrum are of tremendous diagnostic value, since they provide us with a large set of qualitatively different model constraints, which allow us not only to determine the physical parameters (magnetic fields, velocities, temperatures, densities, etc.) as functions of height, but also how these parameters vary on small spatial scales that are beyond the resolution of the telescopes.

3.2. *Radiative transfer formulation*

The basic theoretical tool for the quantitative interpretation of Stokes spectral observations is the polarized radiative transfer equation, which determines how the Stokes vector spectrum that we see is formed in the solar atmosphere in the presence of a magnetic field. It was first formulated by Unno (1956), Stepanov (1958), and Rachkovsky (1962a,b). It can be written as

$$\frac{\mathrm{d}\vec{I}_\nu}{\mathrm{d}\tau_c} = (\vec{\eta} + \vec{E})\vec{I}_\nu - \vec{S}_\nu \,, \tag{3.19}$$

where $\vec{I}_\nu$ is the Stokes vector at frequency ν, τ_c is the continuum optical depth, defined by

$$\mathrm{d}\tau_c = -\kappa_c \, \mathrm{d}s \tag{3.20}$$

(κ_c being the continuum opacity), $\vec{S}_\nu$ is the source function vector, which will be specified later, $\vec{E}$ is the 4×4 unity matrix (for which the diagonal elements are unity, the off-diagonal zero), and $\vec{\eta}$ is the line absorption matrix, given by

$$\vec{\eta} = \begin{pmatrix} \eta_I & \eta_Q & \eta_U & \eta_V \\ \eta_Q & \eta_I & \rho_V & -\rho_U \\ \eta_U & -\rho_V & \eta_I & \rho_Q \\ \eta_V & \rho_U & -\rho_Q & \eta_I \end{pmatrix} . \tag{3.21}$$

For a normal Zeeman triplet

$$\begin{aligned} \eta_{I,Q,U,V} &= \eta_0 H_{I,Q,U,V} \,, \\ \rho_{Q,U,V} &= 2\eta_0 F_{Q,U,V} \,. \end{aligned} \tag{3.22}$$

Here

$$\eta_0 = \kappa_0/\kappa_c \,, \tag{3.23}$$

where $\kappa_0 H(a,0)$ is the line absorption coefficient at the center of the line. The Voigt function $H(a,v)$ is defined below. First we define $H_{I,Q,U,V}$ by

$$\begin{aligned} H_I &= H_\Delta \sin^2 \gamma + \tfrac{1}{2}(H_+ + H_-) \,, \\ H_Q &= H_\Delta \sin^2 \gamma \cos 2\chi \,, \\ H_U &= H_\Delta \sin^2 \gamma \sin 2\chi \,, \end{aligned} \tag{3.24}$$

$$H_V = \tfrac{1}{2}(H_+ - H_-)\cos\gamma\,,$$

with

$$
\begin{aligned}
H_q &= H(a, v - q v_H)\,, \quad q = 0, \pm 1\,, \\
H_\Delta &= \tfrac{1}{2}[H_0 - \tfrac{1}{2}(H_+ + H_-)]\,.
\end{aligned}
\tag{3.25}
$$

The corresponding expressions for $F_{I,Q,U,V}$ are obtained if we simply replace H by F. The Voigt function $H(a,v)$ is given by

$$H(a,v) = \frac{a}{\pi}\int_{-\infty}^{+\infty}\frac{e^{-y^2}\,dy}{(v-y)^2+a^2}\,,
\tag{3.26}$$

which when integrated over v has an area of $\sqrt{\pi}$ (so defined to make $H(0,0)=1$). The line dispersion function $F(a,v)$ is given by

$$F(a,v) = \frac{1}{2\pi}\int_{-\infty}^{+\infty}\frac{(v-y)\,e^{-y^2}\,dy}{(v-y)^2+a^2}\,.
\tag{3.27}$$

Further

$$
\begin{aligned}
v &= (\nu_0 - \nu)/\Delta\nu_D\,, \\
q\,v_H &= \Delta\nu_H/\Delta\nu_D\,,
\end{aligned}
\tag{3.28}
$$

while the damping parameter

$$a = \gamma/(4\pi\Delta\nu_D)\,.
\tag{3.29}$$

Here ν_0 is the central frequency and $\Delta\nu_D$ the Doppler width.

To cover the case of anomalous Zeeman splitting we have to replace H_q and F_q with the weighted averages of the π (for $q = 0$) and σ (for $q = \pm 1$) components, with the transition strengths between the respective magnetic sublevels (with $\Delta M = 0$ for the π, $\Delta M = \mp 1$ for the σ components) as the weights.

Let us next turn to the source function vector $\vec{S}_\nu$. It can be written

$$\vec{S}_\nu = S_L\,\vec{\eta}\vec{1} + S_c\vec{1} + \vec{j}_{\rm coh}/\kappa_c\,.
\tag{3.30}$$

S_c is the source function for the continuum (here assumed to be unpolarized), which is usually treated in LTE (local thermodynamic equilibrium), in which case $S_c = B_\nu$, the Planck function. The emission vector $\vec{j}_{\rm coh}$ will be discussed later in connection with the Hanle effect. When the line is assumed to be formed in LTE, then there is no scattering at all, and the line source function S_L is set equal to the Planck function B_ν. The assumption of LTE thus represents an enormous simplification of the problem and works well for weak, photospheric lines, but is inadequate for strong lines, and excludes the Hanle effect from the outset. Figure 12 gives an example of Stokes line profiles calculated under the assumption of LTE.

The expression for S_L depends on the atomic model used. The simplest and commonly used non-LTE case is that of a 2-level model atom, which also provides some insight into the difference between LTE and non-LTE. For the 2-level case

$$S_L = \frac{\epsilon B_\nu + s}{1 + \epsilon}\,.
\tag{3.31}$$

$$\epsilon = \frac{C_{u\ell}}{A_{u\ell}}\left(1 - e^{-h\nu/kT}\right)\,,
\tag{3.32}$$

where $C_{u\ell}$ and $A_{u\ell}$ are the rate coefficients for collisional deexcitation and spontaneous emission, respectively. The parameter s represents the contributions from incoherent,

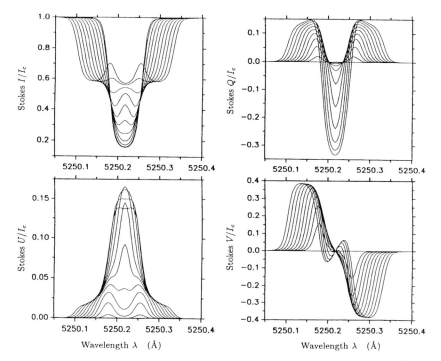

FIGURE 12. Numerical solution of the LTE transfer equations for the Fe I 5250.2 Å absorption line at the center of the solar disk, using a quiet sun model atmosphere with a homogeneous magnetic field of inclination $\gamma = 45°$ and azimuth $\chi = 0$. The ten different curves correspond to field strengths varying from 0.2 to 2.0 kG in steps of 0.2 kG. The Stokes parameters are expressed in units of the intensity I_c of the local continuum. Note that the signal in Stokes U is exclusively due to magnetooptical effects. ¿From Solanki (1993).

non-polarizing scattering and is given by

$$s = \frac{1}{\sqrt{\pi}\Delta\nu_D} \frac{\gamma_c}{\gamma_N + \gamma_c} \int \frac{d\Omega'}{4\pi} \int d\nu' \left(H'_I I_{\nu'} + H'_Q Q_{\nu'} + H'_U U_{\nu'} + H'_V V_{\nu'} \right), \qquad (3.33)$$

where γ_c and γ_N represent the collisional and radiative (natural) damping constants, respectively, and the integrations are done over all directions and frequencies (cf. Stenflo 1994). The first normalization factor is needed because the frequency integral over $H(a, v)$ equals $\sqrt{\pi}\Delta\nu_D$.

¿From the 2-level model expression for S_L one can see that when ϵ becomes $\gg 1$, then we enter into the LTE regime, in which $S_L = B_\nu$. This condition is reached when the collisional deexcitation rate dominates over the spontaneous emission rate. The collisional rate scales with the density, which increases nearly exponentially with depth in the atmosphere. LTE is therefore a good approximation in the lowest regions of the atmosphere (lower photosphere), where the continuum and weak spectral lines are formed. In LTE the source function is determined exclusively by the local temperature. For lines formed at greater heights (upper photosphere, temperature minimum, chromosphere) the scattering term s becomes increasingly dominant. It means that the source function becomes decoupled from the local temperature and instead determined by the radiation field from non-local sources. The non-local s term couples the 3-D atmosphere over distances on the order of the photon mean free path (optical depth unity).

3.3. *Contribution and response functions. Stokes inversion*

3.3.1. *Contribution functions*

The formal solution of the transfer equation can be written as

$$\vec{I} = \int_0^\infty \vec{C}_{\vec{I}}(\tau_c)\, \mathrm{d}\tau_c \,, \tag{3.34}$$

where $\vec{I}$ is the Stokes vector that leaves the atmosphere (at τ_c), and $\vec{C}_{\vec{I}}$ is the *contribution function* for the total Stokes vector $\vec{I}$. With the definition

$$\vec{L}(\tau_c) = \exp\left[\int_0^{\tau_c} (\vec{\eta}(\tau) + \vec{E})\, \mathrm{d}\tau\right] \,, \tag{3.35}$$

$$\vec{C}_{\vec{I}}(\tau_c) = \vec{L}^{-1}\, \vec{S}_\nu \,. \tag{3.36}$$

$\vec{C}_{\vec{I}}$ however does not provide "clean" information on the height of line formation, since the contributions from the line and the continuous spectrum are mixed. In the far line wings these contribution functions mainly tell us where the continuum is formed, not where the line is formed. To obtain something that is of practical use and which exclusively tells us where the line Stokes parameters are formed we need to define the *line depression* contribution functions (Grossmann-Doerth et al. 1988a; Sánchez Almeida 1992)

$$\vec{r} = \int_0^\infty \vec{C}_{\vec{r}}(\tau_c)\, \mathrm{d}\tau_c \,, \tag{3.37}$$

where the line depression $\vec{r}$ is related to $\vec{I}_\nu$ by

$$\vec{r} = \vec{1} - \vec{I}_\nu/I_c \,. \tag{3.38}$$

I_c is the continuum intensity. Let us define

$$\begin{aligned}
\vec{\eta}_r(\tau_c) &= \vec{\eta}(\tau_c) + \vec{E}B_\nu/I_c(\tau_c) \,, \\
\vec{S}_r(\tau_c) &= \vec{\eta}_r(\tau_c)\vec{1} - \vec{S}_\nu(\tau_c)/I_c(\tau_c)
\end{aligned} \tag{3.39}$$

and

$$\vec{L}_{\vec{r}}(\tau_c) = \exp\left[\int_0^{\tau_c} (\vec{\eta}_{\vec{r}}(\tau)\, \mathrm{d}\tau\right] \,. \tag{3.40}$$

It is then found that

$$\vec{C}_{\vec{r}}(\tau_c) = \vec{L}_{\vec{r}}^{-1}\, \vec{S}_r \,. \tag{3.41}$$

3.3.2. *Response functions*

While the contribution functions tell us where the spectrum is formed, it is of more use for the inversion of Stokes data to determine how the emergent Stokes vector $\vec{I}$ responds to perturbations of various physical parameters in the atmosphere. Let $a = a(\tau_c)$ be a scalar physical parameter like temperature, density, velocity, field strength, field angle, or microturbulent velocity. If we perturb a to $a + \delta a$, where δa is a function of τ_c, and restrict ourselves to small perturbations so that only first-order terms in δa, $\delta \vec{I}_\nu$, $\delta \vec{\eta}$, and $\delta \vec{S}_\nu$ need to be retained, the emergent Stokes vector gets changed to $\vec{I}_\nu(0) + \delta \vec{I}_\nu(0)$,

where

$$\delta \vec{I}_\nu = \int_0^\infty \vec{R}_a(\tau_c)\,\delta a(\tau_c)\,\mathrm{d}\tau_c \,. \tag{3.42}$$

This defines the *response function* $\vec{R}_a$ for the Stokes vector $\vec{I}_\nu$ with respect to the parameter a (Landi Degl'Innocenti and Landi Degl'Innocenti 1977; Ruiz Cobo & del Toro Iniesta 1992). One finds that

$$\vec{R}_a = \vec{L}^{-1}\left(\frac{\partial \vec{S}_\nu}{\partial a} - \frac{\partial \vec{\eta}}{\partial a}\vec{I}_\nu\right). \tag{3.43}$$

We can likewise define a response function $\vec{R}_{\vec{r}a}$ for the line depression function $\vec{r}$ by

$$\delta \vec{r} = \int_0^\infty \vec{R}_{\vec{r}a}\,\delta a\,\mathrm{d}\tau_c \tag{3.44}$$

(cf. Grossmann-Doerth et al. 1988a), with

$$\vec{R}_{\vec{r}a} = \vec{L}_r^{-1}\left(\frac{\partial \vec{S}_r}{\partial a} - \frac{\partial \vec{\eta}_r}{\partial a}\vec{r}\right). \tag{3.45}$$

For practical purposes a more useful definition of the response and contribution functions is in terms of $x = \log \tau_c$ as an integration variable instead of τ_c, since x is more closely related to the geometrical height scale and is the variable normally used for numerical integrations of the transfer equations and for tabulations of model atmospheres. The relation between these two definitions is

$$\vec{R}(x) = (\ln 10)\,\tau_c\,\vec{R}(\tau_c) \tag{3.46}$$

for the response functions, and formally the same for the contribution functions (if we substitute $\vec{R}$ by $\vec{C}$).

3.3.3. *Stokes inversion*

The calculation of the emergent Stokes spectrum from a given model atmosphere is the *direct problem*, which can be carried out at various levels of sophistication, depending on the degree of realism introduced in the model (cf. the example in Fig. 13). The problem that one really wants to solve is however the *inverse* one, since it is the emergent Stokes spectrum that is observed, while the model atmosphere is the unknown to be deduced from the Stokes observations.

The relation between the observables (the Stokes line profiles) and the model parameters (the magnetic field and thermodynamic quantities as functions of height) is generally highly non-linear without any analytical formulation. The inverse problem however becomes tractable if it is linearized, which is commonly done in the context of a least-squares fitting procedure, as will be described next.

Let y_i, $i = 1, 2, \ldots, n$ be the observables with errors σ_i, while Y_i are the corresponding synthetic observables that are computed from the model. While there are n observables, the model is characterized by m free parameters. The goodness of the model fit to the observations is then governed by

$$\chi^2 = \frac{1}{n-m}\sum_{i=1}^n \frac{1}{\sigma_i^2}(y_i - Y_i)^2 \,. \tag{3.47}$$

$n - m$ represents the number of degrees of freedom of the problem. If the model is good, we expect that χ^2 is of order unity.

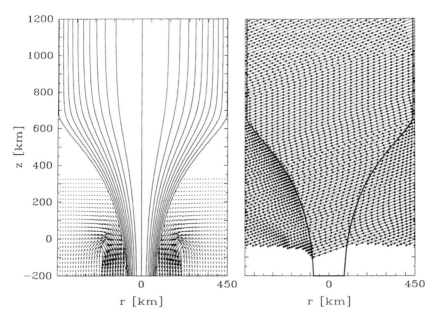

FIGURE 13. Example of an axially symmetric model of a magnetic flux tube with a surrounding convective velocity field. It has been used to successfully reproduce the observed center-to-limb variation of the Stokes V area asymmetry. Left panel: Magnetic field lines and flow lines of the velocity field. Right panel: 20 lines of sight at an angle of $70°$ with the vertical direction, used for the radiative transfer calculations of the synthetic Stokes V spectra. The grid points for the computations (filled circles) have been selected with an adaptive step size method. From Bünte et al. (1993).

The least-squares fitting procedure consists of finding the parameter values for which χ^2 has its global minimum. Let the free model parameters be a_k, $k = 1, 2, \ldots, m$. They may represent any physical parameter at any location, e.g. the temperature gradient at $\tau_c = 10^{-3}$, or the magnetic-field angle at selected depth points. Ideally the a_k parameters, which may be described in terms of an m-vector $\vec{a}$, should represent for each relevant depth point all the physical parameters, i.e., $\vec{a}$ should fully describe the model atmosphere. In practice, however, one needs to limit the number of free parameters and select the set of observables with great care, to ensure stability and uniqueness of the numerical inversion.

χ^2 is a function of $\vec{a}$ and is represented by a surface in the m-dimensional parameter space. Our task is to find the minimum value of χ^2 on this surface. χ^2 has a local minimum when

$$\frac{\partial \chi^2}{\partial a_k} = 0 \qquad (3.48)$$

for all allowed values of k. To find the location of this minimum we linearize the problem by making a Taylor expansion to first order of the synthetic observables.

The location $\vec{a}$ of the χ^2 minimum is determined iteratively, starting with an initial "guess" $\vec{a}_0$. From this starting point on the m-dimensional hypersurface we compute the corrections $\delta\vec{a}$ that should be applied to find the correct location. The new, improved value for the location of the minimum is $\vec{a}_1 = \vec{a}_0 + \delta\vec{a}$, which serves as the starting point for the next iteration. The iterations continue until convergence has been obtained. For more details, see Socas-Navarro (2001).

3.4. *Diagnostics of spatially unresolved magnetic fields*

With improved spatial resolution ever smaller magnetic features have been discovered without an end in sight. One has therefore wanted to answer the question what the intrinsic, resolution-independent magnetic structure of the solar atmosphere would be if we could observe it with infinite resolution. It is possible to give at least partial answers to this question by making use of the differential polarization effects in the solar spectrum. The circumstance that the various spectral lines respond in a non-linear and individual way to the different physical parameters can be used to extract information on the intrinsic atmospheric properties on scales beyond what can be resolved, except that the morphology of the unresolved structures remains undetermined.

3.4.1. *Evidence for intermittency and the flux tube concept*

Line-of-sight magnetograms of the Sun represent maps of the circular polarization recorded in the wing of a spectral line sensitive to the Zeeman effect. The standard calibration of such magnetograms is based on the weak-field assumption that the circular-polarization signal is proportional to the line-of-sight component of the magnetic field (as averaged over the spatial resolution element), with the proportionality factor determined by the effective Landé factor and the gradient $\partial I/\partial\lambda$ of the intensity profile. There are two main reasons why this calibration leads to incorrect values for the average magnetic field: (1) The intensity profile I that is used for the calibration is representative of the average Sun, and can be very different from the intensity profile in the magnetic flux elements due to their very different thermodynamic structures. The intensity profile in the flux elements cannot be used for calibration, since it is not a directly observable quantity (the flux elements are not spatially resolved). (2) Due to the extreme clumping or intermittency of the magnetic flux, the weak-field assumption is not valid. When the Zeeman splitting starts to become comparable to the line width, the relation between polarization and magnetic flux is no longer linear.

Already in the end of the 1960's one started to notice that there were apparent discrepancies between the magnetic flux values measured with different spectral lines. One can for instance record the circular polarization simultaneously in two different spectral lines and make scatter plots of the apparent field strengths derived with the two lines. As an example we illustrate in Fig. 14 such a scatter plot for the field measured simultaneously in the two Fe I 5250.22 and 5247.06 Å lines, which were shown in Fig. 11. They both belong to multiplet no. 1 of iron and have the same line strength and excitation potential. They should therefore be formed in the same way and in the same layers of the atmosphere, and therefore be affected identically by the different thermodynamics. Still the two lines give different values for the magnetic flux.

The only property that is different for this line pair is their Landé factors, 3.0 for the 5250.22 Å line, 2.0 for the 5247.06 Å line. This difference would however not cause a discrepancy between the magnetic flux values unless the intrinsic magnetic field strength is so large that we get Zeeman saturation, i.e., the relation between polarization and field strength deviates from linearity. Since this deviation is larger for the line that has the larger Landé factor, it is the 5250.22 Å line that shows the smaller values of the apparent field strengths. From the amount of the discrepancy the intrinsic field strength can be determined and is found to be 1–2 kG, regardless of what the average field is (Stenflo 1973). Since the typical average field strengths on the quiet sun are on the order of 10 G, the kG field strengths mean that the magnetic flux elements occupy typically only 1 % of the solar surface.

If there were a distribution of intrinsic field strengths on the Sun rather than a strong preference for the kG values, then the region between the regression line and the 45° line

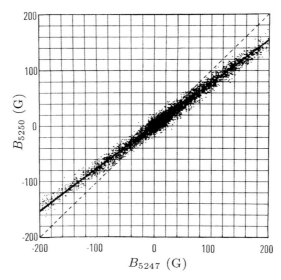

FIGURE 14. Scatter plot of the apparent field strengths B recorded simultaneously in the Fe I 5247.06 and 5250.22 Å lines, for the data set used in Frazier & Stenflo (1978). The slope of the straight regression line corresponds to an intrinsic field strength of 1 kG at the level of line formation.

would be populated with points, which is not seen. This has led to the conclusion that it is a good approximation to consider the flux tube properties as being "unique" (Frazier & Stenflo, 1972, 1978). It means that whatever solar region we look at, the difference between apparently strong and weak "observed" fields B_{obs} is not due to variations of the intrinsic properties of these flux regions, but is instead due to the filling factor or number density of flux elements. These early conclusions based on "old-fashioned" magnetograph observations have later been confirmed by Stokesmeter and FTS polarimetric observations with much smaller instrumental spread, although these observations have also been done with low spatial and temporal resolution.

The nearly unique values of the field strengths and thermodynamic properties of the magnetic flux elements and their small filling factors have been the basis for consistent interpretations with a two-component model and the concept of magnetic flux tubes representing strong-field regions embedded in non-magnetic surroundings. Accordingly the flux tubes occupy on the order of 1 % of the photospheric volume, with the remaining 99 % being field free. This "standard model" has been extremely successful and led to self-consistent interpretations of the Zeeman-effect observations with all their differential effects throughout the solar spectrum (cf. Solanki 1993).

In more recent years polarimetry in the near infrared has allowed us to determine the field strengths more directly. While the widths of spectral lines are to a first approximation proportional to wavelength, the Zeeman splitting increases with the square of the wavelength. Therefore the incomplete Zeeman splitting in the visible part of the spectrum can become complete for certain lines in the infrared. When the splitting is complete, the field strength can be read off directly from the wavelength separation of the sigma components, regardless of the magnitude of the magnetic filling factor.

The near infrared region around 1.6 μm is of special interest, since the solar opacity has a minimum at these wavelengths, which means that the radiation emanates from deeper

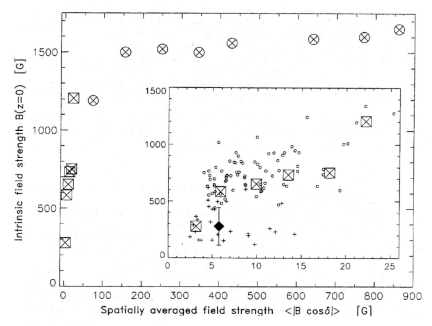

FIGURE 15. Intrinsic field strength of the spatially unresolved magnetic flux tubes as a function of the field strength averaged over the spatial resolution element of the observations. The data are based on observations with the near infrared Fe I 1.5649 μm line, which has a particularly large Zeeman splitting. In regions with magnetic filling factors larger than a few percent the main contributions to the flux comes from flux tubes with field strengths in the kG range, and there is little dependence on the filling factor or average field strength. In regions with little flux, however, the intrinsic field strength drops to much lower values. From Solanki et al. (1996).

layers in the atmosphere than in any other part of the solar spectrum. This enhances the Zeeman splitting still further, since the field strength increases with depth.

The infrared observations have confirmed the kG nature of much of the photospheric magnetic flux, but have also shown the existence of weaker field elements. Except for the smallest magnetic fluxes or filling factors the field strength is kG and increases only slowly with filling factor. In regions of small filling factors, however, the field strengths found are much weaker, which indicates that at the smallest scales the flux elements are intrinsically weak (cf. Fig.15).

Since the flux elements are spatially unresolved, a comparison between the measured magnetic filling factor and the average field strength only gives an *upper limit to the size* of the flux elements. Since we cannot determine the flux morphology we do not know whether the given flux within the spatial resolution element is due to a single element or to many smaller elements.

3.4.2. *3-D flux tube models and numerical simulations*

The flux tube concept has provided a very useful interface between theory and observations. Theoretically the formation of the flux tubes can be explained in terms of the mechanism of "convective collapse". Due to the inhibition of convective energy exchange between the interior and exterior of a flux region because the plasma cannot flow freely across the field lines, the interior is thermally shielded from the exterior, provided that the photon mean free path in the direction perpendicular to the field lines is smaller than the size of the flux element. In this case the motion within the element is more adiabatic

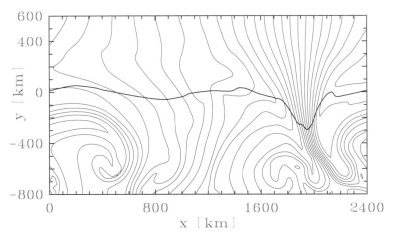

FIGURE 16. Magnetohydrodynamic numerical simulation of the formation of magnetic flux tubes in the solar atmosphere. The magnetic field lines after the formation process are shown. The nearly horizontal wavy line near height zero marks the surface of optical depth unity. It has a depression in the flux tube due to the partial evacuation that is needed to ensure lateral pressure balance. From Grossmann-Doerth et al. (1998).

with a smaller temperature gradient than in the exterior region. This has the result that an initially small downflow in the element leads to an adiabatic cooling with subsequent collapse due to the higher gas pressure in the exterior region (cf. Parker 1978; Spruit & Zweibel 1979).

Once a flux tube has been formed, it is in pressure equilibrium with the surrounding plasma. To contain the large magnetic pressure inside by the external gas pressure, the internal gas pressure must be drastically reduced, which is only possible if the flux tube is partially evacuated. Almost nothing is known from observations about the further fate of the concentrated fields. Their life times with respect to fragmentation by the fluting instability depends on maintaining vortex flows around them. At scales below about one hundred km the convective collapse mechanism will no longer be effective due to radiative exchange, since the photon mean free path in the lower photosphere is on the order of 100 km. Flux on such scales may therefore have intrinsically weaker field strengths and might then not be able to withstand the tangling effect of the turbulent motions. Smaller flux fragments may coalesce to larger structures, for which collapse becomes possible, etc.

Self-consistent magnetohydrodynamic models of magnetic flux tubes at various levels of sophistication have been constructed and used to compute synthetic Stokes spectra to be compared with observations. When viewing the expanding, axially symmetric model flux tubes from the side and simulating spatially unresolved observations, one has to solve the polarized radiative-transfer problem along a large number of lines of sight that pierce the flux tube at different places, and then average the computed spectra (cf. Fig. 13). With Stokes inversion one can use the observed Stokes spectrum to determine the free parameters of the MHD flux tube model. With the complex geometries and many lines of sight, this can be a very demanding task with available computer resources.

Magneto-convection is the underlying process that is involved in most aspects of stellar activity: the dynamo mechanism, flux amplification, transport, destruction, intermittency, etc. Still we are far from a good theoretical understanding of magneto-convection and turbulence in the gravitationally stratified stars. 3-D numerical simulations of

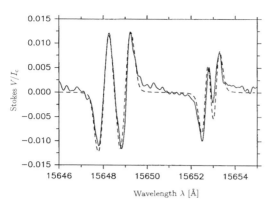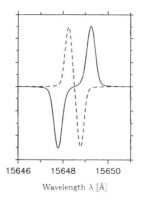

FIGURE 17. Example of anomalous Stokes V profiles requiring more than one magnetic component for their interpretation. Left panel: Observed (solid) and synthetic (dashed) Stokes V/I_c spectrum with unusually looking profiles. I_c is the intensity of the continuous spectrum. A model with two magnetic components has been used to fit the observations. Right panel: The synthetic profiles of the Fe I $1.5649\,\mu$m line for each of the two magnetic components used for the fit. They correspond to a field strength at height $z = 0$ of $1.70\,$kG (solid curve) and $-1.05\,$kG (dashed curve), respectively. From Rüedi et al. (1992).

magneto-convection in the outer layers of the Sun (cf. Nordlund & Stein, 1990) have been successful in reproducing the observed properties of solar granulation, but the computational grid has been too coarse to properly describe such processes as convective collapse, fluting instability, ohmic decay of current sheets, etc. The problem is very complex, since so many length scales are involved, down to the ohmic diffusion scale (on the order of $1\,$km in the photosphere). Nevertheless considerable progress has been made in recent years (cf. Fig. 16).

3.4.3. *"Anomalous" Stokes profiles. Evidence for multi-component atmospheres. Limitations of Zeeman-effect diagnostics*

The two-component model with a strong magnetic (flux tube) component that has a small magnetic filling factor and non-magnetic surroundings has in the past given self-consistent interpretations of the wealth of high spectral resolution polarimetric data obtained in particular with the FTS polarimeter but also with other Stokesmeters or magnetographs. However, these earlier observations were made in the visible and with relatively low spatial resolution (a few arcsec or worse). With the advent of polarimetric observations in the infrared as well as Stokesmeter observations with high spatial and temporal resolution, many instances have been found where the two-component model is inadequate and in need for extension by adding an extra magnetic component, which may have either the same or different polarity as compared with the main magnetic component, but the field strength, Doppler shift, and temperature structure may be different. The need for such an extension becomes obvious for "pathological" Stokes V profiles with a multiple lobe structure (instead of the standard anti-symmetric profiles). Such anomalous profiles are rare in the visible but are frequently seen in infrared Stokes spectra, since the much larger Zeeman splitting in the infrared is able to better separate the components that lie on top of each other in the visible (cf. Fig. 17).

FTS spectra (with low spatial resolution) show that the lobe in the blue wing of Stokes V profiles is larger than the red wing lobe. More recent Stokesmeter spectra with high spatial and temporal resolution show a large variation with frequent extreme values of these Stokes V asymmetries (Sigwarth et al. 1999). Thus it not only happens that one of the Stokes V wing lobes disappears entirely, but many cases are found when the two lobes have the same sign. It has been possible to theoretically show, in terms of a single flux tube model, how such profiles can arise when the line-forming region straddles both sides of a magnetopause or magnetic canopy separating magnetic and non-magnetic regions with different flow fields and temperatures (Grossmann-Doerth et al. 1988b, 2000). In principle such anomalous Stokes V profiles could therefore be used as diagnostic signatures of various types of magnetopause structures (Steiner 2000).

With these amendments and extensions of the two-component model, the basic picture that Zeeman-effect observations have given us is a photosphere with only about 1 % of the volume filled with strong-field kG flux tubes, with indications of many weak-field magnetic elements of mixed polarity in between. The evolution of discrete clumps of apparently weak mixed-polarity fields can be followed in "deep" magnetograms (with long integrations) at scales down to a few arcsec, with hints of many more elements of mixed polarities at smaller scales which presently are not seen, due to cancellation of the opposite polarities within the spatial resolution element (cf. Fig. 5). Due to their weakness or the small fluxes involved, their Zeeman-effect signals are small. Because of the long integrations that are required, the observations are highly sensitive to smearing by atmospheric seeing.

The concept of a two- or multi-component model with a major fraction of the atmosphere being non-magnetic is of course a convenient idealization, which should not be taken to imply that non-magnetic regions do exist on the Sun. The solar plasma has an extremely high electrical conductivity and is in a turbulent state, which means that small-scale magnetic fields are constantly generated by induction effects. In reality we therefore expect the so-called non-magnetic, 99 % of the photosphere, to be fully magnetized and teeming with weak, mixed-polarity fields far smaller than the angular resolution that we can reach. Even if we in some future could reach nearly infinite angular resolution, we would not be able to see mixed-polarity magnetic fields on scales much below the photon mean free path (about 100 km) in the photosphere due to cancellations along the line of sight over the line forming region when there are contributions of mixed signs to the Zeeman effect.

The Zeeman-effect signals are weak if the Zeeman splitting is small as compared with the spectral line width. In this weak-field regime the Zeeman effect only provides information about the magnetic fluxes but not on filling factors and intrinsic field strengths. The field of the magnetic flux tubes expands rapidly with height due to the almost exponential drop of the external gas pressure, which means that the chromospheric magnetic fields should be substantially weaker than the photospheric fields. We would like to find out how much weaker. The few chromospheric lines that we have in the visible and near infrared part of the spectrum are however much broader than the photospheric lines and therefore much less sensitive as diagnostic tools for anything else than the magnetic flux.

In summary, the Zeeman effect observations are nearly "blind" to small-scale magnetic fields of mixed polarity and is not a well suited diagnostic tool for weak magnetic fields and fields above the photosphere. Fortunately there exists another diagnostic tool, the Hanle effect, which has its best sensitivity in the above-mentioned regimes, where the Zeeman effect performs poorly. The two effects are therefore highly complementary to each other. In the following sections we will discuss the diagnostic possibilities offered by the Hanle effect.

4. Hanle diagnostics and coherency effects

4.1. *Observational signatures of the Hanle effect with imaging polarimeters*

Polarization can be produced by radiative scattering, when the incident radiation has some degree of anisotropy. Down in the lower solar atmosphere when a local plane-parallel stratification without horizontal inhomogeneities can be assumed, the anisotropy is due to the limb darkening. For non-magnetic scattering the polarization at disk center is zero for symmetry reasons, but it increases monotonically as we approach the solar limb. The polarization is linear, and in the non-magnetic case the electric vector is usually, with few exceptions, parallel to the nearest limb (perpendicular to the radius vector from disk center).

While the intensity spectrum and the Stokes spectrum formed by the Zeeman effect are governed by many different radiative transfer processes, the scattering polarization is formed exclusively by radiative excitation followed by spontaneous emission. The totality of ways in which the scattering polarization can be modified by the presence of a magnetic field is covered by the term "Hanle effect" (cf. Trujillo Bueno 2001). The magnetic field removes the degeneracy of the magnetic substates of the atomic levels. The corresponding polarization effects depend on the degree of degeneracy removal, i.e., on the ratio between the Zeeman splitting and the damping width (inverse life time) of the atomic states. If these two quantities are comparable in magnitude, we are in the Hanle regime in which the polarization effects depend both on the strength and orientation of the field. For excited states of allowed atomic transitions this occurs for field strengths typically in the 1–100 G range. For ground states of allowed transitions, which have life times that are about two orders of magnitude longer, the Hanle field strength range lies correspondingly lower, typically 0.01–1 G. When the Zeeman splitting is much larger than the damping width, we are in the Hanle saturated regime, in which the polarization effects depend on the orientation of the field but not on its strength. This is the situation for coronal forbidden lines, due to the very long life times of their excited states. If we can observationally separate the Hanle effects in the excited states and ground states from each other, then we may have a tool for diagnosing magnetic fields of all strengths, down to the lower mG range.

Figure 18 illustrates how the linear polarization produced by scattering and by the transverse Zeeman effect have entirely different signatures. The spectrograph slit has been placed parallel to and 2.5 arcsec inside the limb (at $\mu = 0.07$, where μ is the cosine of the heliocentric angle), such that half the slit covers a small facular region, while the other half lies outside it. In the facular region we see the characteristic signatures of the transverse Zeeman effect in practically all the atomic lines, while outside it the only significant signal is the scattering polarization in the Sr I 4607 Å line and occurs only in Q/I (linear polarization parallel to the limb). As we move closer to the facular region, the amplitude of Q/I in the Sr I line decreases, which is a signature of Hanle depolarization by magnetic fields. The Sr I 4607 Å line has been used as a test case for extensive radiative transfer calculations of Hanle depolarization (Faurobert-Scholl 1993; Faurobert-Scholl et al. 1995).

Another observational example of Hanle-effect signatures is given in Fig. 19, which shows the spatial fluctuations along the slit of the polarization amplitude in the Doppler core of the Na I D_2 5890 Å line. While the variations in Q/I and U/I are due to the Hanle effect, V/I is exclusively due to the longitudinal Zeeman effect. Note that the broad Q/I polarization maxima in the wings of the D_2 line are spatially invariant, in accordance with theoretical expectations, since the Hanle effect only operates in the Doppler core (cf. Stenflo 1994, 1998). Note also that there is practically no Q/I signal in the D_1 line,

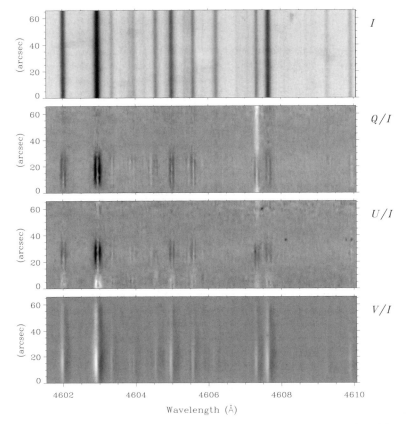

FIGURE 18. Example of the coexistence between scattering polarization (in Q/I in the Sr I 4607 Å line) and the transverse and longitudinal Zeeman effect. The recording was made with ZIMPOL II at NSO/Kitt Peak on March 4, 2000, near the SW limb (at $\mu = 0.07$), where there was some minor facular activity. ¿From Stenflo (2001).

since it can only be produced by lower-level atomic polarization, which gets destroyed by Hanle depolarization in magnetically active regions (see below).

It has long been a common belief that the Hanle effect is only observable in an annular zone close to the limb, where the scattering polarization has sufficient amplitude to be observable. According to this common view, the scattering polarization is largest in the absence of a magnetic field, and the Hanle effect manifests itself by reducing this polarization (depolarization) and rotating its plane. This situation pertains for instance for 90° scattering, which is similar to the scattering geometry near the solar limb. However, in forward scattering, similar to scattering at disk center, the non-magnetic scattering polarization is zero, while a horizontal magnetic field would via the Hanle effect give rise to significant linear polarization (Trujillo Bueno 2001). The Hanle effect is thus not limited to a limb zone but may be used more or less all over the solar disk. It would for instance be a more sensitive tool than the transverse Zeeman effect for the mapping of horizontal magnetic fields in the solar chromosphere.

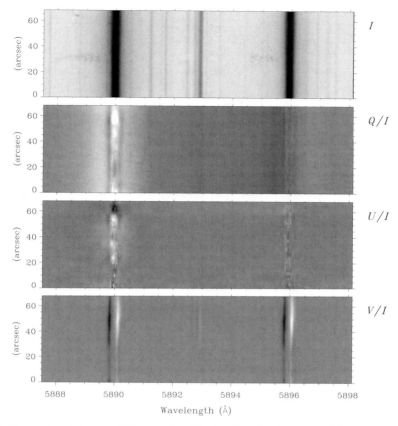

FIGURE 19. Example of the spatially varying Hanle effect in the core of the Na I D_2 line. Q/I shows varying Hanle depolarization, U/I varying Hanle rotation of the plane of polarization, V/I the longitudinal Zeeman effect. The recording was made with ZIMPOL II at NSO/Kitt Peak on October 9, 1999, near the SE limb (at $\mu = 0.1$). From Stenflo et al. (2001).

4.2. *Complementarity between the Zeeman and Hanle effects. Magnetic parameter domains and accessible atmospheric layers*

The discovery that most of the magnetic flux that is recorded in magnetograms (circular polarization maps) has its origin in kG fields with a magnetic filling factor that is typically on the order of 1 % (Stenflo 1973) has led to the concept of discrete kG magnetic flux tubes embedded in a non-magnetic atmosphere. The flux tubes are believed to be formed by convective collapse (Parker 1978; Spruit & Zweibel 1979) and are expected to expand with height to form magnetic canopies that are overlying the field-free atmosphere (Giovanelli 1980; Giovanelli & Jones 1982; Jones & Giovanelli 1983).

This view however has serious deficiencies. No plasma physics theory would ever predict the existence of truly field free regions in such a highly electrically conductive medium as the solar atmosphere, since any initial seed field would immediately be tangled up and amplified by the motions in the solar atmosphere. The absence of observable magnetic flux through a spatial resolution element does not at all imply the absence of magnetic flux, since the Zeeman-effect signals cancel each other for a tangled field. Other deficiencies are that convective collapse is not (yet) an observed phenomenon (cf. Collados

2001), and the canopy concept is based on observations of magnetically active regions (including strong network regions) but has not been verified for very quiet regions.

The reason for this state of affairs is that the Zeeman effect fails to deliver much information about 99 % of the photosphere. Almost all Zeeman-effect diagnostics, with few exceptions, is done with photospheric spectral lines. In chromospheric lines the Zeeman splitting is generally much smaller than the line width and provides information that is mainly limited to magnetic flux. The Zeeman effect is insensitive to magnetic fields that are weak, like the fields between the flux tubes or the fields in the higher layers of the solar atmosphere. It is also nearly blind to magnetic fields of mixed polarities, when the mixing occurs on scales that are smaller than the angular resolution of the observations. Since much of the solar plasma is in a turbulent state, we expect such mixed-polarity magnetic fields to be ubiquitous. Much of the volume between the flux tubes could be filled with such a field, but Zeeman-effect observations are unable to tell.

The Hanle effect on the other hand is practically blind to the flux tube field, since this field has such a small filling factor, and the Hanle effect is insensitive to vertical magnetic fields when the illumination of the scattering particles is axially symmetric. Almost the entire contribution to the Hanle effect comes from the 99 % of the volume to which the ordinary Zeeman effect is almost blind. The Hanle and Zeeman effects therefore ideally complement each other.

Since the flux tube field strength has been found to have an almost unique value of about 1.5 kG in quiet network regions (at the level in the solar atmosphere where the continuum at 5000 Å outside the network is formed), it has been natural to believe that the conjectured turbulent, space-filling magnetic field between the flux tubes should also have a unique field distribution with an rms field strength that is uniquely determined by the kinetic energy spectrum of the solar granulation. This is however not the case. Instead it is found that the appearance of the second solar spectrum changes greatly both spatially and with the phase of the solar cycle (Stenflo et al. 1998) because of large variations in the magnitude of the Hanle effect. This implies that the properties of the field that fills the space between the flux tubes vary and need to be mapped.

Although the Hanle effect opens new diagnostic possibilities that are not available with the Zeeman effect, it has the disadvantage that it does not lend itself to direct mapping of the magnetic field but instead constrains the field properties in more convolved ways. A fundamental reason for this is that the Hanle effect shows up in two observed parameters, Q and U, while the magnetic field vector needs three parameters to fully constrain its three vector components. The field vector is therefore not uniquely constrained by Hanle observations alone, but needs some additional constraint, either from theory or other types of observations (e.g. from the longitudinal Zeeman effect in Stokes V). The Zeeman effect on the other hand can in principle constrain the full vector, the line of sight component from Stokes V, the two transversal components from Q and U (in the case of spatially resolved fields).

4.3. *Extraction of the Hanle depolarization and rotation from the data. Hanle histograms*

Strong resonance lines like the Ca I 4227 Å line that is shown in Fig. 20 most often have scattering polarization (Q/I) profiles that are characterized by a narrow polarization peak in the Doppler core of the line and by broader polarization maxima in the line wings. According to theory the Hanle effect can only operate in the Doppler core, and vanishes in the wings (cf. Stenflo 1994, 1998). The U/I profiles in Fig. 20, which are non-zero due to Hanle rotation of the plane of polarization, therefore only have contributions in the line core, while the wings remain zero.

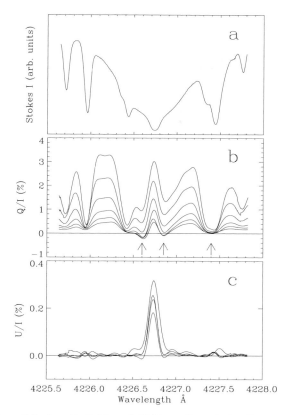

FIGURE 20. Examples of Stokes I, Q/I, and U/I profiles of the Ca I 4227 Å line. **a** Stokes I profile at disk center. **b** Q/I profiles averaged over a sequence of μ windows, to illustrate the center-to-limb variation. **c** U/I profiles averaged over four of the μ windows that were used in the Q/I panel. Only profiles with core amplitudes in excess of 0.05 % have been included in these averages. From Bianda et al. (1999).

While the presence of a U signal reveals the presence of Hanle rotation, it cannot be directly concluded from inspection of the Q/I profile whether Hanle depolarization is present or not. To determine the amount of Hanle depolarization we need to refer to the polarization amplitude that we would have in the absence of magnetic fields. Since theoretical determinations of the non-magnetic polarization in the line core would need to be based on the not well developed theory of polarized partial redistribution and therefore would not provide reliable non-magnetic reference levels for the polarization, we choose to approach the problem empirically in a statistical or differential manner.

The statistical approach is illustrated in Fig. 21, where we present scatter plots of the values read off from the Q/I maxima in the red and blue line wings and in the line core. While the red and blue Q/I wings are highly correlated, the line core values exhibit a large scatter. Such a behavior is expected from Hanle-effect theory, since only the line core is affected by the magnetic fields, not the wings. As the magnetic fields vary from place to place on the solar disk, a scatter in the amount of Hanle depolarization results.

Since we do not know the non-magnetic Q/I reference level from theory, we may use the envelope to the points in the scatter plot diagram as the best empirical estimate of this reference level. As the choice of this envelope is somewhat subjective and affected

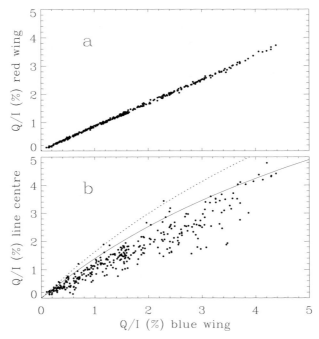

FIGURE 21. Red-wing (**a**) and line center (**b**) Q/I polarization amplitude vs. the Q/I amplitude in the blue line wing, for the Ca I 4227 Å line. The solid and dotted lines are envelope curves, representing estimates of the non-magnetic line center polarization. From Bianda et al. (1999).

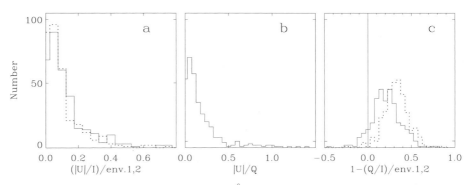

FIGURE 22. Hanle histograms for the Ca I 4227 Å line, showing the distribution of the observed parameters representing Hanle rotation (panels **a** and **b**) and Hanle depolarization (panel **c**). The solid histograms are based on the lower (solid) Q/I envelope in Fig. 21b, the dotted histograms on the upper (dotted) Q/I envelope in Fig. 21b. From Bianda et al. (1999).

by the presence of some instrumental scatter, two possible choices of envelopes have been drawn in Fig. 21. The wing polarization depends uniquely on the center-to-limb distance or μ (cosine of the heliocentric angle), so the abscissa in Fig. 21 may be translated into a μ scale. The height of the envelope for a given μ (or wing polarization) thus represents the non-magnetic reference value with which our observed Q/I and U/I may be normalized, to allow comparison with theoretical predictions for the depolarization and rotation of the plane of polarization.

The distributions of the observed Hanle-effect signals can be represented in the form of *Hanle histograms* as in Fig. 22 for the Ca I 4227 Å line. The histograms of the observed $|U|/I$ and Q/I, normalized to the reference (envelope) values, are shown in the left and right panels, while the middle panel shows the histogram for the ratio $|U|/Q$, which equals $|\tan 2\beta|$, where β is the rotation angle. Such histograms contain information that constrains the intrinsic distributions of the magnetic field vectors on the Sun. However, to limit the degrees of freedom of the problem one needs to introduce some assumptions concerning the shapes of the angular and field strength distributions, and the extent to which the fields are spatially resolved by the observations or not. These assumptions are however not quite ad hoc, since their self-consistency may be constrained and tested.

The histogram for the depolarization values in Fig. 22c is for instance consistent with an isotropic angular distribution of a 5–10 G field. If the field strength were larger, the distribution would crowd more to the right in the diagram. The width of the distribution depends not only on the angular distribution but also on the spread in the field strength values. If we make this spread large, the distribution would become wider than observed.

Hanle depolarization can occur both for spatially resolved and unresolved fields, since the depolarization values for opposite-polarity fields do not cancel each other. The observed non-zero values of U/I, as seen in Figs. 22a and b, however cannot occur for spatially unresolved fields with random angular orientations within the resolution element, since U/I is subject to cancellation effects from fields of opposite directions. The circumstance that U/I signals (due to Hanle rotation) are frequently observed implies that the fields have a net orientation within the spatial resolution element, i.e., the fields are partially resolved.

4.4. *Multi-line approach: Differential Hanle effect*

Interpretations that are based on observations with a single spectral line are sensitive to the model used for the solar atmosphere and the atomic physics, and they need a statistically determined "non-magnetic" reference level as described in the previous section. As in Zeeman-effect diagnostics one can however suppress such model dependence by using the *differential* rather than the absolute polarization effects, which are found from combinations of lines with different sensitivities to the Hanle effect.

Figure 23 provides examples of spectra for the 4886 and 4934 Å wavelength ranges. The top panels show the usual intensity spectra, which hardly vary from region to region on the Sun (as long as the limb distance is kept fixed), while the panels below show the "second solar spectrum" for two solar regions that differ greatly in Hanle depolarization. Inversions with a simplified Hanle-effect model assuming a spatially unresolved turbulent magnetic field with an isotropic angular distribution give turbulent field strengths B_t of 4 and 30 G, respectively, for these two regions.

Regardless of whether one trusts these quantitative values or not, it is apparent from direct visual inspection of the Q/I spectra that the polarizations in the bottom panels are subject to much more Hanle depolarization than the Q/I spectra of the middle panels. One may therefore immediately draw the qualitative conclusion, without any modelling or special assumptions, that the turbulent fields of the solar region that is represented by the bottom panels are much stronger than those of the region that is represented by the middle panels. The inversions merely try to quantify this direct conclusion in terms of G units.

Because of uncertainties in the atomic and radiative transfer physics and in the collisional rates, the effective decay rate to use for a given line transition for computing the Hanle response to different magnetic fields is often not well known and may for many lines need to be treated as a free parameter. The non-magnetic polarization amplitude is in general also unknown. However, even with this many free parameters, the number

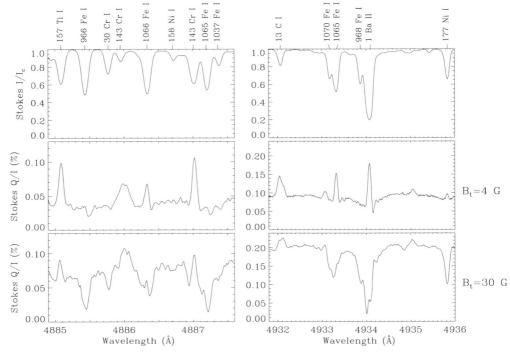

FIGURE 23. Examples of varying Hanle depolarization on the solar disk. For each of the
wavelength ranges 4885–4888 Å and 4932–4936 Å the top panel shows the intensity I (normalized
to the local continuum intensity I_c) with line identifications and multiplet numbers, while the
two lower panels show the fractional linear polarization Q/I for two solar regions (which are
the same for the two wavelength ranges), one with little (middle panel) and one with much
(bottom panel) Hanle depolarization. No attempt has been made here to fix the zero point of
the polarization scale, which is thus arbitrarily chosen, but it is unimportant for the qualitative
comparison of the two solar regions. According to inversions with idealized models the turbulent
field strength B_t has a value of 4 and 30 G for these two regions, which were recorded respectively
on 4-5 April 1995 and 15 September 1996 with the spectrograph slit 5 arcsec inside the limb
at the position angle of geographical north, using ZIMPOL I at the McMath-Pierce facility of
NSO/Kitt Peak. From Stenflo et al. (1998).

of independent observables increases faster than the number of free parameters when
we increase the number of simultaneously used spectral lines with different Hanle sensitivities, and combine recordings made in solar regions with different magnitudes of the
Hanle effect. In the general case we need to sample at least three different solar regions
with at least three different spectral lines to allow an inversion of the observational data
(Stenflo et al. 1998). With more lines and solar regions the problem can be further
constrained, which in principle would allow us to introduce more free parameters, e.g. to
further constrain the shapes of the intrinsic distribution functions for the magnetic field.

4.5. *Superposition of a polarized continuum. Separation of depolarization and intrinsic line polarization*

Note in Fig. 23 that when we have much Hanle depolarization, the polarized spectrum
that otherwise looks like an emission spectrum changes appearance and becomes more
similar to an absorption-line spectrum, since many lines now partially depolarize the
continuum polarization.

Figure 23 illustrates a fundamental problem when trying to extract the intrinsic line polarization: For most spectral lines the continuum polarization is often of the same order of magnitude as the line polarization that we want to determine, except for strong lines like Ca I 4227 Å, Na I D_2–D_1, and a few other lines. The partially polarized continuum opacity combines with the partially polarized line opacity in a complex way that depends on the relative heights of formation of the line and the continuum. Radiative-transfer calculations (Fluri & Stenflo 2001) for instance show that lines formed above the continuum layer do not depolarize in Q/I, while when the line and continuum formation regions overlap, the continuum level in Q/I may be suppressed (in relative units) even more than the suppression of Stokes I by the same absorption line in the intensity spectrum. The interplay between the continuum and line opacities in Q/I needs to be further explored to clarify the proper procedure for extracting the line polarization from the observed polarized spectra.

Another major complication in the quantitative extraction of the intrinsic line polarization is the unknown zero point of the Q/I polarization scale due to instrumental cross talk from Stokes I (instrumental polarization) as well as slight asymmetries in the demodulation process in the detector. Usually the zero point is fixed by globally shifting the Q/I data until Q/I in the continuum agrees with the Q/I that is theoretically predicted from the theory of Fluri & Stenflo (1999). This procedure may however be affected both by deficiencies of the theory and by inaccuracies of the μ value (center-to-limb distance).

5. Extension of the diagnostic range through multi-level effects

5.1. *Evidence for lower-level atomic polarization from differential effects in atomic multiplets*

In the standard view of scattering polarization the emitted radiation gets polarized because of the atomic polarization (alignment) that has been induced in the excited state due to the anisotropy of the incident radiation field. There is however another effect, which has been found to play a fundamental role on the Sun. The spontaneous emission process transfers some of the alignment to the lower state. With many such processes a statistical equilibrium with a polarized lower level is reached. The lower level has thus been optically pumped into an aligned state. Scattering from a polarized initial state produces very different polarization in the emitted radiation as compared with scattering from an unpolarized state.

Our next question is therefore how we can distinguish whether the observed polarization and its accompanying Hanle effect is due to atomic coherences in the excited state or in the ground state. The answer to this question is of greatest significance, since the lower and upper states are sensitive to magnetic fields in entirely different parameter regimes (0.01–1 G and 1–100 G, respectively). To diagnose the magnetic field we need to identify the physical effect that we are dealing with.

For atomic transitions for which lower-level polarization can be neglected the intrinsic polarizability of a scattering transition is determined by the total angular momentum quantum numbers (J, or in the case of hyperfine structure splitting, F) of the lower and upper levels. The observed polarization amplitude should be proportional to this intrinsic polarizability, which is governed by the quantum mechanics of the atom. To test if this is really the case we may compare the observed polarization amplitudes for lines that belong to the same atomic multiplet but have different intrinsic polarizabilities because the total angular momentum quantum numbers are different for the different members of the multiplet.

An example is given by Fig. 24, which shows the three members of multiplet no. 2 of

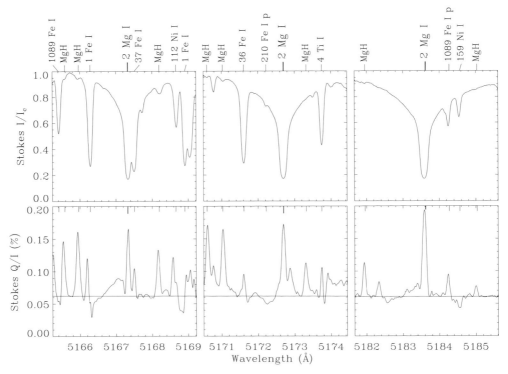

FIGURE 24. Evidence for lower-level atomic polarization in terms of the relative polarization amplitudes of the three strong Mg I lines. Note the prominent polarization peaks due to the molecular MgH lines. The recording was made 5 arcsec inside the solar limb (at $\mu = 0.1$) near the solar north pole on April 4, 1995, with ZIMPOL at NSO/Kitt Peak. ¿From Stenflo et al. (2000b).

Mg I. They have a common upper level with $J = 1$, while for the lower levels J varies from 0 to 1 to 2 when we go from the left to the right diagram in Fig. 24. According to the intrinsic polarizabilities the amplitude of the 5184 Å line should be smaller by between one and two orders of magnitude (when accounting for not only the resonant but also for the fluorescent transitions) as compared with the amplitude of the 5167 Å line, but the observations show that the amplitudes are instead of similar magnitudes. There is no possibility to account for this order-of-magnitude effect by playing around with free parameters in radiative transfer modelling. If one however takes into account lower level atomic polarization produced by optical pumping, then the observed amplitude ratios within the multiplet are reproduced without the need to adjust free parameters (Trujillo Bueno 2001). This result provides convincing evidence for the existence of lower-level atomic polarization and demonstrates that the relative polarization amplitudes within multiplets can be used to unambiguously distinguish it from other physical effects that have entirely different polarization signatures.

Multiplet no. 3 of Ca I (6103, 6122, and 6162 Å) has the same quantum number structure as multiplet no. 2 of Mg I. The relative polarization amplitude ratios for the three members of this multiplet are indeed observed to be the same as for the Mg I case that we have discussed.

A multiplet with an entirely different quantum-number structure is the Ca II infrared triplet at 8498, 8542, and 8662 Å. Also here the predicted amplitude ratios based on

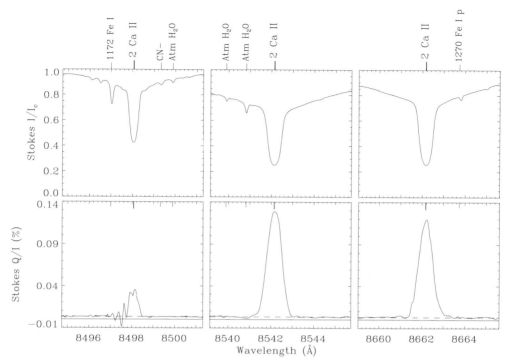

FIGURE 25. Evidence for optical pumping from the relative polarization amplitudes of the Ca II infrared triplet, observed at $\mu = 0.1$ near the solar north pole with ZIMPOL at NSO/Kitt Peak on November 17, 1994. ¿From Stenflo et al. (2000b).

quantum mechanics neglecting optically pumped lower-level coherences deviate from the observed ratios (seen in Fig. 25), not subtly but by order-of-magnitude effects. Inclusion of the optically pumped coherences however again lead to agreement with the observations (Manso Sainz & Trujillo Bueno 2001).

5.2. *Evidence for lower-level atomic polarization from Na I D_1 5896 Å and Ba II 4934 Å*

The scattering polarization across the Na I D_2 and D_1 lines that we showed in Fig. 19 has long remained enigmatic and a fascinating challenge for quantum and radiative-transfer theory. The thin solid curve in the Q/I diagram of Fig. 26 shows a recording made with ZIMPOL in April 1995 in a very quiet limb region (with the slit parallel to and 5 arcsec inside the limb) near the Sun's north pole (Stenflo & Keller 1996, 1997). While the D_2 line transition according to its J quantum numbers should have an intrinsic polarizability $W_2 = 0.5$, the D_1 line, being a $J = \frac{1}{2} \rightarrow \frac{1}{2}$ transition, should be intrinsically unpolarizable ($W_2 = 0$). The observations however show a pronounced narrow polarization peak in the D_1 line core. This peculiar profile shape is not limited to a narrow limb zone but can be seen over a wide range of limb distances (Stenflo et al. 2000a).

The shape of the polarization profile in the line wings, with the remarkable sign reversal of the polarization between the D_2 and D_1 lines, could be explained already two decades ago (Stenflo 1980) in terms of quantum-mechanical interference between the two excited states of different total angular momentum quantum numbers $J = \frac{3}{2}$ and $\frac{1}{2}$, as illustrated by the thick solid curve in Fig. 26. When the quantum interference term is removed,

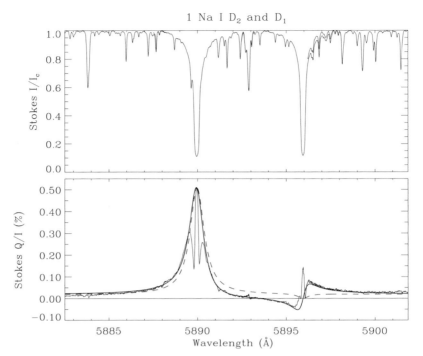

FIGURE 26. The scattering polarization observed in April 1995 with ZIMPOL across the Na I D_2 and D_1 lines (thin solid curves) is modelled taking quantum interference between the $J = \frac{3}{2}$ and $\frac{1}{2}$ excited states into account (thick solid curve), while the dashed curve shows what happens when the interference term is omitted (Stenflo 1997). While this model, which ignores hyperfine structure splitting and lower-level atomic polarization, can reproduce the wing polarization very well, it is unable to account for the narrow polarization peaks in the Doppler cores.

one gets the dashed curve without sign reversal. This model however always gives zero polarization at the center of the D_1 line.

Recently a way was found to obtain scattering polarization in the D_1 line core through a combination of hyperfine structure splitting and optical pumping (Landi Degl'Innocenti 1998, 1999). The $J = \frac{1}{2}$ ground state is not polarizable in principle, even with optical pumping, but when the ground state is split due to coupling between the nuclear spin and the electronic angular momentum, we get a hyperfine structure multiplet with levels that have new total angular momentum quantum numbers (F) and are polarizable by coherency transfer from the excited state (optical depopulation pumping).

Details of the observed shape of the Q/I profile in the Na I D_1 line are shown in Figs. 27 and 28. The polarization peak in the line core can only be explained in terms of lower-level atomic polarization. A very similar profile shape with a core peak and surrounding minima is found for the Ba II 4934 Å line (Stenflo et al. 2000b), which has the same quantum numbers, including hyperfine structure splitting, as the Na I D_1 line. Also here the observed line polarization could not be explained without lower-level atomic polarization.

With ZIMPOL II we are now in a position to collect a wealth of spatially varying Hanle effect signatures. An example was given in Fig. 19, which shows the spatial fluctuations along the slit of the polarization amplitude in the Doppler core of the Na I D_2 5890 Å line. While the variations in Q/I and U/I are due to the Hanle effect, V/I is

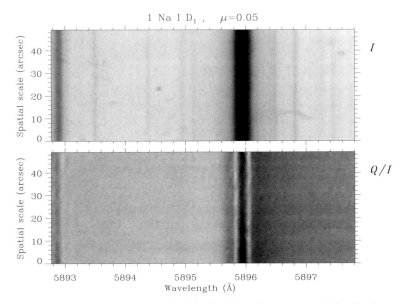

FIGURE 27. 2-D spectra of Stokes I and Q/I for Na I D_1 at $\mu = 0.05$. Darker areas mean stronger polarization (in the positive direction), lighter areas weaker or negative polarization. The slit is parallel to the limb. While there is some spatial variation of the polarized core peak maximum, the wing polarization remains spatially invariant.

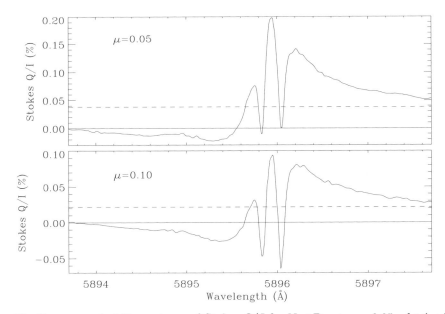

FIGURE 28. Upper panel: 1-D spectrum of Stokes Q/I for Na I D_1 at $\mu = 0.05$, obtained from the corresponding 2-D spectrum in Fig. 27 by averaging along the slit. The horizontal dashed line represents the level of the continuum polarization. Lower panel: The corresponding profile for $\mu = 0.10$.

exclusively due to the longitudinal Zeeman effect. Note that the broad Q/I polarization maxima in the wings of the D_2 line are spatially invariant, in accordance with theoretical expectations, since the Hanle effect only operates in the Doppler core (cf. Stenflo 1994, 1998). Note also that there is practically no Q/I signal in the D_1 line, since it can only be produced by lower-level atomic polarization, which gets destroyed by Hanle depolarization in magnetically active regions.

When one compares the linear polarization fluctuations in the cores of the D_2 and D_1 lines, one finds that they are uncorrelated to a degree that one would expect if the polarizations in the two lines are formed in two entirely different Hanle regimes (Stenflo et al. 2001). This leads us to conclude that while the Hanle effect for the D_1 line is governed by the life time of the ground state and therefore is sensitive to fields of typically 0.01–1 G, the Hanle effect in the D_2 line is governed by the life time of the excited state and therefore responds to fields in the 1–100 G range. In this higher field-strength range the D_1 polarization belongs to the Hanle saturated regime, which is the regime that applies to coronal forbidden lines.

5.3. *Implications for the structure of solar magnetic fields*

In the preceding subsections we have seen how the presence of lower-level atomic polarization has been expressed in the form of the observed relative polarization amplitudes in the multiplets of Mg I at 5167–5184 Å, Ca I at 6103–6162 Å, and the Ca II infrared triplet, as well as in the form of the observed polarization peaks in the Na I D_1 and Ba II 4934 Å lines, which would not polarize in the absence of optical pumping. We therefore now have an array of unambiguous observational signatures of lower-state atomic polarization. However, these signatures are only seen in very quiet regions on the Sun. As soon as we go to magnetically active regions they vanish, which shows that the atomic polarization in the lower level does not survive depolarization by the Hanle effect in these magnetic regions.

Solar magnetic fields are however ubiquitous, even in the most "quiet" solar regions. The convincing observational evidence that we now have for the existence of lower-level atomic polarization in quiet regions provides important constraints on the magnetic field structure there, at height ranging from below the temperature minimum region (Ca I), throughout much of the chromosphere (Ba II, Mg I, Na I D_1, Ca II infrared triplet). If the magnetic field in these regions were stronger than about 0.01 G, then the signatures of lower-state atomic polarization would be severely suppressed, unless the magnetic field everywhere (i.e., with filling factors near unity) throughout the sampled quiet regions is oriented fairly close to the vertical direction. Since we have other grounds to believe that the chromospheric magnetic fields are stronger than milligauss, we are led to the conclusion that the magnetic field in quiet regions is much more vertical than horizontal, at least up to the chromospheric heights where the cores of the Ca II infrared triplet lines are formed. This means that magnetic canopies do not exist there, in contradiction with the previous prevailing view of the Sun's magnetic field.

5.4. *Last scattering approximation and idealized modelling*

A most useful conceptual tool for thinking about and understanding scattering polarization is the concept of the "last scattering approximation". In an optically thick atmosphere we have many absorption and emission processes, but regardless of the number of scattering events, one of them must be the last one before the photon leaves the Sun to travel to the observer. The dominating factor determining the scattering polarization is the degree of anisotropy of the incident radiation field seen by this last scatterer, as well as the external magnetic field at the location of that particle. Although the polarization of the incident radiation influences the polarization of the emitted light, it is a second-order effect, so to a good approximation one can think of the incident radiation field as being unpolarized.

Conceptually one can therefore think in terms of a single scattering event that produces polarization that scales with the intrinsic quantum-mechanical polarizability, diluted by a geometric factor (depending on the anisotropy of the incident radiation field and the scattering angles). In addition we have dilution of the line polarization by continuum photons, which have another degree of polarization. This dilution scales with the relative contributions of the line and continuum opacities to the total opacity, and it is therefore highly wavelength dependent. In this way we can, without the use of radiative transfer, meaningfully model for instance the observed polarization in the Na I D_2 and D_1 lines, as illustrated in Fig. 26. Since the intrinsic polarizability used in this particular model does not account for lower-level atomic polarization, it cannot reproduce the core peak in the D_1 line, but it gives an excellent fit to the wing polarization (although it also neglects the hyperfine structure splitting).

In this highly idealized model the grossly simplifying assumption has been made that the degree of anisotropy of the incident radiation is independent of wavelength within the line, which is certainly invalid for strong non-LTE lines, for which the effects of opacity and partial frequency redistribution (PRD) play major roles. Radiative-transfer calculations for instance indicate that the triplet structure of the D_2 polarization is largely a consequence of PRD effects (Fluri, private communication).

What naturally emerges from concepts and modelling based on the last scattering approximation is the degree of polarization Q/I and *not* Q alone (or Q/I_c, normalized to the local continuum intensity I_c). Q/I is the appropriate quantity to plot and think in terms of. If we would transform Q/I to Q/I_c, the profile shape would be largely dictated by the particular shape of the Stokes I profile, which is due to radiative-transfer effects that have little to do with the processes responsible for Q/I. Furthermore, Q/I is the directly observed quantity that is free from flat-field effects in the detector and spectral features caused by $I \rightarrow Q$ instrumental cross talk, which is not the case for Q/I_c.

5.5. *Outlook*

The magnetic fields leave their "fingerprints" in subtle ways in the polarized spectrum. It is our task to properly meaure and interpret these fingerprints. Since it was first introduced in solar physics by Hale (1908), the Zeeman effect has provided us with powerful diagnostic tools to explore the Sun's magnetic field. With the advent of imaging polarimeters of unprecedented polarimetric sensitivity, the "second solar spectrum" that is produced by scattering phenomena has become accessible to observations. The magnetic-field imprints left in the second solar spectrum via the Hanle effect provide us with a novel diagnostic window to explore hitherto hidden aspects of the Sun's magnetic field.

The observational programs for the scattering polarization, including all the magnetic-field effects that result from the Hanle effect in both the upper and lower atomic states, are still in an exploratory phase. The scattering polarization is however gradually maturing into a powerful tool for magnetic-field diagnostics in regimes that are only poorly, if at all, accessible by Zeeman-effect observations. Examples are diagnostics of the magnetic field in the solar chromosphere, or of the weak and tangled magnetic fields that fill most of the photospheric volume. The polarization signatures of the Hanle effect in the mG regime (lower-level atomic polarization) and the 1–100 G regime can now be unambiguously distinguished from each other, and they are entirely different from the signatures of the Zeeman effect. A recent atlas of the second solar spectrum (Gandorfer 2000, 2001) will help guide us in the selection of suitable line combinations. For overviews of the most recent progress on the topic of solar polarization we refer to the proceedings edited by Nagendra & Stenflo (1999) and by Sigwarth (2001).

As the diagnostic applications of the scattering polarization require polarization accuracies of at least 10^{-4}, we need large photon-collecting telescopes to avoid to have to make too restrictive trade-offs with the spatial, temporal, and spectral resolutions. Since the magnetic field is structured on all spatial scales, we want to push for the highest possible spatial resolution, in future with the help of adaptive optics, to be able to explore the evolution and morphology of the small-scale structures. At this high-resolution end we have to compromise the polarimetric accuracy that can be reached. At the same time, there will always remain an important unresolved tail of the size distribution, which can only be explored by indirect techniques, like multi-line Zeeman observations or Hanle diagnostics, which require much higher polarimetric accuracies and therefore necessitate trade-offs with the angular and temporal resolutions. A comprehensive picture of solar magnetic fields is built from a combination of these various direct and indirect diagnostic methods.

REFERENCES

ALFVÉN, H. & FÄLTHAMMAR, C.-G. 1963 *Cosmical Electrodynamics — Fundamental Principles*, 2nd edition. Oxford Univ. Press.

BIANDA, M., STENFLO, J.O. & SOLANKI, S.K. 1999 *Astron. Astrophys.* **350**, 1060.

BOMMIER, V. 1980 *Astron. Astrophys.* **87**, 109.

BÜNTE, M., SOLANKI, S.K. & STEINER, O. 1993 *Astron. Astrophys.* **268**, 736.

COLLADOS, M. 2001. In *Advanced Solar Polarimetry — Theory, Observation, and Instrumentation* (ed. M. Sigwarth). *ASP Conf. Ser.*, in press.

FAUROBERT-SCHOLL, M. 1993 *Astron. Astrophys.* **268**, 765.

FAUROBERT-SCHOLL, M., FEAUTRIER, N., MACHEFERT, F., PETROVAY, K. & SPIELFIEDEL, A. 1995 *Astron. Astrophys.* **298**, 289.

FLURI, D.M. & STENFLO, J.O. 1999 *Astron. Astrophys.* **341**, 902.

FLURI, D.M. & STENFLO, J.O. 2001. In *Advanced Solar Polarimetry — Theory, Observation, and Instrumentation* (ed. M. Sigwarth). *ASP Conf. Ser.*, in press.

FRAZIER, E.N. & STENFLO, J.O. 1972 *Solar Phys.* **27**, 330.

FRAZIER, E.N. & STENFLO, J.O. 1978 *Astron. Astrophys.* **70**, 789.

GANDORFER, A. 2000 The Second Solar Spectrum, Vol. I: 4625 Å to 6995 Å. ISBN no. 3 7281 2764 7, VdF, Zurich.

GANDORFER, A. 2001. In *Advanced Solar Polarimetry — Theory, Observation, and Instrumentation* (ed. M. Sigwarth). *ASP Conf. Ser.*, in press.

GIOVANELLI, R.G. 1980 *Solar Phys.* **68**, 49.

GIOVANELLI, R.G. & JONES, H.P. 1982 *Solar Phys.* **79**, 267.

GROSSMANN-DOERTH, U., LARSSON, B. & SOLANKI, S.K. 1988a *Astron. Astrophys.* **204**, 266.

GROSSMANN-DOERTH, U., SCHÜSSLER, M. & SOLANKI, S.K. 1988b *Astron. Astrophys.* **206**, L37.

GROSSMANN-DOERTH, U., SCHÜSSLER, M. & STEINER, O. 1998 *Astron. Astrophys.* **337**, 928.

GROSSMANN-DOERTH, U., SCHÜSSLER, M., SIGWARTH, M. & STEINER, O. 2000 *Astron. Astrophys.* **357**, 351.

HALE, G.E. 1908 *Astrophys. J.* **28**, 100.

HANLE, W. 1924 *Z. Phys.* **30**, 93.

IVANOV, V.V. 1991. In *Stellar Atmospheres: Beyond Classical Models* (ed. L. Crivellari, I. Hubeny & D.G. Hummer). Proc. NATO, p. 81. Kluwer.

JONES, H.P. & GIOVANELLI, R.G. 1983 *Solar Phys.* **87**, 37.

LANDI DEGL'INNOCENTI, E. 1998 *Nature* **392**, 256.

LANDI DEGL'INNOCENTI, E. 1999. In *Solar Polarization* (ed. K.N. Nagendra & J.O. Stenflo). *ASSL* **243**, p. 61. Kluwer.

LANDI DEGL'INNOCENTI, E. & LANDI DEGL'INNOCENTI, M. 1977 *Astron. Astrophys.* **56**, 111.

MANSO SAINZ, R. & TRUJILLO BUENO, J. 2001. In *Advanced Solar Polarimetry — Theory, Observation, and Instrumentation* (ed. M. Sigwarth). *ASP Conf. Ser.*, in press.

MORUZZI, G. & STRUMIA, F. (eds.) 1991 *The Hanle Effect and Level-Crossing Spectroscopy.* Plenum.

NAGENDRA, K.N., STENFLO, J.O. (eds.) 1999 *Solar Polarization. ASSL* **243**. Kluwer.

NORDLUND, Å & STEIN, R.F. 1990. In *Solar Photosphere: Structure, Convection and Magnetic Fields* (ed. J.O. Stenflo). *IAU Symp.* **102**, 79.

PARKER, E.N. 1978 *Astrophys. J.* **221**, 368.

PETRASSO, R.D. 1990 *Nature* **343**, 21.

POVEL, H.P. 1995 *Optical Engineering* **34**, 1870.

RACHKOVSKY, D.N. 1962a *Izv. Krymsk. Astrofiz. Obs.* **27**, 148.

RACHKOVSKY, D.N. 1962b *Izv. Krymsk. Astrofiz. Obs.* **28**, 259.

RÜEDI, I., SOLANKI, S.K., LIVINGSTON, W. & STENFLO, J.O. 1992 *Astron. Astrophys.* **263**, 323.

RUIZ COBO, B. & DEL TORO INIESTA, J.C. 1992 *Astrophys. J.* **398**, 375.

SÁNCHEZ ALMEIDA, J. 1992 *Solar Phys.* **137**, 1.

SIGWARTH, M. (ed.) 2001 *Advanced Solar Polarimetry — Theory, Observation, and Instrumentation. ASP Conf. Ser.*, in press.

SIGWARTH, M., BALASUBRAMANIAM, K.S., KNÖLKER, M. & SCHMIDT, W. 1999 *Astron. Astrophys.* **349**, 941.

SOCAS-NAVARRO, H. 2001. In *Advanced Solar Polarimetry — Theory, Observation, and Instrumentation* (ed. M. Sigwarth). *ASP Conf. Ser.*, in press.

SOLANKI, S.K. 1993 *Space Sci. Rev.* **63**, 1.

SOLANKI, S.K., ZUFFEREY, D., LIN, H., RÜEDI, I. & KUHN, J.R. 1996 *Astron. Astrophys.* **310**, L33.

SPRUIT, H.C. & ZWEIBEL, E.G. 1979 *Solar Phys.* **62**, 15.

STEINER, O. 2000 *Solar Phys.* **196**, 245.

STENFLO, J.O. 1973 *Solar Phys.* **32**, 41.

STENFLO, J.O. 1980 *Astron. Astrophys.* **84**, 68.

STENFLO, J.O. 1994 *Solar Magnetic Fields — Polarized Radiation Diagnostics.* Kluwer.

STENFLO, J.O. 1997 *Astron. Astrophys.* **324**, 344.

STENFLO, J.O. 1998 *Astron. Astrophys.* **338**, 301.

STENFLO, J.O. 1999. In *Solar Polarization* (ed. K.N. Nagendra & J.O. Stenflo). *ASSL* **243**, p. 1. Kluwer.

STENFLO, J.O. 2001. In *Advanced Solar Polarimetry — Theory, Observation, and Instrumentation* (ed. M. Sigwarth). *ASP Conf. Ser.*, in press.

STENFLO, J.O. & KELLER, C.U. 1996 *Nature* **382**, 588.

STENFLO, J.O. & KELLER, C.U. 1997 *Astron. Astrophys.* **321**, 927.

STENFLO, J.O., HARVEY, J.W., BRAULT, J.W. & SOLANKI, S.K 1984 *Astron. Astrophys.* **131**, 33.

STENFLO, J.O., KELLER, C.U. & GANDORFER, A. 1998 *Astron. Astrophys.* **329**, 319.

STENFLO, J.O., GANDORFER, A. & KELLER, C.U. 2000a *Astron. Astrophys.* **355**, 781.

STENFLO, J.O., KELLER, C.U. & GANDORFER, A. 2000b *Astron. Astrophys.* **355**, 789.

STENFLO, J.O., GANDORFER, A., WENZLER, T. & KELLER, C.U. 2001 *Astron. Astrophys.*, in press.

STEPANOV, V.E. 1958 *Izv. Krymsk. Astrofiz. Obs.* **18**, 136.

STIX, M. 1989 *The Sun — An Introduction.* Springer.

TRUJILLO BUENO, J. 2001. In *Advanced Solar Polarimetry — Theory, Observation, and Instrumentation* (ed. M. Sigwarth). *ASP Conf. Ser.*, in press.

UNNO, W. 1956 *Publ. Astron. Soc. Japan* **8**, 108.

Polarized Radiation Diagnostics of Stellar Magnetic Fields

By Gautier Mathys

European Southern Observatory, Casilla 19001, Santiago 19, Chile

The main techniques used to diagnose magnetic fields in stars from polarimetric observations are presented. First, a summary of the physics of spectral line formation in the presence of a magnetic field is given. Departures from the simple case of linear Zeeman effect are briefly considered: partial Paschen-Back effect, contribution of hyperfine structure, and combined Stark and Zeeman effects. Important approximate solutions of the equation of transfer of polarized light in spectral lines are introduced. The procedure for disk-integration of emergent Stokes profiles, which is central to stellar magnetic field studies, is described, with special attention to the treatment of stellar rotation. This formalism is used to discuss the determination of the mean longitudinal magnetic field (through the photographic technique and through Balmer line photopolarimetry). This is done within the specific framework of Ap stars, which, with their unique large-scale organized magnetic fields, are an ideal laboratory for studies of stellar magnetism. Special attention is paid to those Ap stars whose magnetically split line components are resolved in high-dispersion Stokes I spectra, and to the determination of their mean magnetic field modulus. Various techniques of exploitation of the information contained in polarized spectral line profiles are reviewed: the moment technique (in particular, the determination of the crossover and of the mean quadratic field), Zeeman-Doppler imaging, and least-squares deconvolution. The prospects that these methods open for linear polarization studies are sketched. The way in which linear polarization diagnostics complement their Stokes I and V counterparts is emphasized by consideration of the results of broad band linear polarization measurements. Illustrations of the use of various diagnostics to derive properties of the magnetic fields of Ap stars are given. This is used to show the interest of deriving more physically realistic models of the geometric structure of these fields. How this can possibly be achieved is briefly discussed. An overview of the current status of polarimetric studies of magnetic fields in non-degenerate stars of other types is presented. The final section is devoted to magnetic fields of white dwarfs. Current knowledge of magnetic fields of isolated white dwarfs is briefly reviewed. Diagnostic techniques are discussed, with particular emphasis on the variety of physical processes to be considered for understanding of spectral line formation over the broad range of magnetic field strengths encountered in these stars.

---

1. General framework

1.1. *Introduction*

The basic physics of the generation of polarized radiation in stars by a magnetic field is studied in the series of lectures presented at this Winter School by Egidio Landi Degl'Innocenti. The way in which it is applied to diagnose stellar magnetic fields has a lot in common with its use for measurements of solar magnetic fields, which is described in the course given here by Jan Stenflo. Many relevant aspects of the instrumentation used for stellar magnetic field studies are covered in Christoph Keller's presentation. The reader is warmly encouraged to refer for additional information to the contributions of these authors appearing in this volume. Yet, the diagnosis of magnetic fields in stars also involves specific aspects which are not addressed by other lecturers. Emphasis in this course is laid on introducing them, and on giving an overview of our current knowledge of stellar magnetism as obtained from (mostly) polarimetric studies.

Studies of stellar magnetic fields have so far been limited by the fact that the disks of the considered stars are not spatially resolved in the observations from which the field is diagnosed. While the progressive development of techniques of optical interferometry opens prospects of overcoming to some extent this limitation, we are nowhere near being able to resolve spatially stellar features of sizes comparable to those of the largest magnetic features of the solar photosphere, the sunspots. Accordingly, a primary difficulty in the interpretation of stellar observations in terms of magnetic fields comes from the fact that the observed signal is integrated over the whole visible stellar hemisphere (or in future interferometric works, a large fraction of it). Of course, this is to some extent true for the observational determination of all the physical parameters of stars (such as temperature, elemental abundances). But while these parameters are scalars and, most often, in good first approximation, do not vary much from place to place on the star, the magnetic field is a vector, whose direction and intensity may typically be very different at different points of the stellar surface. Accordingly, a specific, complex and fundamental aspect of stellar magnetic field studies is to find ways of extracting physically meaningful constraints about the field from disk-integrated observational data.

As an indirect consequence of the disk-integrating process, Doppler effect due to stellar rotation adds its contribution to the observable signal. In some cases, this may hamper the diagnosis of the magnetic field, while in others, this may actually enhance the diagnostic contents of the observations.

Finally, atmospheres of various types of stars are characterized by a wide variety of physical conditions (temperature, density, ...) and of chemical compositions. Not surprisingly, the same magnetic field diagnostics cannot be used in all kinds of stars. This also implies that different field determination methods may have to be applied for stars of different types.

1.2. *Zeeman effect*

The primary physical mechanism leading to observable manifestations of magnetic fields in stars is the Zeeman effect. Let us consider an atomic *level* defined by its energy (in the absence of external perturbation) E_0, its angular momentum quantum number J, and its Landé factor g. In the presence of a magnetic field $\boldsymbol{H}$, this level is split into $(2J+1)$ *states*, characterized by their magnetic quantum number M ($M = -J, -J+1,$ ... , $J-1, J$). They have equally spaced energies $E(M)$ given by:

$$E(M) = E_0 + g\,M\,\hbar\,\omega_{\mathrm{L}}\,, \tag{1.1}$$

where $\omega_{\mathrm{L}} = e\,H/(2\,m_{\mathrm{e}}\,c)$ is the Larmor frequency (m_{e} is the electron mass).

1.3. *Transfer equation*

Interpretation of (spectro)polarimetric observations in terms of magnetic fields is based on the solution of the equation of transfer of polarized light in spectral lines:

$$\mu\,d\mathcal{S}/d\tau = (\mathcal{I} + \boldsymbol{\eta})\,[\mathcal{S} - B(\nu_0, T)\,\mathcal{J}]. \tag{1.2}$$

The state of the radiation is described by the Stokes vector $\mathcal{S} = (I, Q, U, V)^{\mathrm{T}}$. This form of the equation corresponds to a plane-parallel atmosphere, in LTE. The optical depth is denoted by τ; μ is the cosine of the angle θ between the direction of propagation of the light and the normal to the stellar surface. $B(\nu_0, T)$ is the Planck function at temperature T and at the frequency ν_0 of the transition; $\mathcal{I}$ is the 4×4 unit matrix and

$\mathcal{J} = (1,0,0,0)^{\mathrm{T}}$. The Mueller matrix $\boldsymbol{\eta}$ has the usual form:

$$\boldsymbol{\eta} = \begin{pmatrix} \eta_I & \eta_Q & \eta_U & \eta_V \\ \eta_Q & \eta_I & \rho_V & -\rho_U \\ \eta_U & -\rho_V & \eta_I & \rho_Q \\ \eta_V & \rho_U & -\rho_Q & \eta_I \end{pmatrix} , \tag{1.3}$$

with

$$\frac{\eta_I}{\eta_\ell} = \frac{1}{2}\left(\eta_0 - \frac{\eta_+ + \eta_-}{2}\right)\sin^2\gamma + \frac{1}{2}(\eta_+ + \eta_-) , \tag{1.4a}$$

$$\frac{\eta_Q}{\eta_\ell} = \frac{1}{2}\left(\eta_0 - \frac{\eta_+ + \eta_-}{2}\right)\sin^2\gamma\,\cos 2\chi , \tag{1.4b}$$

$$\frac{\eta_U}{\eta_\ell} = \frac{1}{2}\left(\eta_0 - \frac{\eta_+ + \eta_-}{2}\right)\sin^2\gamma\,\sin 2\chi , \tag{1.4c}$$

$$\frac{\eta_V}{\eta_\ell} = \frac{1}{2}(\eta_+ - \eta_-)\cos\gamma , \tag{1.4d}$$

$$\frac{\rho_Q}{\eta_\ell} = \left(\rho_0 - \frac{\rho_+ + \rho_-}{2}\right)\sin^2\gamma\,\cos 2\chi , \tag{1.4e}$$

$$\frac{\rho_U}{\eta_\ell} = \left(\rho_0 - \frac{\rho_+ + \rho_-}{2}\right)\sin^2\gamma\,\sin 2\chi , \tag{1.4f}$$

$$\frac{\rho_V}{\eta_\ell} = (\rho_+ - \rho_-)\cos\gamma . \tag{1.4g}$$

The transition probability between the atomic levels responsible for the line under consideration is accounted for by:

$$\eta_\ell = \frac{1}{\kappa_{\mathrm{c}}}\frac{h\,\lambda_0}{4\,\pi}\,N_1\left[1 - \exp(-h\,\nu_0/k\,T)\right]B_{12} , \tag{1.5}$$

where κ_{c} is the continuum absorption coefficient at the wavelength λ_0 of the line (which is used to define the optical depth scale), B_{12} is the Einstein coefficient of absorption between the lower level 1 and the upper level 2, and N_1 is the population of the lower level. The angle between the magnetic field and the line of sight is denoted by γ, and χ is the azimuth of the magnetic vector with respect to the positive Q direction. Using the notations H_z to represent the component of $\boldsymbol{H}$ along the line of sight, and H_Q and $H_{Q+\pi/2}$ for the components of $\boldsymbol{H}$ resp. parallel and orthogonal to the positive Q direction in the plane perpendicular to the line of sight, one has:

$$\cos\gamma = H_z/H , \tag{1.6a}$$

$$\sin^2\gamma\,\cos 2\chi = (H_Q^2 - H_{Q+\pi/2}^2)/H^2 , \tag{1.6b}$$

$$\sin^2\gamma\,\sin 2\chi = 2\,H_Q\,H_{Q+\pi/2}/H^2 . \tag{1.6c}$$

The form of the expressions in the right-hand sides of Eqs. (1.4) reflects the selection rule that only transitions between states with magnetic quantum numbers such that $M_1 - M_2 = 0$ (known as π components) or $M_1 - M_2 = \pm 1$ ($\sigma_\pm$ components) are allowed. These transitions are described by the coefficients ($q = 0, \pm 1$):

$$\eta_q(\lambda) = \sum_{M_1, M_2} S_q(M_1, M_2)\,\psi[\lambda - \lambda_0 - \Delta\lambda(M_1, M_2)] , \tag{1.7a}$$

which represent absorption and emission processes, and

$$\rho_q(\lambda) = \sum_{M_1, M_2} S_q(M_1, M_2)\,\phi[\lambda - \lambda_0 - \Delta\lambda(M_1, M_2)] , \tag{1.7b}$$

which account for anomalous dispersion effects. The relative strengths of the individual line components are normalized to unity for each set of components corresponding to the same value of q:

$$\sum_{M_1, M_2} S_q(M_1, M_2) = 1 \qquad (q = 0, \pm 1) . \qquad (1.8)$$

Their expression is:

$$S_q(M_1, M_2) = 3 \begin{pmatrix} J_1 & J_2 & 1 \\ M_1 & -M_2 & -q \end{pmatrix}^2 , \qquad (1.9)$$

from which one sees that $S_q(M_1, M_2) \neq 0$ only if $M_1 - M_2 = q$. It is worth noting that the number and relative strengths of the individual magnetic components of a transition depend *only* on the angular momentum quantum numbers of the levels between which it takes place. Their shifts with respect to the nominal wavelength of the line, by contrast, depend on the Landé factors of the involved levels:

$$\Delta\lambda(M_1, M_2) = (g_1 M_1 - g_2 M_2) \Delta\lambda_Z H , \qquad (1.10)$$

where $\Delta\lambda_Z = k \lambda_0^2$ with $k = e/(4\pi m_e c^2) = 4.67 \ 10^{-13}$ Å^{-1}G^{-1}. It is important to note that all the line components corresponding to the same value of q have the same effect on the polarization of the light. On the other hand, all the line components have the same absorption (ψ) and anomalous dispersion (ϕ) profiles. The absorption profile is normalized to unity:

$$\int_{-\infty}^{+\infty} \psi(\lambda - \lambda_0) \, d\lambda = 1 , \qquad (1.11)$$

and $\phi(\lambda - \lambda_0)$ is the corresponding anomalous dispersion profile (see e.g. Mathys 1989 for more details).

The set of relative line strengths $S_q(M_1, M_2)$ and wavelength shifts $\Delta\lambda(M_1, M_2)$ of the components of a transition split by a magnetic field defines its *Zeeman pattern*. Precise determination of stellar magnetic fields depends critically on the knowledge of the correct Zeeman patterns of the diagnostic lines. The main source of uncertainty lies with the Landé factors. While it may be possible to compute them through simple arithmetics from some subset of quantum numbers of the levels between which the transition takes place, such calculations rest on the assumption that those levels pertain to some (almost) pure coupling scheme (*LS* coupling is most frequent). In practice, though, it fairly often happens that this approximation is not particularly good, and the values of the Landé factors that it yields may on occasion be quite poor. As a rule of thumb, experimental values obtained in laboratory studies should be preferred whenever available. When this is not the case, the best alternative is provided by values derived from detailed model atom calculations. In particular, values given in Kurucz's (http://cfaku5.harvard.edu/LINELISTS.html) line lists often prove very good (Mathys 1990b), and their use is recommended (note that for levels for which an experimental value exists, Kurucz gives this value, not a calculated one). Note that, although the issue of availability of correct Landé factors is in principle relevant in both the solar and stellar cases, the problem is exacerbated in the latter context because one is led to consider a wider variety of diagnostic lines, and in particular, may in some instances have no choice but dealing with fairly "exotic" transitions, not necessarily well studied in laboratory or theoretical works.

1.4. *Departures from linear Zeeman effect*

Equation (1.1) describes the linear Zeeman effect. It is valid provided that (*a*) quadratic Zeeman effect is negligible;

(*b*) J is a "good" quantum number.

The first approximation is always valid for the magnetic field strengths encountered in non-degenerate stars, but the contribution of quadratic Zeeman effect has to be taken into account (and may be dominant) for certain magnetic white dwarfs (see Sect. 5.2 for further discussion).

1.4.1. *Partial Paschen-Back effect*

The second condition implies in particular that the magnetic splitting of the level is small with respect to the fine structure separation within the spectroscopic *term* to which this level belongs. Departures from this approximation occur in a number of transitions of great astrophysical interest, such as Fe II λ 6149, an important diagnostic line for stellar magnetic fields (see Sect. 2.4), or the Li I λ 6708 doublet. Splitting of one (at least) of the levels involved in such transitions occurs in a regime of partial Paschen-Back effect: to calculate this splitting, one must take simultaneously into consideration the various levels belonging to the same spectroscopic term. An example of such a calculation is given in Mathys (1990a). The dependence on the magnetic field strength of the energy shifts $\Delta E(M_i)$ of the magnetic states with respect to the unperturbed level is no longer linear, and in most cases, $\Delta E(-M_i) \neq -\Delta E(M_i)$. The expressions of the coefficients η_q and ρ_q remain formally the same as in Eqs. (1.7), but the relative line strengths now depend on the magnetic field intensity. In general, one only of the levels between which the transition takes place belongs to a term for which the fine structure is not large with respect to the Zeeman effect. For illustration, let us consider a transition whose *lower* level belongs to such a term. We assume that this term comprises n levels, which in the absence of a magnetic field, correspond to values $J^{(1)}$, $J^{(2)} = J^{(1)} + 1$, ... , $J^{(n)} = J^{(1)} + n - 1$, of the angular momentum quantum number. The relative strength $S_{q,J^{(\ell)}}(M_1, M_2)$ of the line component corresponding to the transition between magnetic state M_1 of level $J^{(\ell)}$ $(-J^{(\ell)} \leq M_1 \leq J^{(\ell)})$ and magnetic state M_2 of some upper level (formed in pure Zeeman regime) of angular momentum quantum number J_2 can be written as:

$$
S_{q,J^{(\ell)}}(M_1, M_2) = 3 \left[\sum_{J^{(k)}=|M_1|}^{J^{(n)}} a_{J^{(k)}J^{(\ell)}}(M_1, H) \begin{pmatrix} J^{(k)} & J_2 & 1 \\ M_1 & M_2 & -q \end{pmatrix} \right]^2, \quad (1.12)
$$

with the normalization condition:

$$
\sum_{J^{(k)}=|M_1|}^{J^{(n)}} \left[a_{J^{(k)}J^{(\ell)}}(M_1, H) \right]^2 = 1. \quad (1.13)
$$

The strengths of the components corresponding to $M_1 - M_2 = \pm 1$ generally differ from each other. This, combined with the above-mentioned lack of symmetry of the split energy states with respect to the unperturbed level, implies that, in contrast with the Zeeman patterns, the partial Paschen-Back regime line patterns are asymmetric. An example of such a pattern is shown in Fig. 1 for various magnetic field strengths.

1.4.2. *Hyperfine structure*

Some of the most interesting questions in stellar magnetism require the consideration of diagnostic lines with considerable hyperfine structure. This happens, in particular, in studies of Ap stars, whose spectra abound in strong rare earth lines, and may on occasion be so dominated by such lines that the latter become the only usable diagnostics of stellar properties (see e.g. Cowley & Mathys 1998). Most often, the energy perturbations due

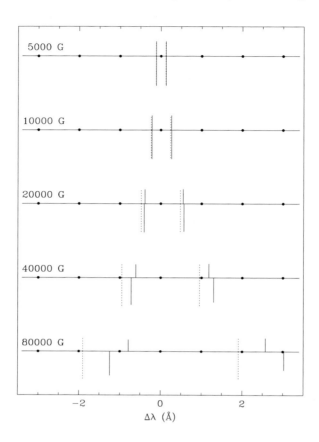

FIGURE 1. Splitting pattern of the line Fe II λ6149.2 in magnetic fields of various strengths. The conventional representation of Zeeman patterns is used, with the line components shown along a wavelength (horizontal) axis as vertical bars of lengths proportional to their relative strengths. Bars above the wavelength axis correspond to π components; bars below it to σ components. The patterns that would be obtained in pure Zeeman effect appear as dotted lines, and the realistic patterns accounting for partial Paschen-Back effect, as solid lines (for a field strength of 5000 G, the two are indistinguishable at the scale of this plot).

to hyperfine structure and to typical magnetic fields are of comparable magnitude, so that they have to be considered simultaneously: the situation is analogous to the partial Paschen-Back effect case discussed in the previous section.

In the presence of hyperfine structure, the total angular momentum of the atom $\boldsymbol{F}$ is defined through vectorial addition of the electronic angular momentum $\boldsymbol{J}$ and of the nuclear spin $\boldsymbol{I}$. The magnetic quantum number f is the projection of $\boldsymbol{F}$ on the direction of the magnetic field. The selection rule $M_1 - M_2 = 0, \pm 1$ is replaced by $f_1 - f_2 = 0, \pm 1$, with π and σ components defined accordingly and playing for polarization the same rôle as in the absence of hyperfine structure. The differences with respect to the latter case come in through the expressions of the wavelength shifts and relative strengths of the individual line components. These expressions, which have been derived by Landi Degl'Innocenti (1975), are significantly more complex than their counterparts of Eqs. (1.10) and (1.9), due to the fact that the hyperfine levels corresponding to different values of F cannot be treated independently from each other (in much the same way as, in the partial Paschen-

Back situation, the various levels of a given spectroscopic term could not be considered individually).

1.4.3. *Combined Stark and Zeeman effects*

In some stars (in particular, fast rotators) hydrogen Balmer lines are the best (or even only) diagnostics of magnetic fields. The physical basis of the interpretation of observations of these lines in terms of properties of the stellar magnetic field is complex, because they are subject to strong linear Stark effect.

Hydrogen atoms in stellar atmospheres permeated by a magnetic field $\boldsymbol{H}$ are subject to external perturbations of three types:

(*a*) linear Stark effect due to interactions with the surrounding charged particles (ions and electrons) of the stellar plasma. The shifts in the energies of the atomic states induced by this effect are of the order of (in *cgs* units):

$$\Delta E_{\mathrm{S}} = 1.875 \ 10^{-9} \, n \, (n-1) \, e \, a_0 \, N_{\mathrm{e}}^{2/3} \,, \tag{1.14}$$

where n is the principal quantum number of the upper level of the transition, a_0 is the Bohr radius, and N_{e} is the electronic density (in cm^{-3}).

(*b*) Zeeman effect due to the magnetic field (assumed to be in the linear regime), whose contribution is of the order of:

$$\Delta E_{\mathrm{Z}} = (n-1) \, \hbar \, \omega_{\mathrm{L}} \,. \tag{1.15}$$

(*c*) Lorentz effect due to the electric field "seen" by the hydrogen atom as a result of its (thermal) motion in the magnetic field. The order of magnitude of the corresponding energy displacements is:

$$\Delta E_{\mathrm{L}} = 6.435 \ 10^{-7} \, T^{1/2} \, H \,, \tag{1.16}$$

where T is the temperature.

The relative contributions of these three effects are shown in Fig. 2 (after Brillant, Mathys & Stehlé 1998). For conditions typical of the atmospheres of non-degenerate stars permeated by a magnetic field, their orders of magnitude are comparable, so that in general, all three of them need to be taken simultaneously into account in the calculation of the atomic states.

This introduces rather severe complications in the expression of the transfer coefficients. While the general form of the Mueller matrix given in Eq. (1.3) remains valid, its elements are no longer given by Eqs. (1.4). Indeed, the plasma and motional electric fields break the cylindrical symmetry about the magnetic field direction, so that the quantum number M is no longer appropriate to characterize unambiguously atomic states within a level (in other words, the hamiltonian of the atom is not diagonal in M). As a consequence, the selection rule $\Delta M = 0, \pm 1$ is not applicable anymore.

Let us open here a parenthesis and note that the situation currently under consideration is quite distinct from the partial Paschen-Back and hyperfine structure cases discussed in Sects. 1.4.1 and 1.4.2. In these two cases, the direction of the magnetic vector defines a privileged direction on which the atom's total angular momentum is projected to define a magnetic quantum number which obeys a selection rule such that its value can only change by 0, +1 or −1 in a radiative transition. The difference between those cases and the "standard" Zeeman case rests with the definition of the total angular momentum itself. The Zeeman effect corresponds to transitions between levels for which the total angular momentum of the atom $\boldsymbol{J}$ is the vectorial sum of the electron orbital momentum $\boldsymbol{L}$ and of the electron spin $\boldsymbol{S}$; M is the projection of $\boldsymbol{J}$ on the magnetic field direction. In the partial Paschen-Back case, one at least of the levels involved in the transition cannot be characterized by a single value of J, but the hamiltonian of the atom remains

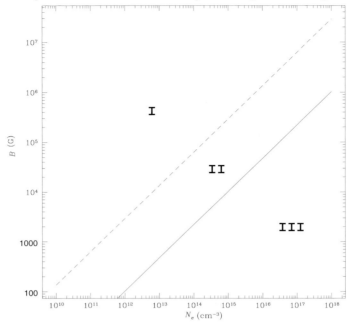

FIGURE 2. Loci of equal Zeeman and plasma Stark effects (*solid line*) and of equal motional and plasma Stark effects (*dashes*), for a temperature of 10^4 K and for a principal quantum number n equal to 2.

diagonal in M: the corresponding atomic states can be expressed as linear combinations of states of different J, but the same M. When hyperfine structure is significant, the hamiltonian is diagonal in f, which is the projection on the magnetic field direction of the total angular momentum of the atom, $\boldsymbol{F} = \boldsymbol{I} + \boldsymbol{J}$. By contrast, for the hydrogen atom in a magnetized plasma, there is no possibility to define, by projection of the total angular momentum of the atom on a specific direction, a "magnetic" quantum number in which the hamiltonian of the atom is diagonal (hence for which a selection rule can be established).

Further complication comes from the existence of a coupling between the Lorentz effect and the thermal Doppler broadening of the line, through the contribution to both of them of the thermal velocity of the atom. As a result, one cannot, in general, apply the "standard" approach, by which the line profiles are first computed for an atom at rest, before being convolved with a velocity distribution function to account for the Doppler effect.

Finally, for levels of astrophysical interest (in particular, the lower level $n = 2$ of the lines of the Balmer series), the order of magnitude of fine structure is not much smaller than the external contributions under consideration. For instance, for $n = 2$, fine structure splitting is of the order of 0.365 cm^{-1} (Bethe & Salpeter 1977); a similar ΔE_{Z} is obtained for a magnetic field of ~ 8 kG.

Further discussion of the theoretical derivation of the expressions of the transfer coefficients entering the Mueller matrix expression of Eq. (1.3) is beyond the scope of these lectures. The most most comprehensive study so far is presented in recent papers of Brillant, Mathys & Stehlé (1998) and of Stehlé, Brillant & Mathys (2000). The new impulse that the work of these authors has given to the subject holds the promise of having soon all the tools necessary to proper interpretation of spectropolarimetric observations of hydrogen lines in stellar atmospheres permeated by magnetic fields. Preliminary conclusions about magnetic field determinations based on the consideration of such lines will be reported in Sect. 2.3.2.

1.5. *Global description of the Zeeman pattern*

For application of some stellar magnetic field diagnostic methods, it proves advantageous to replace the "natural" description of the Zeeman patterns of the transitions in terms of the relative strengths $S_q(M_1, M_2)$ and wavelength shifts $\Delta\lambda(M_1, M_2)$ of their *individual* components by a *global* description that we shall present below.

The *centre of gravity* of the q components ($q = 0, \pm 1$) of a transition is defined as:

$$\lambda_q = \lambda_0 + \sum_{\substack{M_1, M_2 \\ M_1 - M_2 = q}} S_q(M_1, M_2)\, \Delta\lambda(M_1, M_2)\,. \tag{1.17}$$

In particular, the wavelength of the centre of gravity of the π components is the nominal wavelength of the transition in the absence of external perturbations, λ_0. The shift with respect to it of the centre of gravity of the σ_+ components can be written under the form:

$$\lambda_+ - \lambda_0 = \bar{g}\, \Delta\lambda_Z\, H\,. \tag{1.18}$$

The notation $\bar{g}$ represents the *effective Landé factor* of the transition. Its expression in terms of the total angular momentum quantum numbers and of the Landé factors of the levels between which the transition occurs can be derived from Eqs. (1.9) and (1.10):

$$\bar{g} = 0.5\,(g_1 + g_2) + 0.25\,(g_1 - g_2)\,[J_1\,(J_1 + 1) - J_2\,(J_2 + 1)]\,. \tag{1.19}$$

The moment of order n of the q components of the transition with respect to their centre of gravity is then defined as:

$$\mu_q^{(n)} = \sum_{\substack{M_1, M_2 \\ M_1 - M_2 = q}} (g_1\, M_1 - g_2\, M_2 - q\,\bar{g})^n\,. \tag{1.20}$$

Mathys & Stenflo (1987) have shown that the absorption coefficients $\eta_q(\lambda)$ can be Taylor expanded as:

$$\eta_q(\lambda) = \sum_{n=0}^{\infty} C_n^{(q)}\,(\Delta\lambda_Z\, H)^n\, \frac{d^n}{d\lambda^n}\psi(\lambda - \lambda_0)\,, \tag{1.21}$$

with

$$C_n^{(q)} = \frac{(-1)^n}{n!} \sum_{k=0}^{n} \binom{n}{k}\, \mu_q^{(k)}\,(q\,\bar{g})^{n-k}\,, \tag{1.22}$$

provided that the series (1.21) converges. Then, through application of Eqs. (1.4), one can derive expressions for the elements of the Mueller matrix of transfer:

$$\frac{\eta_I}{\eta_\ell} = \frac{1}{2} \sum_{n=0}^{\infty} \left[C_{2n}^{(0)}\, \sin^2\gamma + C_{2n}^{(1)}\,(1 + \cos^2\gamma) \right] (\Delta\lambda_Z\, H)^{2n}\, \frac{d^{2n}}{d\lambda^{2n}}\psi(\lambda - \lambda_0)\,, \tag{1.23a}$$

$$\frac{\eta_Q}{\eta_\ell} = \frac{1}{2}\, \sin^2\gamma\, \cos 2\chi \sum_{n=1}^{\infty} \left[C_{2n}^{(0)} - C_{2n}^{(1)} \right] (\Delta\lambda_Z\, H)^{2n}\, \frac{d^{2n}}{d\lambda^{2n}}\psi(\lambda - \lambda_0)\,, \tag{1.23b}$$

$$\frac{\eta_U}{\eta_\ell} = \frac{1}{2}\, \sin^2\gamma\, \sin 2\chi \sum_{n=1}^{\infty} \left[C_{2n}^{(0)} - C_{2n}^{(1)} \right] (\Delta\lambda_Z\, H)^{2n}\, \frac{d^{2n}}{d\lambda^{2n}}\psi(\lambda - \lambda_0)\,, \tag{1.23c}$$

$$\frac{\eta_V}{\eta_\ell} = \cos\gamma \sum_{n=0}^{\infty} C_{2n+1}^{(1)}\,(\Delta\lambda_Z\, H)^{2n+1}\, \frac{d^{2n+1}}{d\lambda^{2n+1}}\psi(\lambda - \lambda_0)\,. \tag{1.23d}$$

Similar expressions can be obtained for ρ_q and $\rho_{Q,U,V}$. The numerical values of the coefficients $C_n^{(q)}$ can be computed from the Landé factors and total angular momentum numbers of the levels involved in the transition using formulae given by Mathys & Stenflo (1987).

1.6. *Approximate solutions of the transfer equation*

In the most general case, the transfer equation (1.2) must be solved numerically. Various methods have been developed to obtain such solutions. They will not be discussed here: for a good starting point for more information, see Kalkofen (1987). Hereafter, two approximate analytical solutions of great practical interest are briefly introduced. Both are based on a so-called Milne-Eddington atmosphere, that is, on the following set of assumptions: lines are formed in LTE; the magnetic field is constant in the line formation region; scattering is neglected; the ratio of the line transfer coefficients η_q and ρ_q to the continuous absorption κ_c are constant in the line formation region†; the radiation at the bottom of the atmosphere is continuous and unpolarized; and the source function $B(\nu_0, T)$ depends linearly on the optical depth:

$$B = B_0 \left(1 + \beta_0 \, \tau\right), \tag{1.24}$$

where B_0 and β_0 are constant. Under these assumptions, a general analytical solution of Eq. (1.2) can actually be obtained, in which the Stokes parameters of the light emerging from the atmosphere are expressed in terms of B_0, β_0, and the absorption and anomalous dispersion coefficients, $\eta_{I,Q,U,V}$ and $\rho_{Q,U,V}$ (e.g., Landi Degl'Innocenti 1982). The approximate solutions introduced below are derived from this general solution, as described by Mathys (1989). In all cases, in the continuum, the emerging radiation is unpolarized: the only nonzero Stokes parameter is I, for which the solution of the transfer equation is:

$$I_c(\tau = 0, \mu) = B_0 \left[1 + \beta_0 \, \mu \, (1 - \eta_I)\right]. \tag{1.25}$$

1.6.1. *The weak line limit*

For weak lines, i.e., for lines for which $\eta_\ell \ll 1$, the Stokes parameters of the light emerging from the stellar atmosphere are:

$$I(\tau = 0, \mu) \approx B_0 \left[1 + \beta_0 \, \mu \, (1 - \eta_I)\right], \tag{1.26a}$$

$$Q(\tau = 0, \mu) \approx -B_0 \, \beta_0 \, \mu \, \eta_Q, \tag{1.26b}$$

$$U(\tau = 0, \mu) \approx -B_0 \, \beta_0 \, \mu \, \eta_U, \tag{1.26c}$$

$$V(\tau = 0, \mu) \approx -B_0 \, \beta_0 \, \mu \, \eta_V. \tag{1.26d}$$

In other words, the line depressions in the various Stokes parameters are proportional to the corresponding absorption coefficients $\eta_{I,Q,U,V}$.

1.6.2. *The weak field limit*

For weak magnetic fields, that is, when the magnetic splitting of the line is small with respect to its intrinsic width in the absence of a magnetic field, the emergent Stokes profiles from the general solution can be replaced by approximate expansions limited to the first order in H, based on Eqs. (1.23):

$$I(\tau = 0, \mu) \approx B_0 \left[1 + \frac{\beta_0 \, \mu}{1 + \eta_\ell \, \psi(\lambda - \lambda_0)}\right], \tag{1.27a}$$

† Thus, implicitly, the considered solutions are valid only provided that η_q and ρ_q are defined: this excludes the case of the hydrogen lines discussed in Sect. 1.4.3.

$$Q(\tau = 0, \mu) \approx 0 , \qquad (1.27b)$$

$$U(\tau = 0, \mu) \approx 0 , \qquad (1.27c)$$

$$V(\tau = 0, \mu) \approx -\bar{g} \, \Delta\lambda_Z \, H_z \, \frac{dI}{d\lambda} . \qquad (1.27d)$$

In other words, for weak fields, V is to the first order the only Stokes parameter to depend on the magnetic field. For I, Q, and U, the contribution of the magnetic field to the line profile is a second-order effect. Analytic expressions of the expansions of the four Stokes parameters up to terms of the second degree in H can be found in Mathys (1989).

1.7. *Disk integration: the contribution of rotation*

Let us introduce the line depression in the Stokes parameter X ($X = I, Q, U, V$):

$$r_{\mathcal{F}_X} = (\mathcal{F}_{X_c} - \mathcal{F}_X)/\mathcal{F}_{I_c} . \qquad (1.28)$$

The notations $\mathcal{F}_X$ and $\mathcal{F}_{X_c}$ represent the integral over the visible stellar disk of the emergent intensity in the considered Stokes parameter, resp. in the line and in the neighbouring continuum†:

$$\mathcal{F}_X = \int_{-1}^{+1} dx \int_{-\sqrt{1-x^2}}^{+\sqrt{1-x^2}} X(\tau = 0, x, y) \, dy , \qquad (1.29a)$$

and

$$\mathcal{F}_{X_c} = \int_{-1}^{+1} dx \int_{-\sqrt{1-x^2}}^{+\sqrt{1-x^2}} X_c(\tau = 0, x, y) \, dy , \qquad (1.29b)$$

where x and y are the coordinates of a point on the visible stellar disk, in a reference system where the z axis is parallel to the line of sight, the y axis lies in the plane defined by the line of sight and the stellar rotation axis, the origin is the centre of the star, and the unit length is the stellar radius. The use of such a cartesian system is particularly appropriate for the treatment of stellar rotation. The latter can generally be neglected for the continuum; for the lines, it can be introduced in Eq. (1.29a) by specifying explicitly the dependences of the integrand, as follows:

$$\mathcal{F}_X = \int_{-1}^{+1} dx \int_{-\sqrt{1-x^2}}^{+\sqrt{1-x^2}} X[\tau = 0; x, y; \lambda - \lambda_0 - \Delta\lambda_R \, x; \boldsymbol{H}(x, y)] \, dy . \qquad (1.30)$$

In this equation, $\Delta\lambda_R = \lambda_0 \, (v_e/c) \sin i$, where v_e is the projected equatorial velocity of the star, and i, the angle between its rotation axis and the line of sight. Equation (1.30) will be extensively used as a starting point for discussion of stellar magnetic field diagnostic methods in the rest of this course.

2. Ap stars: an ideal laboratory for stellar magnetic field studies

2.1. *Introduction to Ap stars*

Ap and Bp stars (hereafter referred to collectively as Ap stars) are main-sequence A and B stars in the spectra of which lines of a number of elements (such as He, Si, Sr, and lanthanide rare earths) appear abnormally strong or weak with respect to the bulk of "normal" dwarf stars of the same temperature. These peculiarities reflect the existence of departures from the solar abundance pattern in the chemical composition of the stellar surface. It is widely accepted that these abundance anomalies, which may be quite

---

† Note that, in practice, for non-degenerate stars, $\mathcal{F}_{X_c}$ does not, in general, differ significantly from 0 for the Stokes parameters Q, U and V.

extreme (up to 5-6 dex), are confined to the stellar outer layers. Thus the observed composition does not represent the outcome of the nuclear processing within the star of the original interstellar mix from which it formed. Rather, the non-solar abundances result from elemental segregation produced by the action of various competing hydrodynamic processes. In fact, they are one of the most readily observable manifestations of the latter, and as such a powerful tool for improved understanding of stellar hydrodynamics — an essential but still poorly known part of stellar physics.

Most Ap stars also have large-scale organized magnetic fields of kG order. Moreover, they exhibit variations of brightness, spectral line intensities and magnetic field. These variations all occur with the same periodicity, and are interpreted within the framework of the *oblique rotator model*. According to the latter, the magnetic field has a structure which in first, gross approximation, resembles a single dipole at the scale of the whole star. The axis of this dipole makes a nonzero angle with respect to the stellar rotation axis. The magnetic field induces inhomogeneities of brightness and of abundances of various elements over the star's surface, whose distribution is somehow related to the field structure, hence which are not symmetric about the rotation axis. The changing aspect of the visible portion of the surface of the star as it rotates is responsible for the observed photometric, spectroscopic and magnetic variations. Accordingly, the period of these variations is the period of rotation of the star. Rotation periods of Ap stars range from half a day to over 70 years (the longest ones have not been observed over a full cycle yet). On the other hand, no *intrinsic* variations of the magnetic fields of Ap stars have so far been detected, although some of them have been studied for about 50 years.

One of the interests of the study of Ap stars is that, apart from the sun, they are the stars whose magnetic field is most readily accessible to observation. As a matter of fact, until a few years ago, they were the only non-degenerate stars in which magnetic fields had been definitely observed through spectropolarimetry. This privileged situation makes them uniquely suited to study the effect of a magnetic field on stellar atmospheres other than the sun's.

2.2. *First stellar magnetic field detection*

Babcock (1947) reported the first detection of a magnetic field in a star other than the sun. His discovery was based on the observation of a shift between the wavelengths of the spectral lines of the Ap star 78 Vir between spectra simultaneously recorded in right circular polarization (RCP) and left circular polarization (LCP). In hindsight, Babcock's work was remarkably representative of major trends that would predominate in most studies of stellar magnetic fields over the next half century. Indeed, the majority of determinations of stellar magnetic fields achieved to this date are still based on observations of circular polarization in spectral lines. Until a few years ago, the only non-degenerate stars in which the presence of magnetic fields had been undisputedly established were Ap stars. Today still, Ap stars represent the vast majority of the stars where magnetic fields have been measured. Yet, while nowadays, we find logical to expect Ap stars to harbour strong magnetic fields because we assume that such fields must play a key rôle in the generation of their anomalous abundance patterns, Babcock's selection of candidate targets for his attempts to detect stellar magnetic fields was driven by other assumptions, which have since been disproved: namely, that sharp-line A stars must be fast rotators seen almost pole-on, therefore lending themselves best to detection of magnetic fields, the presence of which he believed to be related with fast rotation. Babcock would subsequently contribute to correcting these early misconceptions and establishing the foundations of our current understanding of magnetism in Ap stars, through his extensive and systematic programme of observations of stellar magnetic fields. In particular, his catalogue of magnetic field measurements (Babcock 1958) still remains a primary source of data of this type.

2.3. *Mean longitudinal magnetic field*

For a long time, almost all works about magnetic fields of Ap stars have been devoted to the derivation of their *mean longitudinal magnetic field.* Almost all determinations of this parameter have been achieved through application of either of two polarimetric methods, which will be described in this section.

2.3.1. *The photographic technique*

The *photographic technique* of stellar magnetic field diagnosis draws its name from the fact that, historically, it was developed for the interpretation of observations recorded on photographic plates. In this approach, spectra corresponding to incoming stellar light of opposite circular polarizations are recorded simultaneously on a detector. In the past, this detector was a photographic plate; today, CCDs are used.

Let us denote by $r_{\mathcal{F}_R}$ ($r_{\mathcal{F}_L}$) the depression of a line in the RCP (LCP) spectrum, defined as:

$$r_{\mathcal{F}_R} = (\mathcal{F}_{R_c} - \mathcal{F}_R)/\mathcal{F}_{R_c}\,, \tag{2.31}$$

where $\mathcal{F}_R$ and $\mathcal{F}_{R_c}$ are the disk-integrated emergent intensities in RCP, resp. in the line and in the neighbouring continuum. By definition of the Stokes parameter V, one has:

$$\mathcal{F}_R = (\mathcal{F}_I + \mathcal{F}_V)/2 \tag{2.32a}$$

$$\mathcal{F}_L = (\mathcal{F}_I - \mathcal{F}_V)/2 \tag{2.32b}$$

and, if the continuum is unpolarized,

$$\mathcal{F}_{R_c} = \mathcal{F}_{L_c} = \mathcal{F}_{I_c}/2\,. \tag{2.32c}$$

The wavelength λ_R of the centre of gravity of the line in the RCP spectrum is defined as:

$$\lambda_R = \frac{\int r_{\mathcal{F}_R}(\lambda)\,\lambda\,d\lambda}{\int r_{\mathcal{F}_R}(\lambda)\,d\lambda}\,; \tag{2.33}$$

a similar definition holds for the wavelength λ_L of the centre of gravity of the line in the LCP spectrum. The integrals, in this equation, extend to the whole line (i.e., in theory, they range from $-\infty$ to $+\infty$; in practice, the integration range is limited by the finite signal-to-noise of the data and by the neighbouring lines).

We shall assume that the local emergent V profile at any point of the stellar surface is antisymmetric about its centre. Landi Degl'Innocenti & Landi Degl'Innocenti (1991) have shown that this is true in fairly general conditions. However, departures from antisymmetry can be observed in some cases of practical interest, including those of the lines formed in the regime of partial Paschen-Back effect or with significant hyperfine structure. For a more detailed discussion, see Mathys (1995a). Under this assumption, one has:

$$\int r_{\mathcal{F}_R}(\lambda)\,d\lambda = \int r_{\mathcal{F}_L}(\lambda)\,d\lambda = \int r_{\mathcal{F}_I}(\lambda)\,d\lambda = W_\lambda\,, \tag{2.34}$$

where W_λ denotes the equivalent width of the line. Accordingly, the wavelength shift of the centre of gravity of the line between RCP and LCP can be written as:

$$\lambda_R - \lambda_L = -\frac{2}{W_\lambda\,\mathcal{F}_{I_c}} \int \mathcal{F}_V(\lambda)\,\lambda\,d\lambda\,, \tag{2.35}$$

or, more explicitly, by application of Eq. (1.30):

$$\lambda_R - \lambda_L = -\frac{2}{W_\lambda \, \mathcal{F}_{I_c}} \int_{-1}^{+1} dx \int_{-\sqrt{1-x^2}}^{+\sqrt{1-x^2}} dy \int V[\tau = 0; x, y; \lambda - \lambda_0 - \Delta\lambda_R \, x; \boldsymbol{H}(x,y)] \, \lambda \, d\lambda \, .$$

(2.36)

Performing the change of variable $\lambda \to \lambda + \Delta\lambda_R \, x$, one can see that, thanks to the assumption that V is antisymmetric about its centre, the expression of $\lambda_R - \lambda_L$ is independent of stellar rotation.

At this stage, two approximations are made:

(*a*) the weak line approximation: the emergent V is give by Eq. (1.26d);

(*b*) that the thermodynamic structure of the stellar atmosphere and the distribution of the element responsible for the considered line are homogeneous over the stellar surface: thus the quantities B_0, β_0, and η_ℓ are constant over the star.

Then, V can be expressed as:

$$V[x, y; \lambda - \lambda_0; \boldsymbol{H}(x,y)]$$
$$= -\frac{3}{4\pi} \, W_\lambda \, \mathcal{F}_{I_c} \, \sqrt{1 - (x^2 + y^2)} \, \{\lambda_+[H(x,y)] - \lambda_-[H(x,y)]\} \, \cos\gamma(x,y) \, . \quad (2.37)$$

Taking into account the definition of the effective Landé factor [Eq. (1.18)] and the expression of the line of sight component of the magnetic field H_z [Eq. (1.6)], one finally finds:

$$\lambda_R - \lambda_L = 2 \, \bar{g} \, \Delta\lambda_Z \, \langle H_z \rangle \, , \quad (2.38)$$

where the *mean longitudinal magnetic field* $\langle H_z \rangle$ (also referred to, in short, as the longitudinal field) is the average over the visible stellar disk of the component of the magnetic vector along the line of sight, weighted by the local emergent line intensity:

$$\langle H_z \rangle = \frac{3}{2\pi} \int_{-1}^{+1} dx \int_{-\sqrt{1-x^2}}^{+\sqrt{1-x^2}} H_z(x,y) \, \sqrt{1 - (x^2 + y^2)} \, dy \, . \quad (2.39)$$

Equation (2.38) is the cornerstone of a vast fraction of the published measurements of magnetic fields of Ap stars.

Let us point out that the photographic method of determination of the mean longitudinal magnetic field is qualified as an *integral* method, because it relies on the consideration of observational quantities (the wavelengths of the centres of gravity of the lines) which are obtained by integration over the whole line profile.

In practice, the wavelength shift $\lambda_R - \lambda_L$ is determined for a sample of lines of the studied star, and a least-squares fit (forced through the origin) of these differences as a function of $2 \, \bar{g} \, \Delta\lambda_Z$ is performed to derive $\langle H_z \rangle$. The standard error $\sigma(\langle H_z \rangle)$ of the longitudinal field that is derived from this least-squares analysis is used as an estimate of the uncertainty affecting the obtained value of $\langle H_z \rangle$ (assuming that systematic errors are negligible — this point will not be discussed here, since it depends on the characteristics of the instrumental configuration used to record the polarized spectra). This approach can be refined to take into account the fact that $\lambda_R - \lambda_L$ can be more accurately determined for some lines than for others (e.g., depending on their profiles). To achieve this, the fit of $\lambda_R - \lambda_L$ vs. $2 \, \bar{g} \, \Delta\lambda_Z$ is *weighted* by the inverse of the mean-square error of the $\lambda_R - \lambda_L$ measurements for the individual lines, $1/\sigma^2(\lambda_R - \lambda_L)$. The way in which $\sigma(\lambda_R - \lambda_L)$ is evaluated has been described by Mathys (1994).

An illustration of the relation actually observed between $\lambda_R - \lambda_L$ and $\bar{g} \, \Delta\lambda_Z$ is shown in Fig. 3. The error bars correspond to $\pm\sigma(\lambda_R - \lambda_L)$, and the dashed line is the weighted least-squares fit described above. One can see that the linear dependence predicted by Eq. (2.38) is indeed observed, and that the scatter of the individual measurements of

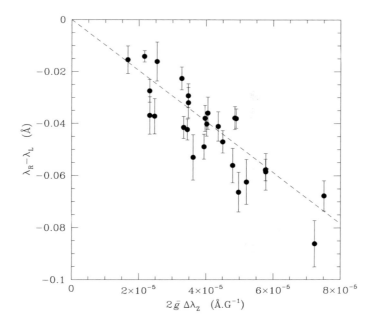

FIGURE 3. Wavelength shift $\lambda_R - \lambda_L$ between observations in opposite circular polarizations of spectral lines of the Ap star HD 201601, plotted against $2\,\bar{g}\,\Delta\lambda_Z$. The dashed line is a weighted least-squares fit to the data by a straight line forced through the origin. The longitudinal field derived through this procedure is $\langle H_z \rangle = (-980 \pm 39)$ G.

$\lambda_R - \lambda_L$ about the best fit straight line is consistent with the size of the error bars. One can check that the residuals with respect to the fit show no dependence on the equivalent width. This is a good indication of the quality of the weak line approximation for the type of analysis performed.

2.3.2. *Balmer line photopolarimetry*

While, as shown in the previous section, determinations of the mean longitudinal field through the photographic method are, in principle, independent of stellar rotation, in practice, the method is not suitable for stars that rotate too fast. Indeed, with increasing rotation, stellar lines become too wide and too shallow for accurate measurement; furthermore, blending with neighbouring lines quickly becomes an insuperable limitation. A workaround is to use hydrogen Balmer lines as magnetic field diagnostics. Indeed, the intrinsic width of these lines is typically much larger than their rotational broadening, even in fairly fast-rotating stars. These considerations are at the origin of the use of *Balmer line photopolarimetry* to determine the mean longitudinal field.

The technique was first introduced by Angel & Landstreet (1970) to diagnose magnetic fields in white dwarfs. It was subsequently adapted by Landstreet *et al.* (1975) for application to studies of non-degenerate stars, and in particular of Ap stars, for which it is the second widely used method, beside the photographic technique. In practice, portions of the wings of a Balmer line of hydrogen (most often Hβ) are observed through a narrow band interference filter, alternatively in RCP and in LCP. The sum and the difference of the intensities observed in both polarizations, which are recorded with a photoelectric photometer, are formed to obtain the observed values in the considered part of the line of the disk-integrated emergent intensities in the Stokes parameters I and V, $\mathcal{F}_I$ and $\mathcal{F}_V$.

The interpretation applied to derive the mean longitudinal magnetic field from these quantities, which has been originally presented by Landstreet (1982), is described hereafter. It rests on the weak field approximation to the solution of the radiative transfer equation. Integrating Eq. (1.27d) over the visible stellar disk, one gets:

$$
\mathcal{F}_V(\lambda) = \bar{g}\,\Delta\lambda_Z \int_{-1}^{+1} dx \int_{-\sqrt{1-x^2}}^{+\sqrt{1-x^2}} H_z(x,y)\, \frac{B_0\,\beta_0\,\eta_\ell}{[1 + \eta_\ell\,\psi(\lambda - \lambda_0 - \Delta\lambda_R\,x)]^2}
$$

$$
\times \frac{d}{d\lambda}\psi[\lambda - \lambda_0 - \Delta\lambda_R(x,y)]\,\sqrt{1-(x^2+y^2)}\,dy. \tag{2.40}
$$

For Balmer lines, it is generally a good approximation to consider the rotational Doppler effect as small with respect to the intrinsic width of the lines, hence that $\psi[\lambda - \lambda_0 - \Delta\lambda_R(x,y)] \approx \psi[\lambda - \lambda_0]$. Assuming furthermore, as in the photographic technique, that B_0, β_0, and η_ℓ are constant over the stellar surface, one can rewrite Eq. (2.40) under the form:

$$
\mathcal{F}_V = -\bar{g}\,\Delta\lambda_Z\,\langle H_z\rangle\,\frac{d\mathcal{F}_I}{d\lambda}\,, \tag{2.41}
$$

with the expression of the mean longitudinal magnetic field $\langle H_z\rangle$ given in Eq. (2.39).

All the measurements of circular polarization in the wings of Balmer lines of hydrogen obtained so far have been interpreted in terms of stellar longitudinal magnetic field by application of Eq. (2.41). Since this interpretation rests on the consideration of the derivative of $\mathcal{F}_I$ with respect to the wavelength, the method is called *differential*. However, it has some shortcomings. The most fundamental one rests with the use of the weak field solution (1.27) as a starting point, which as mentioned in Sect. 1.6, is not valid for hydrogen lines†.

Preliminary results of work in progress towards defining a more correct interpretation of the Balmer line photopolarimetric measurements have recently been published by Mathys *et al.* (2000). They are based on calculations performed using recent theoretical developments on the formation of hydrogen lines in dense magnetized plasmas (Brillant, Mathys & Stehlé 1998; Stehlé, Brillant & Mathys 2000), however provisionally neglecting the contributions of the Lorentz effect and of the fine structure. In these conditions, V is still found to be proportional to $dI/d\lambda$ in the near wings of the hydrogen lines, but Eq. (1.27d) must be replaced by:

$$
V(\tau = 0, \mu) = -0.8\,\bar{g}\,\Delta\lambda_Z\,H_z\,\frac{dI}{d\lambda}\,. \tag{2.42}
$$

The numerical value 0.8 of the "corrective" factor introduced in this equation results from the physics of the combined Stark and Zeeman effects. It could have been different if the other relevant effects had been included; in particular, as a result of the Lorentz effect, it might depend on the magnetic field itself. The fact that the nature and magnitude of differences between the longitudinal field values obtained through the photographic technique and through Balmer line photopolarimetry [using Eq. (2.41)] vary from star to star may possibly be regarded as supporting this latter suggestion.

---

† However, it should be emphasized that, while values of the longitudinal field derived through Balmer line photopolarimetry are, often, not exactly equal to those derived with the photographic technique, both are of the same order of magnitude.

2.4. *Resolved magnetically split lines in Ap stars*

Locally, at a given point of the surface of an Ap star, the profile of a spectral line is split into several magnetic components. Due to rotational Doppler effect, those components are broadened in disk-integrated observations, often to the extent that they can no longer be distinguished from each other. However, a minority of stars have a slow enough projected equatorial velocity, so that the splitting of some at least of their spectral lines can be resolved in high-dispersion spectra. Such cases are particularly interesting because they offer an opportunity to obtain a qualitatively different kind of information about the magnetic fields of the Ap stars, and because, for some lines at least, this information can be derived in a very straightforward and mostly approximation-free manner. While it is in principle possible to exploit this advantageous situation for observations in all 4 Stokes parameter, the additional diagnostic potential allowed by individual line component observations has been exploited only for Stokes I. In this section, we shall discuss the interpretation of such observations. Even though this method of magnetic field diagnosis is not, strictly speaking, polarimetric, its presentation within the framework of these lectures appears justified by its great practical importance and the impact that it has had on our current understanding of the magnetic fields of Ap stars.

2.4.1. *The mean magnetic field modulus*

Best advantage of the possibility to resolve magnetically split lines can be taken for lines that have specific, simple Zeeman patterns. Two configurations are particularly interesting:

- the Zeeman triplet,
- the Zeeman doublet.

Zeeman triplets, also known as *normal* Zeeman patterns, arise from transitions between two levels having the same Landé factor, or between a split level and a level for which $J = 0$. Zeeman doublets are the simplest type of *anomalous* Zeeman pattern. Such a pattern is observed for transitions which split into two π components, one σ_+ component, and one σ_- component, where the shift of each of the $\sigma_\pm$ components with respect to the line centre is equal to the shift of one of the π components. Such a transition can occur only between two levels having both $J = 1/2$, one of which has a zero Landé factor. Among the LS coupling terms, the only level with $J = 1/2$ and $g = 0$ is $^4D_{1/2}$: this strongly restricts the number of existing doublets.

The observed splitting of a triplet or of a doublet can be interpreted in terms of stellar magnetic field in a virtually approximation-free manner. Here we illustrate this for the case of the doublet, which has been, by far, most used in practice. The developments, for a triplet, are rather similar: for details, see Mathys (1989).

For a fully split doublet, radiative transfer occurs independently in each of the two components, to which, for simplicity, we shall refer to as $+$ and $-$. The absorption coefficient in the π component, η_0, can be separated into two contributions, $\eta_{0,+}$ and $\eta_{0,-}$, with self-explanatory notations. The $+$ component of the line results from the transfer of radiation in the superimposed σ_+ and π_+ components. The Mueller matrix of transfer in this component can be written as:

$$\boldsymbol{\eta} = \frac{1}{2}\, \eta_\ell\, \psi(\lambda - \lambda_+) \begin{pmatrix} 1 & 0 & 0 & \cos\gamma \\ 0 & 1 & 0 & 0 \\ 0 & 0 & 1 & 0 \\ \cos\gamma & 0 & 0 & 1 \end{pmatrix}$$

$$+ \eta_\ell \, \phi(\lambda - \lambda_+) \begin{pmatrix} 0 & 0 & 0 & 0 \\ 0 & 0 & \cos\gamma & 0 \\ 0 & -\cos\gamma & 0 & 0 \\ 0 & 0 & 0 & 0 \end{pmatrix}. \tag{2.43}$$

One can see that, in this case, the transfer in the Stokes parameters I and V is uncoupled from the transfer in the Stokes parameters Q and U. For the former two parameters, only the first term of the right-hand side contributes to the transfer. This term appears as the product of a factor symmetric about λ_+ and of a factor independent of the wavelength. Accordingly, under the usual assumption of continuous, unpolarized radiation at the bottom of the atmosphere, the emergent profiles of the $+$ component in Stokes I and V are symmetric about λ_+. (Under the same assumption, $Q = U = 0$ in the emergent line.)

Let $\langle \lambda_+ \rangle$ be the *observed* wavelength of the $+$ component:

$$\langle \lambda_+ \rangle = \frac{\int r_{\mathcal{F}_+}(\lambda) \, \lambda \, d\lambda}{\int r_{\mathcal{F}_+}(\lambda) \, d\lambda}, \tag{2.44}$$

where

$$r_{\mathcal{F}_+}(\lambda) = \frac{\mathcal{F}_{I_c} - \mathcal{F}_+(\lambda)}{\mathcal{F}_{I_c}} \tag{2.45}$$

is the observed profile (in the Stokes parameter I) of the relative depression of the $+$ component of the line. The integral in Eq. (2.44) extends to the $+$ component. Assuming that $v_e \sin i = 0$ (which is reasonable in the situation under consideration), one can express $\mathcal{F}_+$ as:

$$\mathcal{F}_+(\lambda) = \int_{-1}^{+1} dx \int_{-\sqrt{1-x^2}}^{+\sqrt{1-x^2}} I_+[\tau = 0; x, y; \lambda - \lambda_+; \boldsymbol{H}(x,y)] \, dy. \tag{2.46}$$

The same reasoning can be made for the $-$ component of the line, so that, through application of Eq. (1.18), one finds that:

$$\langle \lambda_+ \rangle - \langle \lambda_- \rangle = 2 \, \bar{g} \, \Delta\lambda_Z \, \langle H \rangle, \tag{2.47}$$

where

$$\langle H \rangle = \frac{2}{W_\lambda \, \mathcal{F}_{I_c}} \int_{-1}^{+1} dx \int_{-\sqrt{1-x^2}}^{+\sqrt{1-x^2}} H(x,y) \, dy$$
$$\times \int \{ I_c(\tau = 0, x, y) - I_+[\tau = 0; x, y; \lambda - \lambda_+; \boldsymbol{H}(x,y)] \} \, d\lambda, \tag{2.48}$$

or the equivalent expression in terms of the $-$ component. $\langle H \rangle$ is the *mean magnetic field modulus* (or, in short, field modulus), that is, the average over the visible stellar disk of the modulus of the magnetic vector, weighted by the local emergent line intensity. It may be noted that its expression, as given by Eq. (2.48) is more general than the expression of the mean longitudinal magnetic field in Eq. (1.6). Indeed the latter reflects, in particular, the specific properties of the Milne-Eddington atmosphere, while the field modulus was obtained under much less restrictive assumptions. Actually, the *only* assumptions underlying its derivation are that rotational Doppler effect is small with respect to the intrinsic width of the (local) line components, and the general approximation of LTE, introduced in Sect. 1.3. These are hardly restrictive at all in practice, so that the value of the mean field modulus derived in the described manner appears as a very reliable and accurate measure of the stellar magnetic field.

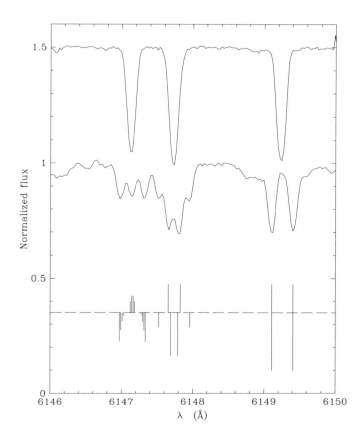

FIGURE 4. Portion of the spectra of HD 133792 (shifted in intensity by 0.5; unresolved lines) and of HD 94660 containing the lines Cr II λ 6147.1, Fe II λ 6147.7, and Fe II λ 6149.2, and Zeeman patterns of those lines. For the sake of clarity, the wavelengths in the stellar spectra have been reduced to the laboratory reference frame.

2.4.2. *Observations of Ap stars with resolved magnetically split lines*

The first star where resolved magnetically split lines have been observed is HD 215441, also known as Babcock's star (Babcock 1960). This star still has nowadays the strongest mean magnetic field modulus observed in an Ap star. Although one cannot exclude the presence of magnetic fields of comparable strength in some fast-rotating Ap stars (where splitting is smeared out by rotational Doppler effect), HD 215441 no doubt stands as one of the most strongly magnetized Ap stars. By 1972, a total of nine stars Ap with magnetically resolved lines were known. The field modulus of four of them had been repeatedly measured throughout their rotation period (Preston 1971 and references therein; see also Huchra 1972); the total number of published measurements of $\langle H \rangle$ was close to 80. These stars received little further attention until the end of the 1980 decade. By that time, resolved magnetically split lines had been observed in three additional Ap stars: these findings were mostly serendipitous, and without follow up. Ten years later, the situation has changed dramatically. At the time of writing, 45 Ap stars with magnetically resolved lines are known, and more than 1000 measurements of their mean field moduli have been obtained. This breakthrough results primarily from a large coordinated project of exten-

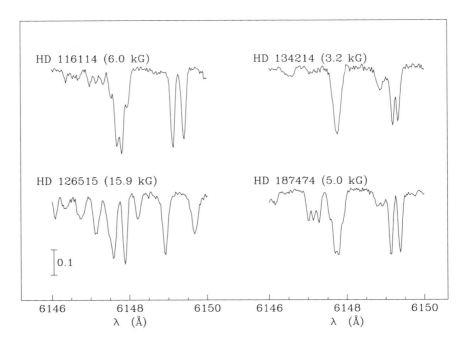

FIGURE 5. Same spectral region as in Fig. 4, for four Ap stars identified next to each tracing. The mean magnetic field modulus at the plotted phase is indicated between parentheses. Note the profile difference between the blue and red components of the line Fe II λ6149.2, and the asymmetry of the latter. Compare with the computed splitting patterns of Fig. 1. (For the sake of clarity, the wavelengths in the stellar spectra have been reduced to the laboratory reference frame.)

sive and systematic search and study of such stars. This project, and its results obtained until August 1995, are described by Mathys *et al.* 1997. Its outcome, which represents a vast fraction of all the measurements of magnetic fields of Ap stars, and one of the largest sets of homogeneous data of that type, is an essential complement to the polarimetric observations of these stars, of which it enhances considerably the diagnostic potential.

All the measurements of $\langle H \rangle$ in this project (hence the vast majority of all published $\langle H \rangle$ data) are based on consideration of a single line, Fe II λ6149.2. The main interest of this line lies in the fact that it has a doublet pattern, in which the split level has a large Landé factor, 2.70. This is illustrated in Fig. 4, where the line, as observed in the Ap star HD 94660, is shown together with neighbouring lines Fe II λ6147.7 (a pseudo-quadruplet) and Cr II λ6147.1 (a pseudo-triplet). The Zeeman patterns of all three lines is also represented, as well as, for comparison, the same portion of the spectrum of HD 133792, an Ap star with sharp, unresolved lines. Additional advantages of the line Fe II λ6149.2 include the facts that it is observed in almost all Ap stars, generally without too strong blends, and that the distribution of iron over the surface of Ap stars is usually rather homogeneous. Although Eq. (2.48) is valid even in case of inhomogeneous distribution of the element responsible for the diagnostic line over the stellar surface, a (fairly) homogeneous distribution ensures that the derived value of $\langle H \rangle$ is representative primarily of the distribution over the star of the magnetic field, rather than of the element abundance distribution, which enters as a weighting function in the mean field modulus definition.

A drawback of the line Fe II $\lambda 6149.2$ is that, for magnetic field strengths typical of Ap stars, it is not formed in a regime of pure Zeeman effect, but rather in a regime of partial Paschen-Back effect. Indeed, the separation between its lower level, $b^4 D_{1/2}$, and the next level within the $b^4 D$ term of Fe II, $b^4 D_{3/2}$, is only 4.01 cm^{-1}. For a magnetic field of 10 kG, the energy difference between two contiguous magnetic states of the latter level would be 0.56 cm^{-1} in the pure Zeeman case. This difference is not negligible with respect to the fine structure separation between the levels $b^4 D_{1/2}$ and $b^4 D_{3/2}$. The impact on the splitting pattern of the line Fe II $\lambda 6149.2$ has been illustrated in Fig. 1. Consistently with this theoretical result, observations show that the two components of the split doublet are different: the blue one is sharper and deeper, while the red one is broader and asymmetric, with a steeper blue edge and a more extended wing on the red side. This is particularly visible in the stars where the magnetic field is stronger; some examples are shown in Fig. 5. However, the impact of this departure from pure Zeeman effect on measurements of the mean field modulus is small: as shown by Mathys (1990a), for fields up to a few tens kG, Eq. (2.48) keeps giving an excellent approximation of the wavelength separation of the split components.

3. Exploitation of line profile information

Early studies of stellar magnetic fields were based on spectra recorded on photographic plates. Because only limited signal-to-noise ratio was achievable, magnetic field diagnosis was to a large extent restricted to consideration of those quantities that could be derived from measurements of wavelengths of entire lines (or line components): this explains the historical importance of mean longitudinal field and mean field modulus determinations. Yet, several authors had already realized that a lot of additional valuable information was contained in the line *profiles* (some examples are mentioned below). Exploitation of this information became feasible with the advent of CCDs as astronomical detectors. This led to the development of new diagnostic techniques, which are reviewed in this section.

3.1. *The moment technique*

The moment technique allows one to determine moments of the magnetic field over the stellar disk from consideration of the observed moments of line profiles recorded in the various Stokes parameters, under certain approximations. It has originally been developed for intensity and circular polarization (Mathys 1988), and subsequently generalized for application to all four Stokes parameters (Mathys 1989).

Let λ_I be the wavelength of the centre of gravity of a spectral line observed in the Stokes parameter I. The moment of order n of this line about its centre, in the Stokes parameter X ($X = I, Q, U, V$), is defined as:

$$R_X^{(n)}(\lambda_I) = \frac{1}{W_\lambda} \int r_{\mathcal{F}_X} (\lambda - \lambda_I) (\lambda - \lambda_I)^n \, d\lambda. \qquad (3.49)$$

The integral extends to the whole observed line: the limits must be far enough in the line wings with respect to the largest wavelength shift of any line component as a result of the combination of the Zeeman effect and of the rotational Doppler effect at any point of the visible stellar hemisphere (see Mathys 1988 for details). Applying Eq. (1.30) and performing the change of variable $\lambda \to \lambda + \Delta\lambda_R x$ in the wavelength integral, one can write a more explicit expression of $R_X^{(n)}(\lambda_I)$:

$$R_X^{(n)}(\lambda_I) = \frac{1}{W_\lambda \mathcal{F}_{I_c}} \sum_{m=0}^{n} \binom{n}{m} \Delta\lambda_R^{n-m} \int_{-1}^{+1} x^{n-m} \, dx \int_{\sqrt{1-x^2}}^{\sqrt{1+x^2}} dy$$

$$\times \int \{X_c(\tau = 0; x, y) - X[\tau = 0; x, y; \lambda - \lambda_0; \boldsymbol{H}(x, y)]\} (\lambda - \lambda_0)^m \, d\lambda. \quad (3.50)$$

This equation is valid quite generally. In the moment technique, interpretation of the observed line profile moments in terms of moments the stellar magnetic field rests on the following two approximations:

(*a*) the weak-line approximation (see Sect. 1.6.1): thus the emergent X is assumed to be proportional to η_X, the absorption coefficient in the Stokes parameter X;

(*b*) that the distribution of the element responsible for the considered line and the thermodynamic structure of the atmosphere are not significantly inhomogeneous over the stellar surface. Thus η_X only depends on the coordinates x and y through the magnetic field, and the coefficient of proportionality linking X and η_X is the product of a constant (independent of x and y) and of the limb-darkening factor.

Then, through application of Eqs. (1.26) and (1.23), one obtains, after some lengthy but straightforward algebra, the following expressions of the n-th order moments about the line centre λ_I of the line profiles in the various Stokes parameters:

$$R_I^{(n)}(\lambda_I) = n! \sum_{m=0}^{n} \frac{1 + (-1)^{n-m}}{4} \frac{\Delta\lambda_R^m}{m!} \sum_{\ell=0}^{m} \frac{\Psi^{(n-m-\ell)}(\lambda_0)}{(n-m-\ell)!} \Delta\lambda_Z^\ell$$

$$\times \left(S_\ell \langle x^m H^\ell \rangle + D_\ell \langle x^m H^{\ell-2} H_z^2 \rangle \right), \quad (3.51a)$$

$$R_Q^{(n)}(\lambda_I) = n! \sum_{m=0}^{n} \frac{1 + (-1)^{n-m}}{4} \frac{\Delta\lambda_R^m}{m!} \sum_{\ell=0}^{m} \frac{\Psi^{(n-m-\ell)}(\lambda_0)}{(n-m-\ell)!} \Delta\lambda_Z^\ell$$

$$\times D_\ell \langle x^m H^{\ell-2} (H_Q^2 - H_{Q+\pi/2}^2) \rangle, \quad (3.51b)$$

$$R_U^{(n)}(\lambda_I) = n! \sum_{m=0}^{n} \frac{1 + (-1)^{n-m}}{2} \frac{\Delta\lambda_R^m}{m!} \sum_{\ell=0}^{m} \frac{\Psi^{(n-m-\ell)}(\lambda_0)}{(n-m-\ell)!} \Delta\lambda_Z^\ell$$

$$\times D_\ell \langle x^m H^{\ell-2} H_Q H_{Q+\pi/2} \rangle, \quad (3.51c)$$

$$R_V^{(n)}(\lambda_I) = n! \sum_{m=0}^{n} \frac{1 + (-1)^{n-m}}{2} \frac{\Delta\lambda_R^m}{m!} \sum_{\ell=0}^{m} \frac{\Psi^{(n-m-\ell)}(\lambda_0)}{(n-m-\ell)!} \Delta\lambda_Z^\ell$$

$$\times C_\ell^{(-1)} \langle x^m H^{\ell-1} H_z \rangle, \quad (3.51d)$$

where

$$S_\ell = C_\ell^{(-1)} + C_\ell^{(0)}, \quad (3.52a)$$

$$D_\ell = C_\ell^{(-1)} - C_\ell^{(0)}, \quad (3.52b)$$

and $\Psi^{(k)}(\lambda_0)$ is the moment of order k about λ_0 of the absorption profile $\psi(\lambda - \lambda_0)$:

$$\Psi^{(k)}(\lambda_0) = \int \psi(\lambda - \lambda_0)(\lambda - \lambda_0)^k \, d\lambda. \quad (3.53)$$

Equations (3.51) show that, in the weak line limit, the moments of the line profiles observed in the four Stokes parameters can be expressed as linear combinations of moments of various orders ($m \geq 0$) about the plane defined by the line of sight and the stellar rotation axis, of products of powers ($\ell \geq 0$) of the local magnetic field modulus H by powers ($0 \leq k \leq 2$) of its components parallel or perpendicular to the line of sight H_j

(where j stands for any of z, Q, or $Q + \pi/2$):

$$\langle x^m \, H_z^\ell \, H_j^k \rangle = \frac{3}{2\pi} \int_{-1}^{+1} dx \int_{-\sqrt{1-x^2}}^{+\sqrt{1-x^2}} x^m \, H_z^\ell(x,y) \, H_j^k(x,y) \, \sqrt{1 - (x^2 + y^2)} \, dy \, . \quad (3.54)$$

In the stars where $\Delta\lambda_R = 0$, only the zero-order moments (in other words, the disk averages) of the powers of the magnetic field can be determined from observation. Explicit expressions of the moments of the line profiles up to fourth order, in the four Stokes parameters, are given in Table IV of Mathys (1989). One can note that the explicit form of Eq. (3.51) for the first order moment of the line profile observed in Stokes V is in fact equivalent to Eq. (2.38), which is used to derive the mean longitudinal magnetic field from the wavelength shift of lines between RCP and LCP. In other words, the photographic technique of determination of the mean longitudinal field appears as a particular case of the more general moment technique of stellar magnetic field diagnosis.

In practice, Eqs. (3.51) appear idealized and unrealistic. Indeed, $\Psi^{(k)}(\lambda_0)$ should account for all the effects contributing to the observed line profile, besides those of the magnetic field and of stellar rotation, which have been explicitly treated to establish those equations. Yet, even in cases when the lines under study are sufficiently weak to fully justify the weak-line approximation, there are some such contributions, always present in the observations, that are not adequately represented by expression (3.53) of $\Psi^{(k)}(\lambda_0)$, in terms of only the intrinsic absorption profile $\psi(\lambda - \lambda_0)$. These include:

- *local* (as opposed to rotational) Doppler effect due to thermal motion (and possibly, other velocity fields, such as microturbulence);
- convolution with the instrumental profile.

These contributions can be taken into account by replacing in Eqs. (3.51) the moments $\Psi^{(k)}(\lambda_0)$ of the absorption profile $\psi(\lambda - \lambda_0)$ by moments $\Phi_0^{(k)}(\lambda_0)$ of more realistic profiles corresponding to the convolution of $\psi(\lambda - \lambda_0)$ with functions describing the local Doppler broadening and the instrumental profile. These functions are, generally, symmetric with respect to the line centre to a very good approximation, and so is the absorption profile $\psi(\lambda - \lambda_0)$. Accordingly, only the moments of even order are non-zero, and they can be expressed as (Mathys 1988):

$$\Phi_0^{(2k)}(\lambda_0) = \sum_{j=0}^{k} \binom{2\,k}{2\,j} \mathcal{F}^{(2k-2j)}(\lambda_0)$$

$$\times \sum_{i=0}^{j} \frac{(2\,j)!}{i! \, (2\,j - 2\,i)!} \left(\frac{\Delta\lambda_D}{2} \right)^{2i} \Psi^{(2j-2i)}(\lambda_0) \, . \quad (3.55)$$

In this equation, $\mathcal{F}^{(k)}(\lambda_0)$ represents the moment of order k of the instrumental profile at wavelength λ_0, and $\Delta\lambda_D = (\zeta_0/c) \, \lambda_0$, where ζ_0 is the most probable line-of-sight velocity of the ion responsible for the considered line, in a reference frame co-rotating with the star. In addition to thermal motion, it may contain a contribution from microturbulence.

As can be seen, in the moment space, the contribution of various broadening agents is accounted for via a linear combination of a number of terms. This suggests that it may be possible to generalize the approach to include other contributions, possibly of effects that are not a priori identified, by replacing $\Psi^{(k)}(\lambda_0)$ in Eqs. (3.51) by:

$$\Phi^{(k)}(\lambda_0) = \Phi_0^{(k)}(\lambda_0) + a_0^{(k)} + \sum_{r=1}^{R(k)} \sum_{s=1}^{S(r)} a_{rs}^{(k)} \, Q_r^s \, . \quad (3.56)$$

In this expression, the quantities denoted by Q_r can be any parameter, or combination of parameters, characterizing the observed line and the transition from which it originates, such as e.g. the equivalent width, the excitation potential of the lower level, the product of both, etc. The notations $a_0^{(k)}$ and $a_{rs}^{(k)}$ represent numerical coefficients, to be determined as a result of the analysis of the observed lines profiles. How many and which parameters Q_r should be used to achieve the best description of the behaviour of the moments of the observed line profiles is then determined semi-empirically by trials and errors, from the analysis of a statistical sample of lines. This approach, which is inspired from the one originally used by Stenflo & Lindgren (1977) to study the magnetic field of the sun, is illustrated below in Sect. 3.1.2. Note that it also allows one to deal with departures from the weak-line approximation, through study of possible dependences of the observed moments on the line equivalent width.

Hereafter, we shall discuss the application of the moment technique to the analysis of the second order moments in Stokes V and I, the two cases of greatest practical importance (leaving aside the determination of the longitudinal field through the photographic technique).

3.1.1. *Crossover*

One can note that $R_V^{(2)}(\lambda_I)$ is, except for a multiplicative factor, the difference of between the second-order moments of the RCP and LCP line profiles about their centre of gravity. The latter, by analogy with the statistical variance, can be seen as characterizing the *spread* of the RCP and LCP line profiles about their respective centre. Accordingly, their difference appears as a measurement of the difference in the width of a spectral line as observed in RCP and LCP. That such width differences between spectral line observations in opposite circular polarizations exist has long been recognized. The effect had already been seen in photographic spectra of Ap stars. It was first detected in HD 125248 by Babcock (1951), who called it the *crossover effect*. The name comes from the fact that the effect is usually largest close to the phases when the mean longitudinal field reverses its sign, or "crosses over" from one polarity to the other.

The explicit form of Eq. (3.51d) for $n = 2$ is:

$$R_V^{(2)}(\lambda_I) = 2\,\bar{g}\,\Delta\lambda_Z\,\Delta\lambda_R\,\langle x\,H_z\rangle. \tag{3.57}$$

Through application of this relation, one can derive the *crossover*, $v_e \sin i \langle x\,H_z\rangle$. If $v_e \sin i$ has been determined independently, one can then obtain the *mean asymmetry of the longitudinal magnetic field*, $\langle x\,H_z\rangle$. The latter is the first-order moment about the plane defined by the stellar rotation axis and the line of sight of the component of the magnetic vector parallel to the line of sight.

The difference of line width between opposite circular polarizations that is measured by the second-order moment of line profiles in Stokes V finds its origin in a correlation between the rotational Doppler shift of the contributions to the observed (disk-integrated) line coming from different parts of the stellar disk and the different Zeeman shifts of their RCP and LCP components, corresponding to the local magnetic field strength and orientation†. In other words, from the point of view of the radiative transfer at a given point of the stellar surface, the relevant effect to be interpreted is the *global shift* of a line between RCP and LCP (as opposed to differences in the *profile* of the local emergent line as observed in opposite circular polarizations), that is, the same effect as sampled by the

---

† Inhomogeneous distribution of the element responsible for the observed line may furthermore contribute, but such inhomogeneities are not *required* to generate crossover. If they exist, they do, actually, complicate the interpretation of the observations, as it becomes necessary to untangle their contribution from that of the magnetic field structure. This problem has, so far, not been addressed within the framework of the moment technique (but there is no fundamental reason why this could not be done).

first-order moment of the Stokes V line profile. Accordingly, the level of approximation in the interpretation of $R_V^{(2)}(\lambda_I)$ is the same as for $R_V^{(1)}(\lambda_I)$ (for more details, see Mathys 1995a).

In practice, the derivation of the value of the crossover from the observation is quite similar to that of the mean longitudinal magnetic field. The second-order moment in the Stokes V parameter is measured for a sample of lines. Then, in application of Eq. (3.57), $v_e \sin i \langle x H_z \rangle$ is determined from a least-squares fit of $R_V^{(2)}(\lambda_I)$ as a function of $2\,\bar{g}\,\Delta\lambda_Z\,(\lambda_0/c)$, forced through the origin. This fit is weighted by the inverse of the mean-square error of the $R_V^{(2)}(\lambda_I)$ measurements for the individual lines, $1/\sigma^2[R_V^{(2)}(\lambda_I)]$.

3.1.2. *The mean quadratic magnetic field*

The second-order moment of line profiles in the Stokes parameter I characterizes the line widths in unpolarized light. Overall line width is defined by the combined effect of various physical processes. One of them is magnetic broadening, which differs from line to line, according to their Zeeman pattern. The first attempt to exploit differential magnetic broadening of spectral lines in the Stokes parameter I to diagnose stellar magnetic fields has been made by Preston (1971). Line width differences between selected spectral lines of high and low magnetic sensitivity, measured in Ap stars with resolved magnetically split lines, were used to establish an empirical relation between mean field modulus and differential line broadening. This relation was then applied to diagnose the magnetic field modulus of stars in which spectral line splitting was not resolved (because their field was weaker and/or because they were rotating somewhat faster). As a matter of fact, we shall see below that the quantity that can be derived from consideration of differential line broadening is not the mean field modulus, but the mean quadratic magnetic field. But at the level of accuracy achievable from spectra recorded on photographic plates (such as used by Preston), the difference between these two quantities is hardly significant for most Ap stars.

Application of Eq. (3.51a) to the specific case $n = 2$ yields:

$$R_I^{(2)}(\lambda_I) = \Phi^{(2)}(\lambda_0) + \Delta\lambda_R^2/5 + \Delta\lambda_Z^2\left(S_2\,\langle H^2 \rangle + D_2\,\langle H_z^2 \rangle\right). \tag{3.58}$$

By application of this equation, one should in principle be able to derive two moments of the magnetic field, the *mean square field modulus* $\langle H^2 \rangle$, and the *mean square longitudinal field* $\langle H_z^2 \rangle$. However, in a typical sample of lines, most transitions do not have very anomalous Zeeman patterns, so that there is in general a fairly strong correlation between S_2 and D_2. Therefore, in most cases, the contributions of $\langle H^2 \rangle$ and $\langle H_z^2 \rangle$ cannot be untangled. The following workaround (Mathys 1995b) proves to work well in practice: one *assumes* that

$$\langle H_z^2 \rangle = \langle H^2 \rangle/3\,, \tag{3.59}$$

and one applies Eq. (3.58) to derive a single quantity, the *mean square magnetic field*, $\langle H^2 \rangle + \langle H_z^2 \rangle$, or its square root (which has the dimension of a magnetic field), the *mean quadratic magnetic field*. The assumption $\langle H_z^2 \rangle = \langle H^2 \rangle/3$ corresponds to a random orientation of the field vector over the visible stellar hemisphere. It is intermediate between the extreme (and very improbable) configurations $\langle H_z^2 \rangle = \langle H^2 \rangle$ (purely longitudinal field) and $\langle H_z^2 \rangle = 0$ (purely transversal field), and should accordingly be closest to any real field configuration. In all the cases treated so far, the difference between the value of the mean quadratic magnetic field derived under the assumption of random field, on the one hand, and of either purely longitudinal or purely transversal field, on the other hand, is smaller than the uncertainty of these determinations.

Let us note in passing that the mean square magnetic field $\langle H^2 \rangle + \langle H_z^2 \rangle$ is always greater

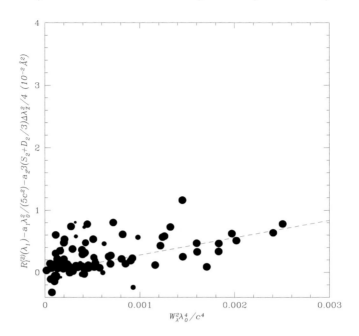

FIGURE 6. Contribution of the intrinsic part of profiles of the lines of Cr II observed in HD 94660 to their second-order moment about their centre (*see text*). The dashed line is the term of the least-squares fit of the observations corresponding to that contribution. The relative weights of the different lines in the regression analysis yielding this best fit are illustrated by the sizes of the dots representing them. The abscissa scale corresponds to the following choice of units: W_λ in mÅ, λ_0 in Å, and c in km s^{-1}.

than the sum of the squares of the mean field modulus and of the mean longitudinal field, $\langle H \rangle^2 + \langle H_z \rangle^2$). This results from the fact that, quite generally, $\langle H \rangle^2 \leq \langle H^2 \rangle$ and $\langle H_z \rangle^2 \leq \langle H_z^2 \rangle$. The difference between both sides of these inequalities reflects the dispersion of the values of the modulus (resp. of the component along the line of sight) of the field vector across the stellar disk. This dispersion is generally small for the field modulus (see Sect. 4.1). But it can be considerably larger for the line-of-sight component, as the field can reverse its polarity over the visible stellar hemisphere.

The form of $\Phi^{(2)}(\lambda_0)$ can be derived from Eqs. (3.56) and (3.55)

$$\Phi^{(2)}(\lambda_0) = \Psi^{(2)}(\lambda_0) + \mathcal{F}^{(2)}(\lambda_0) + \Delta\lambda_D^2/2 + a_0 + \sum_{r=1}^{R(2)} \sum_{s=1}^{S(r)} a_{rs}^{(2)} Q_r^s . \qquad (3.60)$$

Mathys & Hubrig (in preparation) have analyzed sample of lines of several ions in various Ap stars to determine their mean quadratic magnetic field through application of Eq. (3.58), together with Eqs. (3.59) and (3.60). In the latter, they have tried to introduce various parameters or combination of parameters Q_m, such as W_λ, $W_\lambda \lambda_0/c$, χ_e, $\chi_e W_\lambda$, $\chi_e W_\lambda \lambda_0/c, \ldots$ (χ_e is the excitation potential of the lower level of the transition). They finally concluded that the best representation of the observed behaviour of $R_I^{(2)}(\lambda_I)$ is given by an equation of the following form:

$$R_I^{(2)}(\lambda_I) = a_1 \frac{1}{5} \frac{\lambda_0^2}{c^2} + a_2 \frac{1}{4} \left(S_2 + D_2/3 \right) \Delta\lambda_Z^2 + a_3 W_\lambda^2 \frac{\lambda_0^4}{c^4} . \qquad (3.61)$$

It should not *a priori* be expected that this relation is the best one for the analysis of

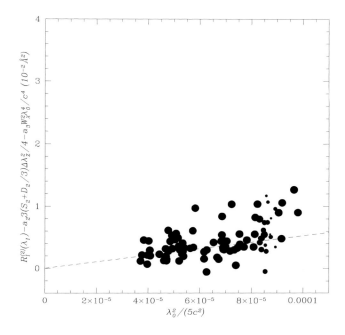

FIGURE 7. Contribution of the Doppler part of the profiles of the lines of Cr II observed in HD 94660 to their second-order moment about their centre (*see text*). The dashed line is the term of the least-squares fit of the observations corresponding to that contribution. The relative weights of the different lines in the regression analysis yielding this best fit are illustrated by the sizes of the dots representing them. The abscissa scale corresponds to the following choice of units: λ_0 in Å, and c in km s^{-1}.

observations of any type of star obtained with any instrument. But it has so far proved adequate in all cases in which it has been applied, that is, for Ap stars over a wide range of temperatures, observed with several different instrumental configurations, so that it appears rather robust.

In practice, $R_I^{(2)}(\lambda_I)$ is measured for a sample of lines, and the coefficients a_1, a_2, and a_3 are determined through a linear regression of these measurements as a function of the parameters appearing in the various terms of the right-hand side of Eq. (3.61). This regression is weighted by the inverse of the mean-square error of the $R_I^{(2)}(\lambda_I)$ measurements for the individual lines, $1/\sigma^2[R_I^{(2)}(\lambda_I)]$. It is forced through the origin: there is no constant term, say a_0, in Eq. (3.61). This is another empirically derived result: observation has shown that, in practical applications carried out so far, if a free a_0 term is introduced in the regression equation, its derived value never significantly differs from zero.

The results of the above-described regression analysis can be used to illustrate each of the dependences appearing in Eq. (3.61), by plotting the difference between $R_I^{(2)}(\lambda_I)$ and the best-fit values of two of the terms of the right-hand side against the independent variable of the third term. This is done in Figs. 6 to 8, which show respectively the intrinsic, Doppler (and related), and magnetic contributions to the second-order moment of the lines in the Stokes parameter I. The same ordinate scale has been used for all three figures, so as to allow the reader to visualize readily the relative importance of each contribution. In the considered case of a very slowly rotating star (the rotation

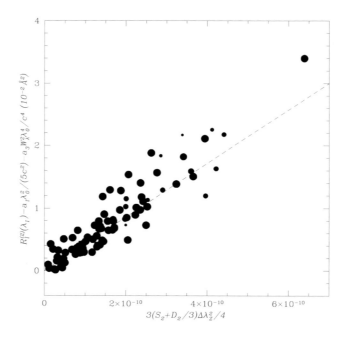

FIGURE 8. Contribution of the magnetic part of the profiles of the lines of Cr II observed in HD 94660 to their second-order moment about their centre (*see text*). The dashed line is the term of the least-squares fit of the observations corresponding to that contribution. The relative weights of the different lines in the regression analysis yielding this best fit are illustrated by the sizes of the dots representing them. The units on the abscissa scale are $\text{Å}^2 \text{G}^{-2}$.

period is approximately 7.7 years), the main broadening agent is the magnetic field. It is particularly significant that the term proportional to the squared equivalent width, which accounts for departures from the weak-line approximation, is small compared to the magnetic term: this demonstrates that the error on the derived magnetic field introduced by the use of this approximation must be small. That it is small for Stokes I implies that it must be even smaller for the other Stokes parameters, where it can be expected to cancel out to a large extent when taking the difference between two mutually orthogonal polarization states.

3.2. *Zeeman-Doppler imaging*

The purpose of the Zeeman-Doppler imaging (ZDI) method is to take advantage of the enhancement of detectability and diagnostic contents of magnetic signatures in polarized line profiles resulting from differential wavelength shifts due to rotation Doppler effect of contributions to the observed lines coming from various locations of the stellar disk, characterized by different magnetic field strengths and orientations. This approach is particularly useful for stars with tangled fields, such as late-type stars, where in the absence of rotation, the net circular polarization signal in spectral lines would be zero due to mutual cancellation of the contributions of regions of opposite field polarity. But as has been seen in Sect. 3.1.1, correlations between Doppler and Zeeman effects on the surface of Ap stars also allows one to derive constraints about the structure of their magnetic fields, and ZDI is one of the methods that can be used for that purpose.

Ultimately, ZDI, which has been developed originally developed by Semel (1989), is a field mapping technique. However, it is also a powerful magnetic field detection tool, which has indeed been successfully used to detect e.g. weak magnetic fields on Ap stars (Donati, Semel & del Toro Iniesta 1990) or magnetic spots on active late-type stars (Donati *et al.* 1990). The present description is restricted to the basic principles underlying the ZDI method.

The interpretation of observations of spectral line profiles through ZDI rests on the weak-field solution of the radiative transfer equation. From Eq. (1.27a), one can see that, in this approximation, the Stokes I line profile emerging at any point of the stellar surface is unaffected by the magnetic field. Furthermore, one can rewrite Eq. (1.27d) in the form:

$$V(\tau = 0, \mu, \lambda) = -\bar{g} \, \Delta\lambda_Z \, H_z \, \mu \, \frac{dI_0(\lambda)}{d\lambda}, \tag{3.62}$$

where $I_0 = I(\tau = 0, \mu = 1)$ is the local emerging intensity in the direction normal to the star's surface. Under the assumption that I_0 is the same at all points of the stellar surface, the following expression of the observed line profile in the Stokes parameter V can be derived through application of Eq. (1.30):

$$r_{\mathcal{F}_V}(\lambda) = \frac{\bar{g}}{\mathcal{F}_{I_c}} \frac{\Delta\lambda_Z}{\Delta\lambda_R} \int_{-1}^{+1} \frac{d}{d\lambda} I_0(\lambda - \lambda_0 - \Delta\lambda_R \, x) \, dx \int_{-\sqrt{1-x^2}}^{+\sqrt{1-x^2}} \mu(x,y) \, H_z(x,y) \, dy. \tag{3.63}$$

Contributions to the observed line profiles of points of different abscissae x undergo different Doppler shifts: this can be reflected by performing the change of variables $x \to \Delta\lambda/\Delta\lambda_R$. Then $r_{\mathcal{F}_V}$ can be rewritten as a convolution:

$$r_{\mathcal{F}_V}(\lambda) = \frac{\bar{g}}{\mathcal{F}_{I_c}} \frac{\Delta\lambda_Z}{\Delta\lambda_R} \int_{-\Delta\lambda_R}^{+\Delta\lambda_R} \bar{H}_z(\Delta\lambda) \frac{d}{d\lambda} I_0(\lambda - \lambda_0 - \Delta\lambda) \, d\Delta\lambda, \tag{3.64}$$

where $\bar{H}_z(\Delta\lambda) \, d\Delta\lambda$ is the magnetic flux along the line of sight, integrated over all points of the stellar surface with rotational Doppler shift between $\Delta\lambda$ and $\Delta\lambda + d\Delta\lambda$:

$$\bar{H}_z(\Delta\lambda) = \int_{-\sqrt{1-(\Delta\lambda/\Delta\lambda_R)^2}}^{+\sqrt{1-(\Delta\lambda/\Delta\lambda_R)^2}} \mu\left(\frac{\Delta\lambda}{\Delta\lambda_R}, y\right) H_z\left(\frac{\Delta\lambda}{\Delta\lambda_R}, y\right) dy. \tag{3.65}$$

The emergent line profile in the Stokes parameter I can similarly be expressed as a convolution:

$$r_{\mathcal{F}_I}(\lambda) = \frac{1}{\mathcal{F}_{I_c}} \frac{1}{\Delta\lambda_R} \int_{-\Delta\lambda_R}^{+\Delta\lambda_R} A(\Delta\lambda) \, I_0(\lambda - \lambda_0 - \Delta\lambda) \, d\Delta\lambda, \tag{3.66}$$

where $A(\Delta\lambda)$ is the projected fractional area of the region of the star where the rotational Doppler shift is between $\Delta\lambda$ and $\Delta\lambda + d\Delta\lambda$:

$$A(\Delta\lambda) = \int_{-\sqrt{1-(\Delta\lambda/\Delta\lambda_R)^2}}^{+\sqrt{1-(\Delta\lambda/\Delta\lambda_R)^2}} \mu\left(\frac{\Delta\lambda}{\Delta\lambda_R}, y\right) dy. \tag{3.67}$$

Then the expression of the Fourier transform of $r_{\mathcal{F}_V}(\lambda)$ is:

$$\tilde{r}_{\mathcal{F}_V}(u) \equiv \int_{-\infty}^{+\infty} e^{iu\lambda} \, r_{\mathcal{F}_V}(\lambda) \, d\lambda$$

$$= -i \, u \, \frac{\bar{g}}{\mathcal{F}_{I_c}} \frac{\Delta\lambda_Z}{\Delta\lambda_R} \tilde{H}_z(u) \, \tilde{I}_0(u), \tag{3.68}$$

where $\tilde{H}_z$ and $\tilde{I}_0$ are the Fourier transforms of, resp., $\bar{H}_z$ and I_0. Similarly, the Fourier transform of $r_{\mathcal{F}_I}(\lambda)$ can be written:

$$\tilde{r}_{\mathcal{F}_I}(u) = \frac{1}{\mathcal{F}_{I_c}} \frac{1}{\Delta\lambda_R} \tilde{A}(u) \, \tilde{I}_0(u), \tag{3.69}$$

with evident notations. Combining Eqs. (3.68) and (3.69), the Fourier transform $\tilde{H}_z(u)$ can be derived:

$$\tilde{H}_z(u) = \frac{i\,\tilde{A}(u)}{u\,\bar{g}\,\Delta\lambda_{\rm Z}}\,\frac{\tilde{r}_{\mathcal{F}_V}(u)}{\tilde{r}_{\mathcal{F}_I}(u)}\,, \qquad (3.70)$$

and from there, the magnetic flux in any band $(\Delta\lambda, \Delta\lambda + d\Delta\lambda)$ of the visible stellar disk, $\bar{H}_z(\Delta\lambda)\,d\Delta\lambda$. One can, of course, achieve better precision by combining the information from several lines, possibly with some appropriate weighting.

3.3. *Least-squares deconvolution*

Both the ZDI method and the moment technique (and, before it, the photographic technique of mean longitudinal magnetic field determination) rely on the analysis of the profiles of limited samples of selected lines. While these approaches have definite advantages, they use only (an often small) part of the diagnostic information contained in the observations. They are, therefore, not ideal to achieve the ultimate limit in the detection of weak magnetic signatures. For the latter purpose, one would of course wish to exploit the information contents of the observations as completely as possible. As has since long been realized, this can be achieved by application of cross-correlation techniques. Early steps towards this for stellar magnetic field studies include the development of the MSHIFT method by Weiss, Jenkner & Wood (1978) for the analysis of RCP and LCP spectra of Ap stars recorded on photographic plates, and the use of a modified version of the radial velocity spectrometer CORAVEL in attempts to detect magnetic fields in late-type stars by Borra, Edwards & Mayor (1984). These had, until recently, remained isolated efforts. However, in the last few years, a new approach, *least-squares deconvolution* (LSD), has been developed (Donati *et al.* 1997), and has started to be applied to various stellar contexts, showing considerable promise, in particular, for systematic investigation of the occurrence of stellar magnetism across the Hertzsprung-Russell diagram.

The weak-field solution of the transfer equation for Stokes V, from Eq. (1.27d), can be rewritten in terms of the velocity coordinate $v = c\,(\lambda - \lambda_0)/\lambda_0$:

$$V(\tau = 0, \mu, v) = -\bar{g}\,k\,\lambda_0\,c\,H_z\,\mu\,\frac{dI_0(v)}{dv}\,. \qquad (3.71)$$

The LSD approach rests on the assumption that, in Stokes I, all the spectral lines in the considered stellar spectrum have the same *shape* $I_1(v)$, hence that the expression of $I_0(v)$ takes the form:

$$I_0(v) = d\,I_1(v)\,, \qquad (3.72)$$

where d is the local line central depth (which is generally different from line to line, but does not depend on v). Then one can write:

$$V(\tau = 0, \mu, v) = -\bar{g}\,\lambda_0\,d\,\mu\,K_H(v)\,, \qquad (3.73)$$

where $K_H(v)$ is a *line-independent* proportionality function. Applying Eq. (1.30) to carry out the disk integration, one gets:

$$r_{\mathcal{F}_V}(v) = \bar{g}\,\lambda_0\,d\,Z(v)\,, \qquad (3.74)$$

where the *mean Zeeman signature* (or *LSD profile*)

$$Z(v) = -\frac{1}{\mathcal{F}_{I_{\rm c}}}\int_{-1}^{+1} dx \int_{-\sqrt{1-x^2}}^{+\sqrt{1-x^2}} \mu(x,y)\,K_H[v - v_R(x); H_z(x,y)]\,dy \qquad (3.75)$$

is the same for all the lines. In this equation, $v_R(x) = x\,\Delta\lambda_R/\lambda_0$. Let us introduce the (line-dependent) scaling factor $w = \bar{g}\,\lambda_0\,d$. If we consider a sample of $N_{\rm line}$ lines (characterized by the index $i = 1, \ldots, N_{\text{line}}$), the corresponding line pattern function is defined as:

$$M(v) = \sum_{i=1}^{N_{\text{line}}} w_i \, \delta(v - v_i),$$ (3.76)

where v_i is the position, in the velocity space, of spectral line i of this sample. Then the Stokes V spectrum can be described by a linear system of equations (the index $j = 1, \ldots, N_{\text{pixel}}$ refers to the pixels in the observed spectrum):

$$r_{\mathcal{F}_V}(v_j) = \sum_{i=1}^{N_{\text{line}}} w_i \, \delta(v_j - v_i) \, Z(v_j),$$ (3.77)

or, in matrix form:

$$\mathsf{V} = \mathsf{M}\,\mathsf{Z}.$$ (3.78)

This involves the implicit assumption that line intensities add up linearly. This is, of course, wrong, especially in the case of strongly saturated lines. To avoid inconsistencies, in practical applications, one imposes the condition that the sum of normalized depths of neighbouring lines does not exceed 1. From the observed spectrum V and the *reference line pattern* (or *line mask*) M, one then derives the mean Zeeman signature Z by a least-squares solution (deconvolution). This can be written as (e.g., Press *et al.* 1992):

$$\mathsf{Z} = (\mathsf{M}^{\mathrm{T}} \mathsf{S}^2 \mathsf{M})^{-1} \mathsf{M}^{\mathrm{T}} \mathsf{S}^2 \mathsf{V},$$ (3.79)

where S denotes a diagonal matrix whose element S_{jj} is the inverse standard error of pixel j of the spectrum, $1/\sigma_j$ (σ_j can be determined as part of the procedure of extraction of the spectrum).

Note that in the right-hand side of Eq. (3.79), $\mathsf{M}^{\mathrm{T}} \mathsf{S}^2 \mathsf{V}$ represents the (weighted) cross-correlation of the observed spectrum with the line mask: thus the LSD technique is in its principle very similar to cross-correlation methods previously used for magnetic field diagnosis. In practice, the line mask is built by computation of a synthetic spectrum based on a model atmosphere corresponding approximately to the temperature of the studied star. One can at will build masks for various selected line subsamples. This is useful, e.g., in studies of Ap stars where different chemical species have different inhomogeneous distributions over the stellar surface: by using masks corresponding to lines of different chemical elements, one samples the magnetic field in a different manner, hence once gains insight into its spatial structure, and the correlation of the latter with abundance inhomogeneities.

3.4. *Multiline techniques: two different philosophies*

The moment technique and the LSD method are both based on the simultaneous consideration of a multiline sample for enhanced magnetic field diagnostic potential. Yet, besides the fact that they rely on somewhat different sets of assumptions and approximations, they also reflect quite different measurement philosophies, which will be briefly reviewed below.

In a way, both methods can be regarded as applying a filter to the observations so as to overcome the limitation of the observational noise affecting profiles of individual lines. In LSD, filtering is achieved simply by adding up signals from a large number of lines. This is, of course, very effective to get rid of random noise. The drawback, though, is that any intrinsic line-to-line difference is also wiped out in the process. In particular, all transitions are treated as if they had the same Zeeman pattern. This introduces some degree of uncertainty in the interpretation of the LSD profile. This profile, in first approximation, resembles that of a single line in the presence of a magnetic field: but which atomic parameters should be used to translate it into a magnetic field measurement? Numerical simulations (Landstreet, private communication) suggest that careful definition of an "average transition" may allow one to achieve reliable results in the analysis of Stokes V spectra — but interpretation of Stokes Q and U observations may be much less straightforward (see Sect. 3.5).

By contrast, the moment technique combines noise filtering at two levels. First, by characterizing line shapes by low-order moments, one effectively filters out high frequency noise in individual profiles. Then, in the simultaneous consideration of several lines, further noise reduction is achieved. Contrary to LSD, combination of information from several lines is not carried out by averaging, but via regression analysis. In this way, the specific properties of each transition (in particular, the Zeeman pattern) are still fully taken into account (through independent variables in the regressions). This does also, to some extent, allow one to overcome in an empirical manner, the limitations of the approximations made for the treatment of the radiative transfer (as illustrated in Sect. 3.1.2). Furthermore, deviating behaviours, such as due e.g. to mistaken line identifications or wrong atomic parameters, can be readily identified. Accordingly, the derivation of constraints about the stellar magnetic field properties through the moment technique rests on a well defined and, mostly, physically sound description of the diagnostic transitions.

On the other hand, the moment technique cannot properly handle line blends (which translate into non-matching behaviours in the regression analysis). Thus, while rotational Doppler effect is duly taken into account in the interpretation of the observations, the technique can only be applied to stars with moderate rotation (otherwise, Doppler broadened line all tend to blend with each other). This restriction does not exist for LSD, which deals naturally with blends as part of the line pattern function. This implies that studies based on LSD can generally use (many) more diagnostic lines than works relying on the moment technique: hence the former are better suited to pushing the detection limit as far as possible.

In summary, LSD is the method of choice for *detection* of weak magnetic signatures, while the moment technique generally lends itself better to physically meaningful diagnosis of stellar magnetic properties. This tendency is strengthened by the fact that the assumptions underlying LSD include the weak-field approximation, which is not a significant restriction for detection work, while strong magnetic fields can be properly handled through the moment technique.

3.5. *Linear polarization*

Circular polarization in spectral lines is primarily determined by the component of the magnetic field along the line of sight. Accordingly, observations in linear polarization are necessary to derive constraints about the field components in the plane perpendicular to the line of sight. Yet, to first order in the field strength, only Stokes V is sensitive to the effect of a magnetic field [see Sect. 1.6.2, and also Eqs. (1.23)]. Linear polarization is a second order effect in the magnetic field: in practice, this implies that magnetic signatures in Stokes Q and U are considerably smaller than in Stokes V. This is further strengthened by the fact that mutual cancellation of the polarization contributions from different regions of the stellar surface is much more effective for linear than for circular polarization: it occurs in Stokes Q and U for field elements with azimuth angles differing by $90°$, while for Stokes V, polarization contributions subtract out only between opposite field polarities.

As a result, until recently, spectropolarimetric studies of stellar magnetic fields had been almost exclusively restricted to Stokes V. Only within the last few years have new state-of-the-art instrumentation and developments of analysis methods finally started to make possible systematic exploitation of Stokes Q and U signals in spectral lines towards magnetic field diagnosis. Studies based on observations in linear spectropolarimetry hold the promise of bringing a major breakthrough in our knowledge of stellar magnetic fields. Yet, at present, this area of research is still very much in its infancy. Here, we shall just outline the main directions in which work has so far developed.

Let us also point out that, until now, direct detection of linear polarization Zeeman signatures in individual lines has been achieved only in very high signal-to-noise ratio (typically > 400) spectra of strong, magnetically sensitive lines of a very small number of strongly magnetic Ap stars (Wade *et al.* 2000). In other words, such detections still remain exceptions. In general, to detect linear polarization of magnetic origin, the use of advanced analysis procedures and tools is required.

3.5.1. *Broad-band linear polarization*

In view of the difficulty of detecting linear polarization in spectral lines, it has in some cases proved advantageous to use an alternative approach based on the measurement of linear polarization through broad band filters, or *broad band linear polarization* (BBLP). The technique has been successfully applied to a number of Ap stars, mostly by J.-L. Leroy and his collaborators (see Leroy 1995 and references therein).

Magnetic fields of Ap stars are not strong enough to generate any significant amount of continuum polarization. What BBLP measurements detect, instead, is the cumulative effect of differential magnetic intensification of all the spectral lines contained in the filter passband. The net linear polarization of weak lines is zero. But for stronger lines, desaturation by a magnetic field is different for the π and σ components. This differential effect is, qualitatively, similar for all lines, so that in broad band observations, the contributions of all the lines add up. The observed values of Q/I and U/I may typically reach up to a few 10^{-4}. Although these are very small degrees of polarization, the precision required to detect them can be achieved in broad band observations thanks to the high throughput allowed by integration of the signal over a wide passband, compared to high-resolution spectra.

The interpretation of the observations is not straightforward. Landolfi *et al.* (1993) have developed a canonical model that provides reasonably simple analytical results. The absolute value of the polarization degree depends on a number of poorly defined parameters, such as the line density and the average line strength in the spectral range of interest. Accordingly, it is an ill-suited diagnostic. By contrast, its relative variation as the star rotates, as well as the variation of the direction of polarization, lend themselves much better to the derivation of meaningful constraints. For this purpose, it is convenient to build a polarization diagram from the observational data, by putting in the $(Q/I, U/I)$ plane the points corresponding to linear polarization measurements performed at various phases of the rotation cycle of the star, and connecting them in phase order, to draw the path that the star describes. These paths may have very diverse shapes, from one single, almost circular loop, to a complex path intersecting itself in various places. An example is shown in Fig. 9.

In practice, constraints on the magnetic field properties are derived from such observational data is based on the following considerations (Bagnulo, Landi Degl'Innocenti & Landi Degl'Innocenti 1996). The observed broad-band linear polarization can be written:

$$Q = Q_0 + Q_{\rm F}\,, \tag{3.80a}$$

$$U = U_0 + U_{\rm F}\,, \tag{3.80b}$$

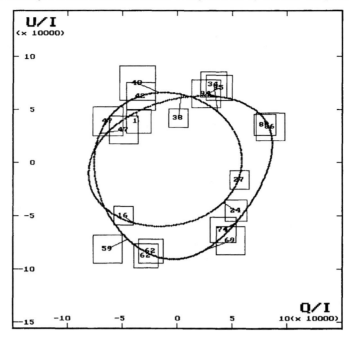

FIGURE 9. Linear polarization measurements of the Ap star HD 118022 in the B band, shown in the $(Q/I, U/I)$ plane. The individual measurements are represented by error boxes labeled with the rotation phase at which they have been obtained (fraction of the rotation cycle multiplied by 100). The solid curve represents the path that the star describes in the plane; it is based on fits of the curves of variations of Q/I and U/I against phase by third-order Fourier expansions. (From Leroy 1995.)

where

$$Q_0 \propto \langle H_Q^2 - H_{Q+\pi/2}^2 \rangle, \tag{3.81a}$$

$$U_0 \propto \langle 2\, H_Q\, H_{Q+\pi/2} \rangle, \tag{3.81b}$$

and

$$Q_F \propto \langle -2\, H_Q\, H_{Q+\pi/2}\, H_z \rangle, \tag{3.82a}$$

$$U_F \propto \langle (H_Q^2 - H_{Q+\pi/2}^2)\, H_z \rangle. \tag{3.82b}$$

One can note that, not too surprisingly, Q_0 and U_0 are sensitive to the same moments of the magnetic field as the second-order moments of line profiles in the Stokes parameters Q and U [see Eqs. (3.84) below]. By contrast, Q_F and U_F have no equivalent in the moment technique. This comes from the fact that these contributions to the BBLP correspond to the anomalous dispersion effects. Due to the weak-line approximation underlying the moment technique, these effects do not contribute to the moment expressions. Yet, they are significant for strong lines, and it is important to take them into account in the BBLP method, since the fact that this method works at all is due to line saturation.

3.5.2. *Linear polarization in spectral lines*

Even though linear polarization signatures may not be detectable in individual lines, it is generally possible to detect them, and to extract valuable diagnostic information from them, through application of multiline techniques. The same multiline techniques that are used for circular polarization data analysis can be and have been applied in linear polarization works: LSD and the moment technique.

The LSD approach is still based on the weak field solution of the radiative transfer problem. Yet, since Stokes Q and U are not sensitive to the magnetic field to first order, the solution must be expanded to the second order. One can describe the Stokes Q and U spectra by expressions similar to Eq. (3.77):

$$r_{\mathcal{F}_{Q,U}}(v_j) = \sum_{i=1}^{N_{\text{line}}} w_i^{(Q,U)} \, \delta(v_j - v_i) \, Z(v_j)\,, \tag{3.83}$$

where the scaling factor is now $w^{(Q)} = w^{(U)} = \bar{g}^2 \lambda_0^2\, d$ (Wade *et al.* 2000). The weakness of this approach lies in the fact that the approximation that a mean Zeeman signature adequately represents all the lines is much less good for linear than for circular polarization, as demonstrated by numerical simulations (Landstreet, private communication). As a consequence, the signal-to-noise ratio amplification by cross-correlation is significantly less effective for Stokes Q and U than for Stokes V, and the interpretation of the LSD profiles in terms of magnetic field properties is more ambiguous.

The first non-zero moments of the line profiles in Stokes Q and U are those of second-order. Their expression can be derived from Eqs. (3.51):

$$R_Q^{(2)}(\lambda_I) = -\Delta\lambda_Z^2\, D_2 \left\langle H_Q^2 - H_{Q+\pi/2}^2 \right\rangle\,, \tag{3.84a}$$

$$R_U^{(2)}(\lambda_I) = -\Delta\lambda_Z^2\, D_2 \left\langle 2\, H_Q\, H_{Q+\pi/2} \right\rangle\,. \tag{3.84b}$$

However, this form proves not to be ideally suited to the analysis of real data. An equivalent, but more appropriate to practical applications, set of equations is:

$$\left\{ \left[R_Q^{(2)}(\lambda_I) \right]^2 + \left[R_U^{(2)}(\lambda_I) \right]^2 \right\}^{1/2} = \Delta\lambda_Z^2\, D_2 \left\langle H_\perp^2 \right\rangle\,, \tag{3.85a}$$

$$R_U^{(2)}(\lambda_I) / R_Q^{(2)}(\lambda_I) = \tan(2\,\langle\chi\rangle)\,, \tag{3.85b}$$

where $\langle H_\perp^2 \rangle^{1/2}$ is the *mean transverse magnetic field*. Its square is defined by an expression similar to Eq. (3.54), where the magnetic term in the integrand is $(H_Q^2 + H_{Q+\pi/2}^2)$. The second quantity which is derived, $\langle\theta\rangle$, is called the *mean angle of the transverse magnetic field*.

4. Magnetic geometries and structures

4.1. *Magnetic fields of Ap stars: general properties*

Exploitation of the diagnostics described in Sects. 2 and 3 allow one to derive a number of conclusions about the generic properties of the magnetic fields of the Ap stars. A number of these properties are reviewed below, with emphasis on how they are established. This presentation is not intended to give an exhaustive review of our current knowledge of Ap star magnetic fields. Rather, its primary purpose is to illustrate the use of the diagnostic methods previously introduced.

From consideration of the measurements of $\langle H_z \rangle$ and $\langle H \rangle$ in Ap stars, one can derive a general overview of the properties of their magnetic fields.

As can be seen in Fig. 5, the resolved components of magnetically split lines are sharp and well defined: this implies that the range of field strengths at the stellar surface is fairly narrow. Furthermore, when the splitting is sufficient, the profile of the line Fe II λ 6149.2 goes back all the way to the continuum between the two components: this shows that the magnetic field covers the entire stellar surface. This conclusion receives further support from the fact that, in all Ap stars where a magnetic field is detected, it is observed throughout the whole rotation cycle.

The mere fact that non-zero longitudinal fields are observed at all indicates that the magnetic fields of Ap stars have a fairly simple large scale organization, in strong contrast with the magnetic field of the sun. The latter is intermittent, confined in many small elements of alternating polarity, so that the polarization that these elements induce locally mostly cancels out in disk-integrated observations.

Various considerations converge to support the view that, to first order, magnetic fields of Ap stars must include a sizeable dipole-like contribution. On the one hand, the polarization generated by toroidal fields, or multipolar fields of higher orders, cancels out to a large extent in the disk-integration process. Therefore, a roughly dipolar structure is the only one that can account for longitudinal fields of the observed order of magnitude without implying values of the mean field modulus so high that they would generally distort the Stokes I line profiles well beyond the limits allowed by observation. For instance, the order of magnitude of the ratio between the extrema of the mean longitudinal field and of the field modulus is typically $\langle H_z \rangle / \langle H \rangle \sim 0.3$ for a dipole, and $\langle H_z \rangle / \langle H \rangle \sim 0.05$ for a quadrupole (assuming in both cases that the multipole is at the centre of the star).

On the other hand, the curve of variation of the longitudinal field with rotation phase is in most cases closely sinusoidal (see e.g. Mathys 1991). The simplest model consistent with such variations is one in which the stellar field has a purely dipolar structure, with its centre at the centre of the star, and its axis inclined at some angle with respect to the stellar rotation axis.

However, this model does not stand closer scrutiny. For a fraction of the Ap stars, the extrema of the longitudinal field are of opposite sign. This implies that both magnetic poles of these stars alternatively come into view. If their magnetic field was a centred dipole, their field modulus should have two maxima and two minima per rotation cycle (since for a centred dipole, magnetic field strength is maximum at the magnetic poles and minimum at the magnetic equator). However, of all the Ap stars with resolved magnetically split lines for which $\langle H \rangle$ has so far been repeatedly determined throughout a rotation cycle, only for one does the curve of variation of this field moment show two maxima and two minima (Mathys *et al.* 1997). This represents a strong statistical evidence that the magnetic fields of Ap stars are generally not centred dipoles, since one expects the fraction of stars with reversing $\langle H_z \rangle$ to be considerably higher. This view receives further support from a recent systematic study of the longitudinal fields of Ap stars with magnetically resolved lines (Mathys, Manfroid & Wenderoth, in preparation). To the extent that the variations of both $\langle H_z \rangle$ and $\langle H \rangle$ can be adequately represented by sinusoids with their extrema in phase, models such as a dipole offset from the star centre along its axis, or the superposition of a centred dipole and centred quadrupole with the same axis, prove convenient. Indeed, they provide a good match between the number of independent observables (four: the maximum and minimum of $\langle H_z \rangle$, and the maximum and minimum of $\langle H \rangle$), and the number of free parameters:

• for the decentred dipole: the angle i between the rotation axis and the line of sight, the angle β between the magnetic and rotation axes, the polar field strength H_d, and the offset a (expressed as a fraction of the stellar radius);

• for the superposition of collinear centred dipole and quadrupole: i, β, and the polar field strengths H_d and H_q corresponding resp. to the dipolar and quadrupolar components.

Such models, have accordingly, been popular in a number of past studies. Actually, one can show that they are equivalent for small departures from a centred dipole (i.e., for a

or H_q/H_d small compared to 1). For more details on this subject (mostly of historical interest), see e.g. Landstreet (1980).

The possibility to extract line profile information is, of course, providing additional constraints, which lead to further refinement of our knowledge of the structure of Ap star magnetic fields. Full exploitation of state-of-the-art observational data requires the use of elaborated modelling techniques, involving considerable amounts of numerical computation, the principles of which will be briefly described in Sect. 4.2. Development of the relevant methods and tools has been going on for a few years. A number of magnetic field maps have been derived. But for the time being, these results are mostly limited to a few individual stars. It may still take a while before enough stars have been analyzed to allow general conclusions to be drawn. Ability to derive as detailed and realistic information as possible of the structure of magnetic fields of some stars is only one approach to achieve progress in our understanding of the origin and of the physics of magnetic Ap stars. But focusing attention on the detailed properties of each tree does not necessarily reveal the overall layout of the forest. Important insight can also be gained from consideration of statistical samples of stars, whose properties may possibly be diagnosed only in a more approximate manner, through application of simpler methods.

An illustration of the potential of such an approach is given by the recent work of Landstreet & Mathys (2000). These authors have shown that a simple model consisting of the superposition of a dipole, a quadrupole, and an octupole, all centred at the centre of the star and with the same axis, can generally give an adequate, though not exact, description of the variations of the longitudinal field, the crossover, the quadratic field and the field modulus. This model is, obviously, a generalization of the long popular dipole plus quadrupole model. While the dipolar component accounts primarily for the longitudinal magnetic field and the quadrupole gives the field strength contrast between the magnetic poles, the octupole allows one to adjust the equator-to-pole contrast of the field strength. Deriving such models for a sample of stars, Landstreet and Mathys showed that in Ap stars with rotation periods longer than approx. one month, the angle β between the magnetic and rotation axes is generally small, unlike stars with shorter periods, where this angle is predominantly large. Another indication of correlation between magnetic properties and stellar rotation had been found by Mathys *et al.* (1997). Namely, no stars with rotation period in excess of 150 d have a mean magnetic field modulus in excess of 7.5 kG, whereas among stars with shorter rotation periods, more than 50% have a field modulus larger than that value. Mathys *et al.* also reported that there apparently exists a sharp cutoff a the low end of the distribution of $\langle H \rangle$: no star with a field modulus (averaged over the stellar rotation period) smaller than 2.8 kG has been found so far. This does not appear to be due to an observational bias: there are strong arguments supporting the view that weaker fields should be easily detectable with observations of the resolution and quality achieved in the published studies of Ap stars with magnetically resolved lines.

Currently, it is widely accepted that the magnetic fields of Ap stars are *fossil*, that is, that they have been acquired by the stars at the time of their formation. However, many details remain unclear, such as the exact origin of the fields, and how and when they become observable at the stellar surface. Progress in the theoretical understanding of these aspects depends critically on the availability of additional observational constraints. Therefore, results of statistical nature about the strength and structure of Ap star magnetic fields, such as those sketched about, are particularly valuable. At present, their implications have not been fully worked out yet, but they undoubtedly open new perspectives for our understanding of the origin and evolution of these fields.

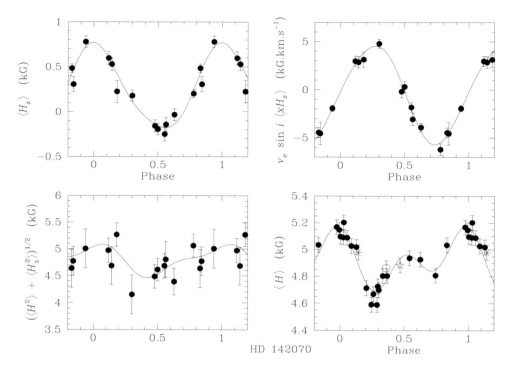

FIGURE 10. Mean longitudinal magnetic field (*top left*), crossover (*top right*), mean quadratic magnetic field (*bottom left*), and mean magnetic field modulus (*bottom right*) of the Ap star HD 142070, against rotation phase. Curves are fits of the data by a sine wave and its first harmonic.

Yet, it has become increasingly evident in recent years that, beyond their first-order dipole-like structure, magnetic fields of Ap stars have geometries significantly more complex than had long been believed. In particular, departures from cylindrical symmetry of the magnetic field about an axis passing through the centre of the star (though inclined with respect to the rotation axis), which in the past could seem to be restricted to a small number of particularly exotic objects, are now recognized as a common property of the majority of Ap stars. This is indicated in particular by the observation that, in most stars, one at least of the curves of variation with phase of $\langle H_z \rangle$, $\langle x\,H_z \rangle$, $(\langle H^2 \rangle)^{1/2}$, or $\langle H \rangle$ does not show mirror symmetry about some phase (Mathys 1993). An example of typical variation curves of the various field moments is shown in Fig. 10. In this specific case, the curves of variation of $\langle H_z \rangle$ and $\langle x\,H_z \rangle$ depart only marginally from sinusoids, but the variations of $(\langle H^2 \rangle)^{1/2}$ and $\langle H \rangle$ are strongly asymmetric. Precise modelling of such magnetic fields requires the application of the methods sketched in the next section.

4.2. Numerical mapping of stellar magnetic fields

4.2.1. Full numerical inversion of line profile

Ultimately, one would wish to be able to derive from observations a map of the magnetic field over the stellar surface, with as little *a priori* assumptions as possible. The formal description of this problem is straightforward. Let us assume that spectra of a star have been obtained in all four Stokes parameters at N_{phase} rotation phases φ_i. These spectra can be used to build an observation matrix:

$$\mathcal{O}(\lambda_j, \varphi_i) = \{I(\lambda_j, \varphi_i), Q(\lambda_j, \varphi_i), U(\lambda_j, \varphi_i), V(\lambda_j, \varphi_i)\}, \qquad (4.86)$$

where the index $j = (1, \ldots, N_{\text{pixel}})$ refers to the individual wavelength bins (pixels) of the observed spectra.

Let $\boldsymbol{H}(x_m, y_n)$ $(m = 1, \ldots, N_x; n = 1, \ldots, N_y)$ represent the magnetic vector at a grid of points over the stellar surface (the coordinates x and y are defined in some suitable, generally non-cartesian, reference system). For some assumed magnetic field distribution $\boldsymbol{H}(x_m, y_n)$ (and for a given model atmosphere), one can solve the radiative transfer equation to derive a computed spectrum in the four Stokes parameters. Repeating this calculation at the N_{phase} rotation phases φ_i, one can build a computed matrix $\mathcal{C}(\lambda_j, \varphi_i)$, with a structure similar to the observation matrix $\mathcal{O}(\lambda_j, \varphi_i)$.

In principle, determination of the geometrical structure of the stellar magnetic field should be achieved by identifying the distribution $\boldsymbol{H}(x_m, y_n)$ such that the computed matrix $\mathcal{C}(\lambda_j, \varphi_i)$ is the best fit to the observation matrix $\mathcal{O}(\lambda_j, \varphi_i)$. In practice, this *inversion problem* is ill-posed: its solution is non-unique. In order to guarantee that a unique solution can be derived, a so-called regularization condition must be imposed. The choice of this condition is arbitrary and it does in fact reflects an implicit assumption about the magnetic field structure. For instance, a frequently used approach involves maximizing some adequately defined entropy function (e.g., Brown *et al.* 1991). This, in practice, corresponds to looking for the "smoothest" magnetic field structure. As another example, the poster paper prepared for this Winter School by one of the students, O. Kochukhov, presents a mapping approach in which the regularization criterion requires the magnetic field structure to be as close as possible to a superposition of multipoles. Other choices exist in the literature; their presentation as well as further discussion of the details of the numerical inversion techniques are beyond the scope of these lectures.

4.2.2. *Generalized multipolar model*

For Ap stars, an approach intermediate between the semi-analytical, axisymmetric models described in Sect. 4.1 and the full numerical inversion of line profiles introduced above has been developed in recent year by S. Bagnulo and collaborators. The general formalism underlying it is presented in Bagnulo, Landi Degl'Innocenti & Landi Degl'Innocenti (1996).

In all applications of this method that have been published until now, it is assumed that the magnetic field of the studied stars can be represented by the superposition of a dipole and of a non-linear quadrupole, located at the centre of the star. The best such model is derived through an inversion procedure based on χ^2 minimization of differences between values of selected observables predicted and observed at phases distributed throughout the stellar rotation period. The observables that have been considered so far are (Bagnulo, Landolfi & Landi Degl'Innocenti 1999):
- the mean longitudinal magnetic field,
- the crossover,
- the mean quadratic magnetic field,

and, more recently (Bagnulo *et al.* 2000):
- the mean magnetic field modulus,
- and the broad-band linear polarization.

Frequently, through this method, it is impossible to derive a unique best model. Instead, there are often two or, more rarely, a small number of models, with fundamentally different geometries, that fit equally well the observables. It may be, in the future, be possible to solve this ambiguity by fitting simultaneously additional observables. Obvious candidates are the magnetic moments derived from higher order moments of the line profiles in the Stokes parameters I and V, as well as information extracted from study of the line profiles observed in linear polarization. Yet, it must be noted that some of the models derived so far have been used to compute synthetic spectra in all four Stokes parameters. These synthetic spectra have been compared to observed spectra (independent from those on the basis of which the model had been established). In some cases at least, a reasonable match between both was observed, even for the Q and U spectra, although none of the original observables used as input for the model had been derived from these spectra (Bagnulo *et al.* 2001). This suggests that the addition of new observables may not always be sufficient to fully constrain the models.

On the other hand, a common feature of most models derived so far is that they involve strong gradients of the magnetic field on the part of the stellar surface that never comes into view. Statistically, it is very improbable that the geometry of observation of all those stars is such that, by coincidence, we never see the regions of their surface where the variation of the magnetic field is steepest. It appears much more plausible that the systematic existence of strong field gradients on the hidden part of the stellar surface reflects an intrinsic weakness of the models. One possible way to try to improve them may be to add to it a non-linear octupolar component.

4.2.3. *Modelling techniques: two different approaches*

Let us briefly discuss the respective advantages and drawbacks of the modelling techniques introduced in the previous two sections.

An obvious limitation of the generalized multipolar model is that, since its observables are derived from consideration of low-order moments of line profiles (or use no line profile information at all, in the case of BBLP measurements), the information contained in the observed spectral line shapes is generally not fully exploited. By contrast, direct numerical inversion uses all the available line profile information. In principle, the latter is advantageous, since it should allow one to derive the most complete constraints on the magnetic field structure. In practice, though, it also implies that any noise in the observations is used as input to the model. Unavoidably, the numerical inversion procedure tries to fit this noise as well as the actual signal. This may, and does, lead to numerical instabilities. Instabilities may also result from insufficient wavelength or phase sampling of the data. Accordingly, numerical line profile inversion is very demanding in terms of signal-to-noise, spectral resolution and phase coverage of the observations, unlike the generalized multipolar model method. The latter is particularly well suited to the analysis of fairly noisy data recorded at mild resolution, thanks in particular to the fact that use of line profile moments quite effectively filters out the observation noise (see Sect. 3.4).

On the other hand, its basic assumption that the stellar field can be represented by a generalized multipolar model implies that it is applicable to Ap stars (and, possibly, white dwarfs), but that it is generally not appropriate for other stars (in particular, active late-type stars). By contrast, direct inversion of line profiles does not assume any *a priori* field geometry — at least, explicitly: the regularization criterion does to some extent involve an implicit assumption on the field structure, and different regularization conditions may be better suited to different types of stars.

From a practical point of view, one can also note that line profile inversion involves much heavier, more time-consuming numerical computations than the generalized multipolar model method. This suggests that it may in some cases be advantageous to take the latter (or, as a matter of fact, one of the simple semi-analytical models discussed in Sect. 4.1) as a first approximation of the actual models, and to use this approxima-

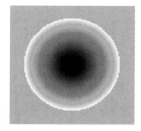 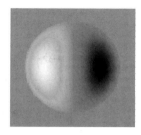

FIGURE 11. Simulated SPIN Stokes V observations of a slowly rotating star with a centred, dipolar magnetic field, with $i = \beta = 90°$. The phases illustrated correspond to those when the magnetic field is seen pole-on (*left*) and equator-on (*right*). (From Rousselet-Perraut *et al.* 2000.)

tion to restrict the space of free parameters allowed for full numerical inversion. While apparently promising, this approach does not seem to have ever been tried.

4.2.4. *The future: spectropolarimetric interferometry*

A promising prospect to gain additional insight into the structure of stellar magnetic fields should come from the use of a spectropolarimeter in combination with an optical interferometric system (such as ESO's VLTI). The potential of this technique, called SpectroPolarimetric INterferometry (SPIN) is under study (see Rousselet-Perraut *et al.* 2000). As an illustration of these theoretical predictions, a simple example is shown in Fig. 11: that of a centred magnetic dipole in a slowly rotating star, with $i = \beta = 90°$, observed in Stokes V. One can note that the magnetic field is detected not only when it is seen pole-on, but also when it is seen equator-on, by contrast with standard spectropolarimetry, where no magnetic signature is seen in Stokes V in the latter case. The technique also has the potential to resolve ambiguities which affect models based on standard spectropolarimetric observations. The first observations applying this technique are planned to take place in 2001 with the GI2T interferometer.

5. Polarimetric diagnostics of magnetic fields in non-Ap stars

5.1. *Non-degenerate stars*

For more than three decades, Ap stars have remained the only non-degenerate stars (besides the sun) in which the presence of magnetic fields was unquestionably established through *direct* observation, in spite of many attempts to detect such fields in stars of various types. A breakthrough was achieved in 1980 when Robinson, Worden & Harvey (1980) convincingly detected a magnetic field in the late-type dwarfs ξ Boo A and 70 Oph A. This generated renewed interest in the search for magnetic fields in active late-type stars, and success was met in a significant number of them in the following years. Yet, all those detections, like the original one of Robinson, Worden & Harvey, were based on observation of differential magnetic broadening of spectral lines in Stokes I spectra. It still took a decade until the first spectropolarimetric diagnosis of a magnetic field in a non-Ap star: a Stokes V magnetic signature was repeatedly observed on the active K component of the RS CVn system HR 1099, which was interpreted, using ZDI, as due to a localized, largely monopolar magnetic spot (Donati *et al.* 1990). The next major step came with the introduction of the LSD technique, and its application to an extensive search for magnetic fields in active stars of various types (Donati *et al.* 1997).

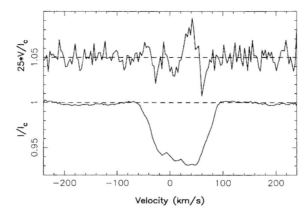

FIGURE 12. LSD Stokes V (*top*) and I (*bottom*) profiles observed in the weak-lined T Tauri star V410 Tau. Note the several sign reversals of the V profile, which reflect the complexity of the magnetic field structure on the stellar surface. (From Donati *et al.* 1997.)

This work definitely established LSD as the technique of choice for *detection* of (weak) stellar magnetic fields. It has since been used to establish the presence of such fields in stars of a variety of types, over a wide range of temperatures. A brief overview of the results recently obtained is given below.

5.1.1. *Pre-main sequence and late-type active stars*

Pre-main sequence and late-type stars have long been known to show activity phenomena qualitatively similar to those observed on the sun, but with different strengths and variability timescales. It is quite natural to attribute them to dynamo mechanisms at work in the convective outer layers of those stars. The most active of them appear therefore as prime candidates for attempts at achieving direct detections of stellar magnetic fields.

Donati *et al.* (1997) have conducted a systematic survey of such stars, as a result of which the presence of a magnetic field has been definitely established in one or several stars of the following types:

- weak-line T Tauri stars,
- pre-main sequence binaries,
- Herbig Ae stars,
- cool stars on the ZAMS,
- dwarf flare stars,
- RS CVn stars,
- FK Com stars.

Stokes V magnetic signatures in these stars are often complex, with several sign reversals throughout the line profile. An example is shown in Fig. 12. This indicates that the field structure is complex, with several small-scale magnetic regions of opposite polarities over a stellar hemisphere.

For some of those stars (e.g., the ZAMS star AB Dor and the RS CVn system HR 1099), observations sampling one or several rotation cycles have been obtained, and they have been used to build maps of the magnetic field over the stellar surface.

5.1.2. *Hot stars*

Over recent years, an increasing number of observations of hot stars have been obtained, which provide increasing *indirect* evidence that they must have magnetic fields. Such observations include, for instance, rotationally modulated winds in O stars, X-ray emission in Be stars, and transient features in the profiles of absorption lines in the visible spectrum of Be stars. Such features can be best explained theoretically by assuming that those stars have a (generally moderate) magnetic field. Other attempts to interpret them without introducing the effect of a such a field must call to exotic combinations of processes and circumstances, which overall seem much less plausible than the presence of a magnetic field.

However, direct detection of magnetic fields in hot stars is particularly challenging, because the density of their spectral lines is much lower than in cooler stars, and because the line profiles are often strongly broadened by fast rotation and other mechanisms, making them difficult to measure accurately. The low line density implies, in particular, that the gain in detection efficiency allowed by use of the LSD technique is much lower than in late-type stars.

Yet, the only definite detection of a magnetic field in a hot star published so far has been achieved through application of LSD. Henrichs *et al.* (2000) have indeed reported the presence of a field in β Cep, a somewhat unusual member of the Be star group, since its projected radial velocity is low: $v_e \sin i = 25$ km s^{-1}. This represents a very favourable circumstance for the detection of a magnetic field. Henrichs *et al.* measured a mean longitudinal magnetic field varying quasi-sinusoidally, with a semi-amplitude of 96 G about a mean value of -10 G, over a period of 12 days. This is also the period of variation of the UV spectral lines of the star. It is interpreted as being its rotation period. In other words, the observed magnetic field variations of β Cep can be satisfactorily interpreted within the framework of the oblique rotator model, and it appears plausible that this Be star has a magnetic field roughly similar to that of the Ap stars.

If the field strength of β Cep is representative of the magnetic fields of Be stars, and of other hot stars, field detection in most of those will be very challenging, given their very high average rotational velocity.

5.2. *White dwarfs*

5.2.1. *Magnetic fields in white dwarfs: overview*

In 1970, Kemp *et al.* (1970) reported the first detection of circularly polarized light from a white dwarf, Grw $+70°8247$. While the details of the processes responsible for generating the circular polarization were not understood at the time, the observation was correctly interpreted as a convincing indication for the presence of a strong magnetic field. Further discoveries of white dwarfs showing circular polarization in their spectra soon followed, so that after Ap stars, white dwarfs became the second group of stars (other than the sun) where the presence of magnetic fields was definitely established.

Before discussing the techniques used to diagnose magnetic fields in white dwarfs, we shall present a brief overview of our current knowledge of their properties. For more detail, the reader should refer to the recent, comprehensive review of Wickramasinghe & Ferrario (2000).

Two types of magnetic white dwarfs are distinguished: isolated ones, of which 65 are currently known, and white dwarfs in magnetic cataclysmic variables, in similar number. The properties of magnetic white dwarfs in cataclysmic variables are affected by the interactions with their companion in the binary system, so that their study involves a number of specific issues which are outside the scope of these lectures. Accordingly, we shall not consider them any further, and we shall discuss only the properties of the isolated magnetic white dwarfs. The latter are estimated to comprise about 5% of all white dwarfs. Observed magnetic field strengths range from $3\,10^4$ G (comparable to the highest fields of magnetic Ap stars) to 10^9 G. The majority of the isolated magnetic white dwarfs are of type DA (i.e., their spectra only show hydrogen lines). The magnetic fields of a fraction of them are found to vary periodically. These variations are interpreted as resulting from stellar rotation, through the same oblique rotator model as for Ap stars. The white dwarfs where no field variations have been observed must either have very long periods (in excess of 100 y), or their magnetic field must be symmetric about their rotation axis. When rotational modulation is present, it can be used to derive constraints about the geometrical structure of the magnetic field. In a first approximation, fields may often be represented by a dipole offset from the centre of the star, along the dipolar axis, but there is strong evidence that more complex structures must be invoked to reproduce adequately the details of the observations. The parallel with the situation for Ap stars is striking. This supports the view that magnetic Ap and Bp stars are the progenitors of the magnetic white dwarfs, since the distribution of the magnetic field strengths in the latter can then be explained as the result of magnetic flux conservation in the post-main sequence evolution of the former. This picture is also consistent with the good match between the rate of occurrence of Ap stars (relative to normal A and B stars), on the one hand, and of magnetic white dwarfs (compared to the whole class of white dwarfs), on the other hand.

5.2.2. *Magnetic field diagnosis in white dwarfs*

The diagnosis of magnetic fields in white dwarfs represents a significantly more complex challenge than for non-degenerate stars. In the latter, as we have seen in the previous sections, the effect of the magnetic field on line formation is generally quite satisfactorily described within the framework of the theory of the linear Zeeman effect. Exceptions to this rule correspond to the use of diagnostic transitions for which, for some reason, other effects reach an order of magnitude comparable to that of the linear Zeeman perturbation. They are not related to any particular characteristic of the magnetic field (or of other physical properties) of some specific star(s). Typically, use of the considered spectral lines for magnetic field diagnosis in any non-degenerate star would involve the same physics.

By contrast, interpretation of the (polarized) spectra of magnetic white dwarfs requires the consideration of vastly different physical processes in different stars, depending on the strength of their magnetic field. Furthermore, in most regimes, the behaviour of the energies of the atomic states, and the related radiative transfer processes, can be dealt with only numerically, and not through the use of (approximate) analytical expressions. As a matter of fact, at the high end of the field strength distribution, the physics of the atoms is still rather poorly understood. Finally, a non-negligible fraction of the magnetic white dwarfs have short rotation periods, ranging from 12 minutes to a couple of days. Since, on the other hand, these stars are intrinsically faint, their observation often requires rather long integration times. The latter can represent a non-negligible fraction of the rotation period, so that the resulting spectrum does not give a snapshot picture of the star's surface (and in particular, of its magnetic field) at a given instant, but rather some average image of the changing aspect of the visible stellar hemisphere as it drifts in and out of the observer's view during the observation. This obviously adds intricacies to the extraction of constraints on the magnetic field from the observations. Note however that the advent of the new generation of 10-m class telescopes allows one to overcome this limitation to a large extent.

Here, we shall just briefly introduce the general principles underlying the measure-

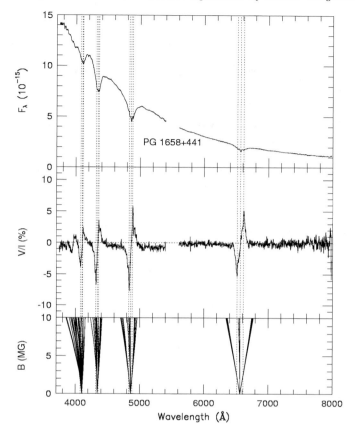

FIGURE 13. Stokes I and V spectra of the magnetic white dwarf PG 1658+441. Note the typical S-shape of the Stokes V line profiles, which is similar to that observed in non-degenerate (Ap) stars. The bottom panel shows the wavelength dependence of the line components on magnetic field strength. The splitting is to first order, similar to a linear Zeeman pattern, with systematic wavelength shifts due to the quadratic effect, increasing for the higher members of the Balmer series. (From Putney 1997, by permission of the AAS.)

ments of white dwarf magnetic fields in different strength ranges. For a more detailed presentation of the physical processes involved, see Wickramasinghe & Ferrario (2000).

For classification of the effects of a magnetic field on atomic structure at various field strengths, it is convenient to refer to the parameter $\beta = H/4.7\,10^9$ G. For $\beta \ll 1$, that is, approximately, for fields not exceeding 1 MG, line formation in white dwarfs is governed by the same physics as in non-degenerate stars, that is, it occurs primarily in a regime of linear Zeeman effect†. Towards the high field end of that range, quadratic Zeeman effect becomes increasingly significant, but its contribution is still a perturbation to the global effect. Accordingly, in first approximation at least, the methods that are used for non-degenerate stars can still be applied to the determination of magnetic fields in

† Strictly speaking, for the lines of hydrogen, which are most often observed in white dwarfs, this is a regime of complete Paschen-Back effect, since for the field strengths of interest, the magnetic perturbation largely exceeds the fine structure splitting. This implies that the magnetic quantum number M referred to in the selection rule $\Delta M = 0, \pm 1$ is the projection on the field direction of the orbital quantum number L, not of the total angular momentum quantum number J, but it has otherwise no impact on the present discussion.

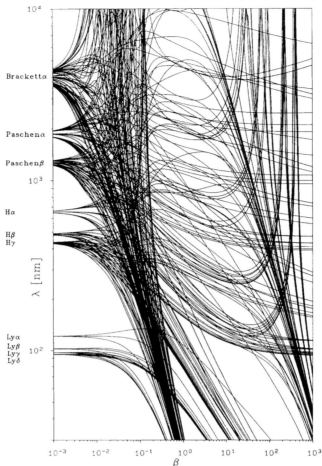

FIGURE 14. Computed wavelengths of hydrogen transitions as a function of magnetic field strength (characterized by the parameter $\beta = H/4.7\,10^9$ G). (From Ruder *et al.* 1994.)

white dwarfs in that field strength range. In particular, one may note that the Balmer line photopolarimetric technique was originally developed to study white dwarf magnetic fields (Angel & Landstreet 1970). Until very recently, the linear Zeeman effect regime was the only one for which the effect of the magnetic field could be adequately taken into account for line of other elements than hydrogen. Figure 13 shows, as illustration, the Stokes I and V spectra of a white dwarf in this regime.

Contrary to linear Zeeman effect, quadratic Zeeman effect cannot generally be treated analytically. Furthermore, at higher field strengths (for $\beta \sim 1$), it becomes comparable to the separation between levels of different principal quantum numbers. Finally, when β significantly exceeds 1, the magnetic term is the dominating one in the hamiltonian of the atom. In those regimes, heavy numerical computation is required to determine the energies of the atomic states. Their dependence on the magnetic field strength is intricate, with virtually no systematic structure. This is illustrated in Fig. 14, where the wavelengths of the allowed transitions are shown as a function of β. Note that some of the line components are dramatically shifted in wavelength: e.g., transitions ascribed to the Paschen series at low field strengths (thus occurring in the infrared) end up in the

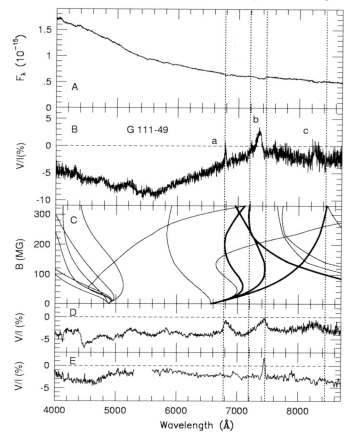

FIGURE 15. Stokes I and V spectra of the magnetic white dwarf G111-49 (top two panels). The third panel shows the wavelength dependence of the line components on magnetic field strength. Note the correspondence between (fairly) narrow features in the Stokes V spectrum and stationary wavelengths of components for field strengths of the order of 100 MG. The bottom two panels show the Stokes V spectra of two other magnetic white dwarfs: Grw $+70°8247$ and G227-35. (From Putney 1997, by permission of the AAS.)

far ultraviolet for $\beta \sim 1$ or greater. Accordingly, mere identification of the transition responsible for an observed spectral feature in a magnetic white dwarf may turn into a major challenge. This is even more so because, as usual, the observable spectrum is the result of the disk-integration of local contributions corresponding to a range of magnetic field strengths and orientations. The fact that, in most cases, component wavelengths vary very strongly with the field intensity makes this averaging process much more critical than in non-degenerate stars (in particular, Ap stars), and the final outcome is often very broad features which do not stand out from the continuum nor can be distinguished from each other. However, it is important to note in Fig. 14 that there are field strengths at which the wavelength variation of some σ_+ components reverses its slope, so that the wavelengths of these components remain quasi-stationary over a range of field strengths around these turning points. Such line components will tend to appear as (fairly) sharp and well defined features in spectra of white dwarfs with magnetic fields in the corresponding range, and can thus very advantageously be used as diagnostics of the field intensity. A practical application of this method is illustrated in Fig. 15.

Many of the studies of white dwarf magnetic fields so far rely almost only on Stokes *I* observations. On the one hand, the magnetic effects are large enough so that they can almost always be *detected* in unpolarized light. On the other hand, the fact that polarization is at most a (fairly small) fraction of the Stokes *I* intensity, hence that reliable measurements require high quality, low noise data, combined with the typical faintness of the sources, makes spectropolarimetric observations challenging. Yet, spectropolarimetry is definitely advantageous, or even required, for a number of purposes:

- to detect fields too weak to produce resolved line splitting in Stokes *I*;
- to derive effective constraints about the geometric structure of the stellar field;
- to detect and measure magnetic fields in binaries where the contribution of the white dwarf to the Stokes *I* spectrum is masked by unpolarized sources of radiation (such as gas streams in the system).

Examples of Stokes *V* spectra can be seen in Figs. 13 and 15. Linear polarization observations are still scarce in the literature. Predictably, the acquisition of spectropolarimetric observations of white dwarfs in the coming years should be boosted by the possibility to take advantage of the collecting power of the 10-m class telescopes.

REFERENCES

ANGEL, J.R.P. & LANDSTREET, J.D. 1970 Magnetic observations of white dwarfs. *Astrophys. J. Letters* **160**, L147–L152.

BABCOCK, H.W. 1947 Zeeman effect in stellar spectra. *Astrophys. J.* **105**, 105–119.

BABCOCK, H.W. 1951 The magnetically variable star HD 125248. *Astrophys. J.* **114**, 1–36.

BABCOCK, H.W. 1958 A catalog of magnetic stars. *Astrophys. J. Suppl. Ser.* **3**, 141–210.

BABCOCK, H.W. 1960 The 34-kilogauss magnetic field of HD 215441. *Astrophys. J.* **132**, 521–531.

BAGNULO, S., LANDI DEGL'INNOCENTI, M. & LANDI DEGL'INNOCENTI, E. 1996 Multipolar magnetic fields in rotating Ap stars: modeling of observable quantities. *Astron. Astrophys.* **308**, 115–131.

BAGNULO, S., LANDOLFI, M. & LANDI DEGL'INNOCENTI, M. 1999 Modelling of magnetic fields of CP stars. II. Analysis of longitudinal field, crossover, and quadratic field observations. *Astron. Astrophys.* **343**, 865–871.

BAGNULO, S., LANDOLFI, M., MATHYS, G. & LANDI DEGL'INNOCENTI, M. 1999 Modelling of magnetic fields of CP stars. III. The combined interpretation of five different magnetic observables: theory, and application to β Coronae Borealis. *Astron. Astrophys.* **358**, 929–942.

BAGNULO, S., WADE, G.A., DONATI, J.-F., LANDSTREET, J.D., LEONE, F., MONIN, D.N. & STIFT, M.J. A study of polarized spectra of magnetic CP stars: predicted vs. observed Stokes *IQUV* profiles for β Coronae Borealis and 53 Camelopardalis. *Astron. Astrophys.*, in press.

BETHE, H.A. & SALPETER, E.E. 1977 *Quantum mechanics of one- and two-electron atoms.* Plenum.

BORRA, E.F., EDWARDS, G. & MAYOR, M. 1984 The magnetic fields of the late-type stars. *Astrophys. J.* **284**, 211–222.

BRILLANT, S., MATHYS, G. & STEHLÉ, C. 1998 Hydrogen line formation in dense magnetized plasmas. *Astron. Astrophys.* **339**, 286–297.

BROWN, S.F., DONATI, J.-F., REES, D.E. & SEMEL, M. 1991 Zeeman-Doppler Imaging of solar-type and Ap stars. *Astron. Astrophys.* **250**, 463–474.

COWLEY, C.R. & MATHYS, G. 1998 Line identifications and preliminary abundances from the red spectrum of HD 101065 (Przybylski's star). *Astron. Astrophys.* **339**, 165–169.

DONATI, J.-F., SEMEL, M. & DEL TORO INIESTA, J.C. 1990 The magnetic field of the Ap star ε UMa. *Astron. Astrophys. Letters* **233**, L17–L20.

DONATI, J.-F., SEMEL, M., REES, D.E., TAYLOR, K. & ROBINSON, R.D. 1990 Detection of a magnetic region on HR 1099. *Astron. Astrophys. Letters* **232**, L1–L4.

DONATI, J.-F., SEMEL, M., CARTER, D.B., REES, D.E. & CAMERON, A.C. 1997 Spectropolarimetric observations of active stars. *Mon. Not. R. Astron. Soc.* **291**, 658–682.

HENRICHS, H.F., DE JONG, J.A., DONATI, J.-F., CATALA, C., WADE, G.A., SHORLIN, S.L.S., VEEN, P.M., NICHOLS, J.S. & KAPER, L. 2000 The magnetic field of β Cep and the Be phenomenon. In *The Be Phenomenon in Early-Type Stars*, IAU Coll. No. 175 (eds. Smith, M.A., Henrichs, H.F. & Fabregat, J.). Astron. Soc. Pacific Conf. Series vol. 214, pp. 324–329.

HUCHRA, J. 1972 An analysis of the magnetic field of 53 Camelopardalis and its implications for the decentered-dipole rotator model. *Astrophys. J.* **174**, 435–438.

KALKOFEN, W. (ed.) 1987 *Numerical radiative transfer.* Cambridge University Press.

KEMP, J.C., SWEDLUND, J.B., LANDSTREET, J.D. & ANGEL, J.R.P. 1970 Discovery of circularly polarized light from a white dwarf. *Astrophys. J. Letters* **161**, L77–L79.

LANDI DEGL'INNOCENTI, E. 1975 Hyperfine structure and line formation in a magnetic field. *Astron. Astrophys.* **45**, 269–276.

LANDI DEGL'INNOCENTI, E. 1982 On the effective Landé factor of magnetic lines *Solar. Phys.* **77**, 285–289.

LANDI DEGL'INNOCENTI, E. & LANDI DEGL'INNOCENTI, M. 1991 Radiative transfer for polarized radiation: symmetry properties and geometrical interpretation. *Il Nuovo Cimento* **62B**, 1–16.

LANDOLFI, M., LANDI DEGL'INNOCENTI, E., LANDI DEGL'INNOCENTI, M. & LEROY, J.-L. 1993 Linear polarimetry of Ap stars. I. A simple canonical model. *Astron. Astrophys.* **272**, 285–298.

LANDSTREET, J.D. 1980 The measurement of magnetic fields in stars. *Astron. J.* **85**, 611–620.

LANDSTREET, J.D. 1982 A search for magnetic fields in normal upper-main-sequence stars. *Astrophys. J.* **258**, 639–650.

LANDSTREET, J.D. & MATHYS, G. 2000 Aligned magnetic and rotation axes of long-period Ap stars. *Astron. Astrophys.* **359**, 213–226.

LANDSTREET, J.D., BORRA, E.F., ANGEL, J.R.P. & ILLING, R.M.E. 1975 A search for strong magnetic fields in rapidly rotating Ap stars. *Astrophys. J.* **201**, 624–629.

LEROY, J.-L. 1995 Linear polarimetry of Ap stars. V. A general catalogue of measurements. *Astron. Astrophys. Suppl. Ser.* **114**, 79–104.

MATHYS, G. 1988 The diagnosis of stellar magnetic fields from spectral line profiles recorded in circularly polarized light. *Astron. Astrophys.* **189**, 179–193.

MATHYS, G. 1989 The observation of magnetic fields in nondegenerate stars. *Fundam. Cosmic Phys.* **13**, 143–308.

MATHYS, G. 1990a Ap stars with resolved Zeeman split lines. *Astron. Astrophys.* **232**, 151–172.

MATHYS, G. 1990b Better Landé factors for iron-period elements. *Astron. Astrophys.* **236**, 527–530.

MATHYS, G. 1991 Spectropolarimetry of magnetic stars. II. The mean longitudinal magnetic field. *Astron. Astrophys. Suppl. Ser.* **89**, 121–157.

MATHYS, G. 1993 Magnetic field diagnosis through spectropolarimetry. In *Peculiar versus normal phenomena in A-type and related stars*, IAU Coll. No. 138 (eds. Dworetsky, M.M., Castelli, F. & Faraggiana, R.). Astron. Soc. Pacific Conf. Series vol. 44, pp. 232–246.

MATHYS, G. 1994 Spectropolarimetry of magnetic stars. III. Measurement uncertainties. *Astron. Astrophys. Suppl. Ser.* **108**, 547–560.

MATHYS, G. 1995a Spectropolarimetry of magnetic stars. IV. The crossover effect. *Astron. Astrophys.* **293**, 733–745.

MATHYS, G. 1995a Spectropolarimetry of magnetic stars. V. The mean quadratic magnetic field. *Astron. Astrophys.* **293**, 746–763.

MATHYS, G. & STENFLO, J.O. 1987 Anomalous Zeeman effect and its influence on the line absorption and dispersion coefficients. *Astron. Astrophys.* **171**, 368–377.

MATHYS, G., HUBRIG, S., LANDSTREET, J.D., LANZ, T. & MANFROID, J. 1997 The mean magnetic field modulus of Ap stars. *Astron. Astrophys. Suppl. Ser.* **123**, 353–402.

MATHYS, G., STEHLÉ, C., BRILLANT, S. & LANZ, T. 2000 The physical foundations of stellar magnetic field diagnosis from polarimetric observations of hydrogen lines. *Astron. Astrophys.* **358**, 1151–1156.

PRESS, W.H., TEUKOLSKY, S.A., VETTERLING, W.T. & FLANNERY, B.P. 1992 *Numerical Recipes in C. The Art of Scientific Computing*. 2nd edition. Cambridge University Press.

PRESTON, G.W. 1971 The mean surface fields of magnetic stars. *Astrophys. J.* **164**, 309–315.

PUTNEY A. 1997 Surveying DC white dwarfs for magnetic fields. *Astrophys. J. Suppl. Ser.* **112**, 527–556.

ROUSSELET-PERRAUT, K., CHESNEAU, O., BERIO, PH. & VAKILI, F. 2000 Spectro-Polarimetric Interferometry (SPIN) of magnetic stars. *Astron. Astrophys.* **354**, 595–604.

ROBINSON, R.D., WORDEN, S.P. & HARVEY, J.W. 1980, *Astrophys. J. Letters* **236**, L155–L158.

RUDER, H., WUNNER, G., HEROLD, H. & GEYER, F. 1994. *Atoms in Strong Magnetic Fields*. Springer.

SEMEL, M. 1989 Zeeman-Doppler imaging of active stars. I. Basic principles. *Astron. Astrophys.* **225**, 456–466.

STEHLÉ, C., BRILLANT, S. & MATHYS, G. 2000 Polarised hydrogen line shapes in a magnetised plasma. *European Phys. J. D* **11**, 491–503.

STENFLO, J.O. & LINDEGREN, L. 1977 Statistical analysis of solar Fe I lines. Magnetic line broadening. *Astron. Astrophys.* **59**, 367–378.

WADE, G.A., DONATI, J.-F., LANDSTREET, J.D. & SHORLIN, S.L.S. 2000 Spectropolarimetric measurements of magnetic Ap and Bp stars in all four Stokes parameters. *Mon. Not. R. Astron. Soc.* **313**, 823–850.

WEISS, W.W., JENKNER, H. & WOOD, H.J. 1978 A statistical approach for the determination of relative Zeeman and Doppler shifts in spectrograms. *Astron. Astrophys.* **63**, 247–257.

WICKRAMASINGHE, D.T. & FERRARIO, L. 2000 Magnetism in isolated and binary white dwarfs. *Publ. Astron. Soc. Pacific* **112**, 873–924.

Polarization Insights for Active Galactic Nuclei

By ROBERT ANTONUCCI[1]

[1]Department of Physics, University of California, Santa Barbara, CA 93106-9530, USA

Optical spectropolarimetry and broadband polarimetry in other wavebands has been a key to understanding many diverse aspects of AGN. In some cases polarization is due to synchrotron radiation, and in other cases it's due to scattering. Recognition of relativistically beamed optical synchrotron emission by polarization was vital for understanding blazars (BL Lacs and Optically Violently Variable quasars), both physically and geometrically. Radio polarimetry of quiescent AGN is equally important, again for both purposes. Scattering polarization was central to the Unified Model for Seyferts, Radio Galaxies and (high ionization) Ultraluminous Infrared Galaxies. It provides a periscope for viewing AGN from other directions. Finally, if we could understand its message, polarization would also provide major insights regarding the nature of the AGN "Featureless Continuum" and Broad (emission) Line Region.

I point out that high ionization ULIRGs have all the exact right properties to be called Quasar 2s. Mid-IR observations generally don't penetrate to the nucleus, greatly reducing their ability to diagnose the energy source. In particular, LINER ULIRGs aren't necessarily starburst-dominated, as has been claimed.

1. Seyfert Galaxies

1.1. Type 2 Seyferts

1.1.1. Polarization alignments and hidden Type 1 Seyfert nuclei

In the 1970s the continua of Seyfert 2s were decomposed into two parts: relatively red light from the old stellar population, and a bluer component modeled satisfactorily with a power law. The latter was called the "Featureless Continuum," in a commendable attempt to avoid prejudice as to its nature. (Unfortunately some of them were later found to have strong features; see below.)

The small ($\sim 1\%$) V-band polarization often seen usually derives from the power law component. Most of these continua are dominated by unpolarized starlight, so the implied "FC" polarization is sometimes intrinsically large. The red starlight also strongly affects the wavelength-dependence of P so it really had to be removed. It was shown that for the brightest and best observed Seyfert 2, NGC1068, the true FC polarization was a surprisingly high 16%, and independent of wavelength (Miller & Antonucci 1983; McLean et al. 1983). The former paper noted the cause might be either scattering above and below an opaque torus, or synchrotron emission; the latter focused on synchrotron radiation as the more likely.

Following the initial discovery of a geometrical relationship between optical polarization and radio axis for quasars (Stockman, Angel, & Miley 1979), Martin et al. (1983) and I (Antonucci 1982a, 1983) sought such patterns in Seyferts and radio galaxies. (Ulvestad and Wilson were discovering tiny, weak, but linear radio sources in many Seyferts, using the new Very Large Array.) The Martin et al. paper presented a lot of data, but didn't find alignment effects. I did claim to see them, and with essentially the same data. I think there were two reasons for the difference: 1) I didn't consider the whole sample statistically, but only the few whose polarization was very likely to be intrinsic to the nuclei, and 2) I divided them into the two spectroscopic classes (Type 1 and 2), considering each separately.

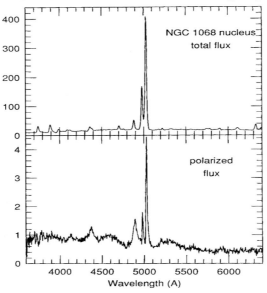

FIGURE 1. Total and Polarized Flux Spectra of NGC 1068.

There was pretty good evidence that the Type 2s tended to be polarized perpendicular to the radio axis, and not so good evidence that the Type 1s were parallel. (I tended to believe the latter though, because of the parallel polarizations of their "cousins"(?), the radio loud quasars.)

At the same time Joe Miller and I (I was the thesis student) were observing NGC1068 and the radio galaxy 3C234 with a new spectropolarimeter. This device had just been built, mainly by J Miller and G Schmidt. (Since I have no talent with instrumentation, my goal was just to avoid breaking it.) I realized pretty quickly that the explanation for the high "perpendicular" polarization of 3C234 was reflection from a quasar hidden inside a torus (Antonucci 1982a, b). We had to puzzle over NGC1068 longer, I suppose because the strong starlight confused us. Because of the wavelength independence of starlight-subtracted P, as well as some other moderately good reasons, we thought the scattering was by free electrons. Miller *et al.* (1991) then presented much better data, and showed the polarized spectrum *as scattered off dust clouds in the host galaxy*. These data indicated that 1) the dust-scattered light is much bluer than the inner, putative electron scattering light; and 2) the broad lines are narrower, suggesting that the line widths are somewhat smaller than those in the electron-scattered polarized flux plot when the reflection is from dust. This confirms that the nuclear scatters are electrons, and provides a temperature estimate of $\sim 300,000$ K. Note that according to the true line widths, NGC1068 would be classified as a "Narrow Line Seyfert 1" by astronomers looking from above. The Miller *et al.* 1991 nuclear data appear as the present Fig. 1.

We now know that many Seyfert 2s are just Seyfert 1s hidden inside opaque tori, and that therefore at least part of the class difference is just orientation. Some early references are Miller and Goodrich 1990; Tran *et al.* 1992. We still don't know if this applies to all the 2s. For a while I thought it probably didn't because Kay (1994) was finding that the FC for many Seyfert 2s have low polarization. However, starting with 3C234, it was becoming clear that some of that blue light was coming from hot stars or some other source, and that there was in fact highly polarized reflected light from hidden

1s in many more (Miller and Goodrich et al 1990, Tran 1995a,b). The high polarization of the actual scattered part of the FC is proven by the polarization of the broad lines, and the normal Seyfert 1 spectra in polarized flux. (The key here is that the broad lines have normal equivalent width in polarized flux.) In most cases we just have very high ~ 10–20% lower limits on the broad line region (BLR) polarization since at least I can't see them at all in the total flux in many objects (P = polarized flux in broad line/total flux in broad line.) Soon afterwards Heckman *et al.* 1995, and Gonzalez-Delgado *et al.* 1998 showed definitive spectroscopic evidence that most of the FC in several famous objects is in fact light from hot stars.

As noted, the general applicability of the hidden-1 model for the 2s is still poorly known. It would be great to select a sample by some nearly isotropic property such as 60μ emission, and observe them all down to a certain level of sensitivity. One (contentious) idea is that the scattering regions are themselves often partially occulted, so that the trick works only for those viewed at relatively small inclinations (Miller and Goodrich 1990). Models of the torus (e.g., Pier and Krolik 1993) indicate that such relatively polar views would expose warm dust to the observer, and it's been argued that we can in fact detect the hidden 1s in all the *warm* Seyfert 2s (Heisler *et al.* 1997).

1.1.2. *Mirror and torus; a 3-D image of an AGN*

These components were entirely hypothetical in the early 1980s. The mirrors (scattering regions) weren't resolved significantly from the ground (but see Elvius 1978 for resolution of the outer dust-scattering part of the NGC1068 mirror). HST however could spatially resolve some of them. Our multiaperture HST UV spectropolarimetry resolved the inner "electron-scattering" mirror in NGC1068 (Antonucci *et al.* 1994), as did the beautiful polarization images by Capetti *et al.* (1995a,b).

The central arcsec or so shows neutral (wavelength-independent) scattering, and for this and other reasons, electron scattering seems to dominate there. The ~ 400km/sec redshift of the broad lines in polarized light indicates polar outflow; recall that the scattering must be polar to explain the position angles (PAs). The dominance of electron scattering means that this gas has lost virtually *all* its dust, probably by travelling inside the sublimation radius ($\lesssim$ 1pc). Finally, as noted above, the gas temperature is thought to be $\sim 300{,}000$K because the electron-scattered versions of the broad lines are somewhat wider than the lines seen scattered off dust clouds in the host galaxy (Miller, Goodrich and Mathews 1991).

The probable physical basis of this whole occultation/reflection scenario was first provided by Krolik and Begelman (1986, 1988). Krolik and collaborators also calculated the theoretical requirements and consequences of the scenario, predicting, for example, that the scattering region should produce a high-ionization Fe K-α line of enormous ($>$ 1keV) equivalent width, as observed. The large $\sim 1" = 75$pc size of the electron scattering mirror was anticipated by the models of Miller, Goodrich and Mathews (1991).

Capetti *et al.* did a fine job analyzing the HST imaging, delineating for example the inner electron-scattering (neutral scattering) mirror, and the outer regions which show dust scattering (strong rise in cross-section with frequency). There was a slight puzzle left over from their analysis: in the inner region the polarization PAs were not quite centrosymmetric as expected for scattering. In principle this means that the hidden source isn't quite pointlike. However, Kishimoto (1999 and p.c., 2000) found that the deviations were entirely traceable to instrumental effects. Also, since we know the polarization phase function perfectly for electron scattering, it's possible to determine the angle between the nucleus and a scattering cloud, relative to the sky plane. For example, if some scattered light has a polarization of 100%, we know it's right-angle scattering, so

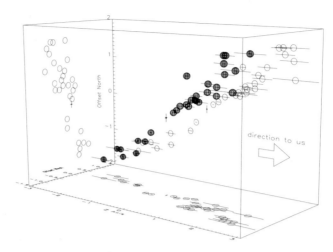

FIGURE 2. Three dimensional inversion of the electron-scattering region in NGC1068.

that the cloud is right in the sky plane. Kishimoto's paper shows the first (I think) 3-D image of an AGN (Fig. 2)! See that paper for some caveats. The overall image is fairly robust I think, and wonderful.

I won't say much about the earliest detections of the torus material since they didn't involve polarimetry. But structures which are acceptable manifestations of it have been observed in the infrared, and in various molecular emission lines. The observed molecular tori, whose "outer radii" are $\sim$ 100pc, are observed in NGC1068, and also in luminous quasars and infrared galaxies (Section 3 has some more information on this). Two points to bear in mind: the polarization and spectral-energy distribution information indicate that the *inner* radius is somewhere near the sublimation radius. Independent of the unified model, the near-IR upturn in the SEDs alone implies a substantial covering factor of hot dust. And remember that the polarimetry indicates a torus only in the sense of something opaque, with holes along the axis so the photons can get out. In the real world there could be (and are) more complicated structures, including bars; it's also possible that a warped thin disk could do the job, but it would have to be extremely warped and "tall" to simulate the obscuring behavior of a torus.

1.2. *Type 1 Seyferts*

1.2.1. *Intrinsic Nuclear Polarization*

Type 1 Seyfert nuclei generally have low intrinsic polarization ($\lesssim$ 2%), so it is often overwhelmed by dichroic or other processes on the $\sim$kpc scale. This is apparent when the narrow line region (NLR) and continuum polarizations agree, and/or both have PAs parallel to the host major axis in a high-inclination galaxy (Thompson and Martin 1988). If in doubt, a polarization image can help determine whether or not the polarization is associated with the nucleus.

There is fairly good evidence that the nuclear polarization PAs tend to align with the axes of the nuclear radio sources. As with the 2s, considerable uncertainty comes from the curved nature of the latter. Many years ago the available small sample seemed to show the alignment (Antonucci 1983). Since then various people have updated it, and I haven't always agreed with what was done. Here is my personal update:

The original criteria were:

(*a*) $P \gtrsim 0.6\%$

(*b*) sky position in a *good* Galactic interstellar polarization zone.

This means it lies within a personal pre-defined sky area with relatively little expected interstellar polarization. I'd chosen these areas using the maps in Mathewson and Ford 1970.

(*c*) linear radio source.

Updates relative to the 1983 paper:

(*a*) I'd drop 3C390.3 on the grounds that it's a radio galaxy; I was young and dumb in 1983 so I was following the classifications in an earlier paper by someone else.

(*b*) Drop NGC3227 because the [O III] data show that the polarization arises in the host galaxy rather than in the nucleus (Thompson *et al.* 1980).

(*c*) Add Mrk509: Polarization PA: ∼ 1.0%, at 130–150 (true variability) degree: see references in Singh and Westergaard 1992. The Radio PA is 130 ± 10 (Singh and Westergaard 1992): nearly parallel

(*d*) Mrk704: Delta PA = 81°! from Martel 1996; also in Goodrich and Miller 1994

(*e*) Mrk1048 Delta PA = 12°! from Martel 1996; also in Goodrich and Miller 1994

(*f*) NGC3516 Polarization ∼ 0.8% at ∼ 2°; Radio 0–10; See Miyaji *et al.* 1992: parallel.

(*g*) NGC5548 Delta PA = ∼ 55: Martel 1996

(*h*) Mrk9 Delta PA = ∼ 28: Martel 1996

(*i*) Mrk304 Delta PA = ∼ 90!: Martel 1996: perpendicular.

(*j*) Mrk957 $P = 0.62 \pm 0.06$ at 43 ± 3 (Goodrich 1989) Radio PA 50 ± 10; see also Ulvestad *et al.* 1995)

While the data are still marginal, an optimistic and plausible summary is this. Most Seyfert 1 nuclei show optical polarization parallel to the radio axis; the exceptions may favor a perpendicular relationship.

1.2.2. *Variety of BLR behavior. Constraints on the nature of the underlying continuum*

In all known cases, the BLR line polarization is less than or about equal to that of the continuum, in magnitude. The action within the line profiles depends on the object. Often both the magnitude and the angle vary rapidly across the profile. It is undoubtedly encoded with lots of information on the nature of the BLR, but it's very difficult to decode. The most heroic and intriguing attempt is Martel's (1998) analysis of the broad H-alpha line in NGC4151.

Here are some more references: There are several c. 1980 papers by the Steward group (R Angel, I Thompson, E Beaver, H Stockman, and probably some others). These authors were the pioneers measuring Seyfert polarization. Later (CCD) data supercede their observations.

NGC4151 was observed by Schmidt and Miller (1980), who found that the integrated broad line polarization was undetectably low with their data quality. There are some quasars that also have undetectably polarized broad lines. This is of great interest because it means that the polarized flux plot looks like a noisy version of the spectrum, but with *the broad lines and small blue bump (Fe II plus Ba continuum from the BLR) scraped off as by a razor!* See e.g., Antonucci 1988 where I present some data "borrowed" from Miller and Goodrich (pc); and Schmidt and Smith 2000. There are a few others.

Thus we can tell what the underlying continuum is doing... at for example, the Ba edge. The answer is: nothing. (Nothing happens at the Ly edge either.) This is not trivial. Models such as accretion disks which are cool enough to match the optical slopes would at least naïvely show the Ba edge in absorption. In one very important regard the

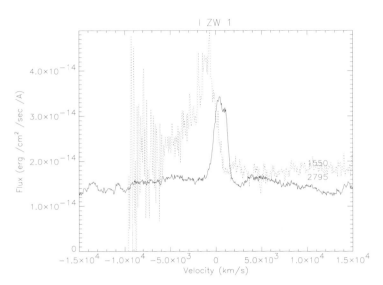

FIGURE 3. Overplot of C IV and Mg II broad line profiles in the Narrow Line Seyfert 1 called I Zw 1, M Kishimoto 2000 p.c. This figure strongly suggests that C IV (and other high-ionization lines) derive from a wind, whereas Mg II (and the other low-ionization lines) do not. For a detailed analysis of I Zw 1, see Laor *et al.* 1997.

Ba edge is more interpretable than the Ly edge: the relativisitic smoothing effects at the Ly edge are much reduced. If there's a feature, we should see it!

We were asked to suggest observing projects for young astronomers, and I think the following is quite good if practical. The high-ionization lines in "Narrow Line" Seyfert 1s are often blueshifted and very broad relative to the low ionization lines and the systemic velocities. (As far as I know this important result was first shown by Rodriguez-Pascual *et al.* (1997), though you wouldn't get that impression from some more recent papers!) To tempt the reader, I show the spectacular example of I Zw 1 in Fig. 3. (Some of those in Leighly 2000 are even better.) This was given to me by M. Kishimoto. I needed no convincing after seeing that plot, that the high-ionization lines are from a wind, in that object and maybe others. Early proponents such as Collin-Souffrin *et al.* (1988) deserve congratulations. It would be great to measure the polarization behavior of the two types of line. There are at present no space spectropolarimeters, so what's needed is a bright object with enough redshift to bring a high ionization line like C IV 1549 into the optical — can we find such an object that shows the difference in the line profiles clearly?

Recovering from that diversion, I just refer you to these papers on Seyfert 1 polarization: Berriman (1989) interpreted his broadband Seyfert 1 survey data as indicating dust scattering for most objects. More data including spectropolarimetry were presented by Brindle (1990). J. Miller and students obtained some fairly large (thesis-size) data sets (Goodrich and Miller 1994; Martel 1996). These papers show in detail the complex behavior inside the broad emission lines.

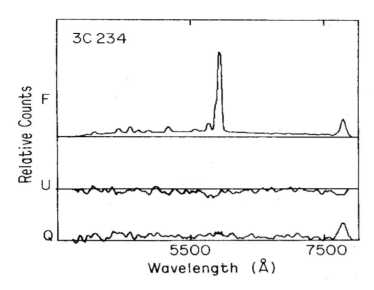

FIGURE 4. The 'polarized fluxes' for the Q and U Stokes parameters, as originally discovered in 1982. These represent the fractional Q and U spectra multiplied by the total flux spectra. For cases such as this, where the position angle is pretty constant, these are like ordinary polarized flux plots, but with a symmetric and unbiased error distribution. Note that the strong permitted line, Hα, shows in polarized flux and the forbidden [O III] λ4959, 5007 do not.

2. Radio Galaxies

2.1. *Unification with quasars*

2.1.1. *Optical properties*

This will be a quick history of the subject as I know it. The usual apologies for omissions [actually, you can still tell me before the printing!]. Relatively complete early references can be found in my 1993 ARAA paper.

Most powerful radio sources have spectra in the same two spectral classes as for the radio quiet ones: Type 2 Radio Galaxies, usually called Narrow Line Radio Galaxies, and often just called radio galaxies if at high redshift; and Type 1, those with strong broad lines in total flux which for historical reasons are called Broad Line Radio Galaxies or radio loud quasars, according to luminosity, basically. I show my original spectropolarimetric data on the first recognized case of a hidden BLR (3C234) in Fig. 4, along with modern Keck data (in a slightly different form) by Tran *et al.* 1995 in Fig. 5 (see Kishimoto *et al.* 2001 for the UV). Excellent data and analysis of many more radio galaxies can be found in Cohen *et al.* 1999.

A great HST picture of a torus is that in 3C270 = NGC 4261, found in Jaffe *et al.* 1996. Note that spectropolarimetry provides only the crudest information on the torus size. I believe in most cases the inner radius is near the sublimation point ($\sim$ 1 pc), because the spectral energy distributions show that all Type 1 objects must have a substantial covering factor of hot dust. This HST image and some molecular mapping of various objects both show a (quasi) outer edge at 100–300pc.

Radio loud AGN can also have a contribution (sometimes dominant) from highly variable, highly polarized synchrotron radiation. From the point of view of the radio-optical Spectral Energy Distribution, the optical synchrotron source is seen to be simply the

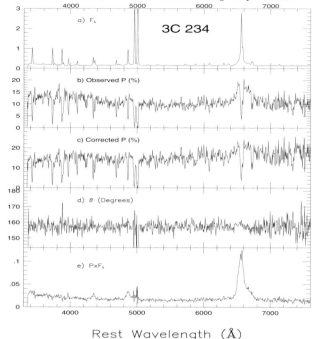

Rest Wavelength (Å)

FIGURE 5. Modern quality spectropolarimetric data on 3C234 from Tran *et al.* 1995.

high-frequency tail of the radio core emission (Landau *et al.* 1986, Impey and Neuge-bauer 1988). These objects are the blazars (Wolfe 1978). Those with very low equivalent width emission lines are historically called BL Lac Objects. However, this subgroup has no physical meaning, and since the objects vary, they alternate their classification over time with this nomenclature! For example, BL Lac is often not a BL Lac object with this definition (Vermeulen *et al.* 1995). Another problem with this nomenclature is that it mixes low-luminosity nearby cases with FR1-level extended radio emission with high redshift very high luminosity objects. Currently some people are saying that BL Lacs are specially oriented FR1 radio galaxies. This is sloppy in the extreme since a significant subset of them are known to be inconsistent with it (e.g., Kollgaard *et al.* 1992).

I've often argued that a better split for the blazars would be according to whether their extended radio emission is consistent with an FR1 or FR2 power level. I think this has more hope of having physical meaning than setting an arbitrary equivalent width limit on the emissions lines on the discovery spectrum, and it would retain the advantages of the latter.

Also, some authors have gone to the extreme of calling something a BL Lac object largely because it doesn't have the strong 4000Å break expected for late-type stars. That seems crazy to me because it includes zillions of faint blue starburst galaxies...like the sources for the blue arc lenses in clusters. It took a long time even to get redshifts for the latter because their spectra are so featureless. High polarization or at least high variability is required when defining "BL Lac."

Next, let's consider some polarized images which resolve the mirrors in Narrow Line Radio Galaxies. Ideally, these are made in the rest-frame near-UV, where dust scattering cross-sections would be large, and where there is reduced confusion with the host galaxy light. Fig. 6 shows our HST image of 3C321 (Hurt *et al.* 1999). The pattern is centro-

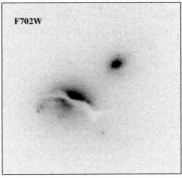

FIGURE 6. The spatially resolved ($\sim$ 8 kpc) mirror in 3C321, by HST in the near UV (F320 FOC filter). The lower plot shows the red (HST WFPC2 F702 filter) total flux image.

symmetric within the noise, and P is locally rather large ($\lesssim$ 50%). Young *et al.* (1996) report scattered broad lines from the hidden quasar. Their data also show that the polarized flux spectrum is indistinguishable from a total-flux quasar spectrum, and this suggests unreddened electron scattering. It doesn't prove electron scattering though, and we hope to look for spatially resolved scattered X-rays to make a strong test. It's very important to establish this with certainty, because of the very remarkable consequences!

Many of the ($z \sim 1$) "aligned" radio galaxies (optical extension parallel to radio) have behavior very similar to that at 3C321 (e.g., Best *et al.* 1997). In order to have sufficient optical depth in electron scattering, given the huge physical sizes of the polarized mirrors, the ionized gas masses are huge. Depending on assumptions, they may be $10^{12} M_\odot$ or considerably more. Also, in cases where the broad lines are clearly seen in polarized flux, the putative scattering electrons cannot be too hot — at X-ray halo temperatures the scattering process would broaden the lines beyond recognition. They must be $\lesssim 10^6$K, which means they aren't typical cooling flows or hot haloes supported hydrostatically. So all that mass would likely be falling in on a dynamical timescale — conceivably forming a central cluster galaxy like M87. A recent discussion can be found in Kishimoto *et al.* 2001.

Incidentally, at even higher redshifts, the tiny sample observed spectropolarimetrically so far does not show polarization, but instead shows absorption lines probably (at least in part) from stellar photospheres. These objects are necessarily observed at shorter rest wavelengths (~ 1500Å) so the comparison with the $Z \sim 1$ objects isn't completely clear.

It does show, however, that almost everyone's theory for the optical radio alignment effect is right somewhere (Dey *et al.* 1997).

2.1.2. *Radio properties*

There are two types of powerful extragalactic radio source, the normal double sources and the core-dominant superluminal sources. (This oversimplifies, of course.) It's rather well established that this is mostly another orientation effect: the latter are seen in the jet direction, and are greatly boosted in flux because of special-relativistic aberration. Indeed, these core-dominant sources generally show *radio* "haloes" consistent with lobes seen in projection around the bright core, if mapped with good dynamic range (Browne *et al.* 1982; Antonucci & Ulvestad 1985; Wrobel & Lind 1990, and many others).

Now the double radio sources themselves can be divided into morphological classes, by whether the lobes are edge-darkened or edge brightened. The former ("FR1") turn out to be basically the low luminosity objects.† They have several other correlated differences which gives this separation some physical significance.

I think that the *most luminous* of the (luminous) FR2 are all just hidden quasars. In many cases the polarized flux spectra already show it. In the 1980s, radio astronomers were making statistical tests of the identification of double radio *quasars* with core-dominant superluminal sources and statistical problems arose, such as finding "too many" fast superluminals relative to the expected beaming solid angle (e.g., Laing 1988, Barthel 1987). These can all be explained by a dearth of objects with axes nearly in the sky plane. The reason is now obvious: quasars in the sky plane do exist, but are called radio galaxies because the obscuring tori block our view of the nuclei.‡

These very powerful radio galaxies have strong narrow emission lines like Seyfert 2s, but many of the weaker FR2 types do not, and *most* of the (low-luminosity) FR1s do not. Let's consider first the FR2 radio galaxies: are the relatively weak ones hidden quasars? The optical-UV continuum is thought by many people to be thermal radiation from optically thick matter falling into a black hole. If there is instrinsically no hidden quasar (BBB), then according to current theory, there can't be much of an accretion flow. In that case only the black hole spin energy would be available to produce and sustain the powerful radio jets. These objects can be called nonthermal AGN (if they exist). It does seem though that at least a few radio galaxies with low-ionization relatively weak narrow lines *do* have pretty good evidence for a hidden quasar (see Sambruna *et al.* 2000 for Hydra A).

"Optically dull" FR2 radio galaxies show very little optical polarized light in general. The reason could be: there's no hidden quasar; there's no mirror; the hidden quasar is completely surrounded by dust; relatively low-column kpc-scale foreground dust lanes block our view of the scattered nuclear light as in 3C223.1 (Antonucci & Barvainis 1990). A more robust test for a hidden quasar would be looking for the inevitable "waste heat" in the mid infrared. D. Whysong are I are trying to make this test at Keck, but it's pretty hard. (A few relevant objects were observed by ISO.)

Now let's consider the FR1 sources: *most* of these are optically dull: the narrow line emission is very weak, and of low excitation. This isn't suggestive of a hidden quasar, but again the inner narrow line region could be occulted, or the nucleus could be completely surrounded by absorbing dust. In a very significant series of papers, Chiaberge *et al.*

---

† For a very interesting refinement of this statement, see Owen & Ledlow (1994).

‡ Also note this interesting pair of papers on the projected linear sizes: Singal (1993); Gopal-Krishna *et al.* (1996). They discuss whether the lower-luminosity FR2s have unexpectedly small projected linear sizes, relative to beam-model predictions.

(1999, 2000), show that archival HST images have nuclear point sources in *most* of the optically dull 3C galaxies of both FR types. They reason that an AGN cannot be hidden in most optically dull radio galaxies because we can seemingly see into the center in order to detect the point sources (thought by the authors to be synchrotron emission from the tail of the radio core spectra). This behavior is unlike that of Seyfert 2s and the very powerful narrow line radio galaxies, and ultraluminous infrared galaxies, all of which show *no* point source in the optical band.

A good example of a somewhat optically dull FR1 (or FR borderline) radio galaxy is M87 (Reynolds *et al.* 1996). It shows a nuclear point source in the optical, and has no powerful thermal IR source. We find that at 11.7 microns there is only a weak ∼ 15mJy point source, and this could easily be explained as nonthermal emission from the base of the jet. *M87 really can't have a hidden AGN with luminosity comparable to the jet power* (Owen *et al.* 2000).

On the other hand, the nearest FR1 is Centaurus A... which has *no* optical point source, lots of thermal dust emission, and considerable polarimetric evidence for a hidden "thermal" optical/UV nucleus (Marconi *et al.* 2000, Capetti *et al.* 2000). Thus the FR1 family is a heterogeneous one. Also regarding Cen A, it's well worth taking a look at the beautiful CO 2-1 torus image, Fig. 2 of Rydbeck *et al.* 1993!

Remember also that the FR1 radio galaxy 3C218 (Hydra A) has (relatively) weak emission lines of low excitation (Ekers & Simkin 1983), yet strong evidence for a hidden AGN (Sambruna *et al.* 2000); it shows that nuclei can have hidden AGN even if they are rather "optically dull".

A related question is whether a quasar or Broad Line Radio Galaxy can have an FR1 radio source. *A few are in fact known* (e.g., see Lara *et al.* 1999). As those authors put it, this makes their source "a nontrivial object from the point of view of current unification schemes." Another recent case is from Sarazin *et al.* 1999: 1028+313.

2.2. *Observing the ionized intergalactic medium*

Radio polarization maps of high-redshift radio galaxies and quasars could supply a key cosmological parameter. We've been trying to detect Thompson-scattered radio halos to detect the ionized intergalactic medium. This was originally suggested by Sholomitskii. The expected baryon density based on nucleosynthesis is ∼ 5% of closure, corresponding to IGM optical depths of ∼ 10^{-3} over the expected halo size (∼ 10^{7-8} light years, based on guessing the AGN lifetime). Sensitivity is needed on large scales and it turns out the Australia Telescope Compact Array is the best choice.

Our first attempt at this was published in Geller *et al.* 2000, providing an upper limit to the IGM density below 100% of closure (you have to start somewhere). We hope to do much better. If we can eventually detect the halos with confidence, we'll learn the quasar lifetime and beaming pattern in principle. More to come!

3. Ultraluminous Infrared Galaxies

3.1. *Spectropolarimetry and Type 2 quasars*

3.1.1. *Quasars of Type 2*

Let's look again at the spectral energy distribution of unobscured AGN (Fig. 7). Obscured AGN are similar except the optical/UV is much lower, and sometimes the X-rays are as well. What is your definition of a "Quasar 2"? I'd expect that a Seyfert 2 with an extremely powerful nucleus would have an extremely powerful infrared bump, absorbing the AGN light, and reradiated in relatively isotropic infrared dust emission.

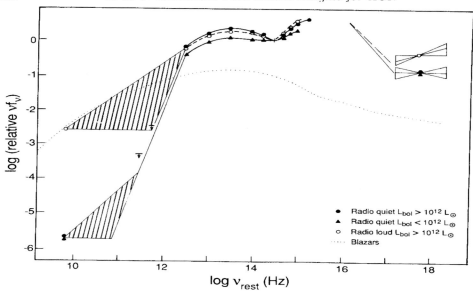

FIGURE 7. Composite spectral energy distribution for AGN (Sanders *et al.* 1989).

Next I'd expect a Quasar 2 to have powerful high excitation narrow emission lines, since the Narrow Line Region is generally outside the torus, and hence it emits rather isotropically. It would not scale with the infrared, simply because empirically the narrow line equivalent widths decrease with luminosity in the objects seen directly (Seyfert 1s and radio quiet quasars). But it should be much more powerful than those in the Seyfert 2s.

I'd also expect, based on the Seyfert 2s and the unified model, that there would be *no* optical/UV point source, and that the hidden Type 1 nucleus appears at good contrast in the polarized flux spectra.

Note that the optical/UV continuum flux would *not* scale with the AGN luminosity, because that continuum in Seyfert 2s is almost always strongly dominated by the light from the host galaxy. Suppose 90% of the Seyfert 2 continua derives from light from the underlying old steller population, a fraction not in great dispute. Then a simple scaling of the AGN power by a factor of 10 would lead to an increase in optical/UV flux of only a factor of two. This is oversimplified given differences among the AGN nuclear regions (reddening, young stars), but I think it's qualitatively correct.

I've just described *exactly* the high-ionization ultraluminous infrared galaxies. I still read that there are no Quasar 2s, or that their existence remains to be demonstrated. If you think that this class remains undetected, please tell me your definition of Quasar 2.

On the radio side, I think there would be little argument that the Type 2s are the powerful narrow line radio galaxies — at least for those with the highest radio luminosity.

3.1.2. *Properties of Quasars of Type 2*

There are a *lot* of high-ionization ultraluminous infrared galaxies. The estimates I've seen indicate that the fraction of all infrared galaxies comprised by those of high excitation rises with infrared luminosity, and reaches half of those more luminous than 1–3×10^{12} Lo.

Fig. 7 shows the generic SED shape for unreddened quasars. It is important that the

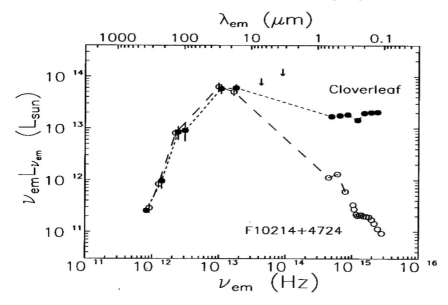

FIGURE 8. Compared with an unobscured quasar, the Cloverleaf is depressed and reddened in the Big Blue Bump region. For the infrared galaxy (ULIRG) *all* of the direct nuclear photons are absorbed.

IR bump has about 30% of the integrated flux. The simple and plausible interpretation is that the "tori" cover $\sim$ 30% of the sky, as seen from the nuclei. This seems at first glance to be consistent with the space density of high ionization ultraluminous galaxies relative to quasars matched in apparent bolometric flux (Gopal-Krishna & Biermann 1998). However, the correct thing to do here is to compare ULIRG and quasar space densities *as matched by far-IR, since that, not* L_{BOL}, *is isotropic*. We can conclude that the (UV-selected) far PG quasars are well below average in dust coverage as deduced by many others.

Fig. 8 shows a comparison of the SED between the first "hyperluminous" infrared galaxy (F10214+4724), and the "Cloverleaf" reddened quasar. Both are lensed and the quantitative close agreement in the far-IR is fortuitous. But it makes the point that even a somewhat-reddened quasar has a much higher fraction of the light in optical/UV. The infrared galaxy has no point source visible in those regions, but there is a little scattered light from the hidden quasar which makes it detectable in the optical/UV. I say this with confidence because the hidden quasar appears in the polarized flux plot, just as in nearby Seyfert 2s (Goodrich *et al.* 1996).

It's also known that the relatively isotropic CO/far-IR ratios are generally the same in quasars as in ultraluminous infrared galaxies, even for those classified as starbursts (Alloin *et al.* 1992; Evans *et al.* 2001).

The IR-warm ultraluminous infrared galaxies start to reveal a pointlike nuclear source in the near-IR (e.g., Surace, Sanders & Evans 2000). These are attributed to penetration of the outer parts of the absorber well enough to see the very warm unresolved nuclear dust. This hot dust can be reached much more easily, apparently, than the nuclei themselves.

Now an extremely important question arises, in terms of understanding the global SEDs. What is the torus column density and the opacity at various wavelengths? Starting with NGC 1068, the lack of a strong variable X-ray source immediately shows that many are "Compton thick," blocking even the $\gtrsim$ 10keV photons. At solar abundances this means the column density is $\gtrsim 10^{25}$ cm^{-2}. Recently it's been possible to measure the column density (or limit) distribution for fairly complete Seyfert 2 samples and the median is almost that high (Maiolino *et al.* 1998; see also Salvati and Maiolino 1988). It immediately follows that in most cases *mid-IR observations reveal the conditions some fraction of the way into the tori unless the dust/gas ratio is extremely low.* Apparently, there is star formation in the tori, since the mid-IR observers often see "starburst" spectra. *Whatever the dust-gas ratio, the point is that the Type I nucleus is seen in the x-ray but not in the mid-IR.*

3.2. *Distinguishing hidden AGN from hidden starbursts in ULIRGs*

3.2.1. *Opacities and Luminosities*

There has been a great deal written on this subject which implicitly or explicitly presumes the tori are optically thin in the mid-IR. I can't put much faith in these papers because we have very strong evidence that this would not be the case in typical Seyferts. Thus the spectra that simply show starburst mid-IR lines are probably just studying conditions inside the tori. I'd be convinced of an important starburst, if some observed spectral features indicated *starburst luminosities* consistent with the far-IR power. However, even in such cases I reject the claim that an object or certain objects get "most" of their energy from starbursts. That language requires the starburst energy contribution to be $75 \pm 25\%$ of the total luminosity. But there is no such precise bolometric luminosity that can be inferred from any emission line feature. Similarly, the best predictor of hidden AGN luminosity is the hard X-rays — for objects in which they get through the torus. It's pretty robust but only works at the factor-of-three level, certainly not sufficient for anyone to say the hidden-AGN luminosity is $75 \pm 25\%$ of the total. As soon as somone makes a statement about energy sources with this kind of precision I tend to stop reading because I've lost faith in my author.

Join my new group: the Militant Agnostics. Our motto is, "I don't know, and you don't either!"

3.2.2. *What are the LINER ULIRGs?*

My *guess* would be that the ULIRGs with high ionization narrow lines are mostly powered by AGN, and likewise for the starbursts. It's just a guess. But what about the large minority of ULIRGS which have LINER (Low Ionization Nuclear Emission Region) spectra? Here we can't even tell if the region producing the lines *that we see* derives from a hidden AGN or a starburst.

A remarkable paper and those surrounding it illustrate the point about LINER ULIRGs. After extensive studies of the mid-IR spectra of ULIRGs, it was claimed by many that their energy source is usually a hidden starburst. Then some mid-IR experts got together with an optical AGN spectroscopist, in part to see what the mid-IR spectra have to say about the optical LINERs (Lutz, Veilleux and Genzel 1999). They showed in a remarkably clean manner that those with optical starburst spectra also have infrared starburst spectra; same for the AGN; but the interesting part is that the many optical LINERs show starburst infrared spectra! This was interpreted as indicating that the LINERs are simply a slightly different manifestation of a hidden energetically dominant starburst. Only trouble is, I think few if any of these mid-IR spectra came from the actual nuclei, as discussed above. Maybe the observations prove that there is a lot of star formation inside the tori, but not that it contributes $75 \pm 25\%$ of the luminosity.

A good example is NGC6240, a famous "prototype" (see Table 1 of Lutz *et al.* 1999). The optical spectral type is listed as LINER, and the mid-IR is listed as starburst. (It is

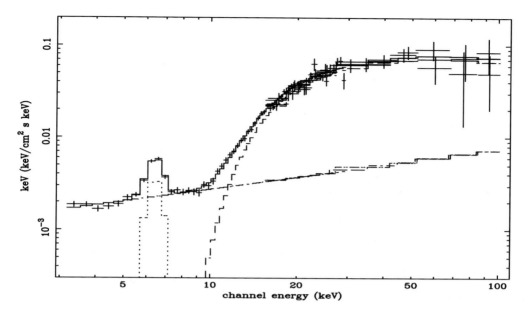

FIGURE 9. Wideband view of a Seyfert 2 with a column density of $\sim 5 \times 10^{24}$ cm^{-2}.

even a "prototype" starburst in Genzel *et al.* 1998.) The simplest conclusion is that it's simply a starburst galaxy, with little AGN contribution. Well, the X-ray opacity at $\gtrsim 10$ keV is a few times smaller than that in the mid-IR, so for a small (but non-negligible) number of cases the X-ray penetrates, though the mid-IR doesn't. Shortly after (or perhaps contemporaneously with) the Lutz *et al.* paper, Vignati *et al.* (1999) published the BeppoSAX spectrum, covering the high energy as well as the low energy X-rays. It clearly shows a column of $\sim 2 \times 10^{24}$ cm^{-2}, so that there would indeed very probably be high opacity in the mid-IR.

The hard X-rays are strong (direct), penetrating and rapidly variable, and really must come from a hidden AGN. The corresponding bolometric AGN luminosity is to within uncertanties just that observed. Even in that case, given the factor of ~ 3 dispersion in the X-rays/bolometric ratios seen in other AGN, it would be too much to claim the AGN luminosity is $75 \pm 25\%$ of bolometric. But it sure isn't negligible either, so the mid-IR spectrum is *not convincing evidence for a dominant starburst for this object and thus potentially for all of them.*

There are many similar cases, but this is getting far afield from polarimetry. I'll just cite and show the case of NGC 4945. It also is a "template" starburst in Genzel *et al.* 1998. Yet it has powerful rapidly-variable ~ 10keV flux, and certainly contains a powerful hidden AGN. The reason I bring this one up is that it has a wonderful wideband X-ray spectrum, and the soft and midrange absorption indicating the very high column couldn't be clearer (Fig. 9, from Madejski *et al.* 2000; see also Eracleus *et al.* 2001 for another example.).

The mid-IR observers aren't convinced though. In a "reply" paper they say that "the starburst may well power the entire bolometric luminosity" of NGC4945 with the caveat that they can't really prove that an AGN doesn't provide "up to 50% of the power". Don't you love science?

The recently-discovered "Scuba", millimeter-selected galaxies are quite similar. They can be detected easily because of a very favorable "K-correction" for objects at significant redshift. We probably know a tenth as much about these compared with the old ULIRGs. Yet at least one of the discovery papers simply *assumes* they're all powered by star formation, and doesn't even mention the AGN possibility. This unjustified assumption has major consequences for the luminosity density history of the universe (Madau Diagram), for the X-ray background (Almaini *et al.* 1999), and also for black hole demographics.

4. Emission Mechanism for the Big Blue Bump Spectral Component

4.1. *Introduction and Polarization Diagnostics*

We have some hope of understanding quasars because the spectral energy distributions are quite generic. Aside from an occasional blazar (radio core, synchrotron) component, they seem to comprise an IR bump (thought by almost everyone to be thermal dust emission), a usually energetically dominant "Big Blue Bump" optical/UV continuum component, and an extremely interesting but less powerful X-ray component. Refer again to Fig. 7. Note that the X-ray component does become competitive with the Big Blue Bump (BBB) in some of the lowest-luminosity nuclei.

Since the BBB is generally energetically dominant, and peaks to order of magnitude in the right spectral region, it is often assumed to be thermal emission from optically thick accreting matter. The models almost all assume the emitter is a standard thin opaque, quasistatic disk converting gravitational potential energy through viscosity into radiation. These are sometimes called "Shakura-Sunyaev" disks. Almost no one would argue that this is qualitatively correct any more. Nevertheless most theorists assume *de facto* that whatever is going on, you get the same spectrum!

Polarization has played a role in testing the accretion disk model as follows. The tale is a bit convoluted. In an optically-thick disk with a scattering atmosphere and with the heat deposited at large optical depths, the polarization should range from 0 to 11.7%, and always lie along the (projected) disk plane. However, it's been known since the world began that this doesn't describe the observations.

The polarization degree might be okay if the edge-on quasars were missing (manifesting as NLRGs or Infrared Galaxies). There are claims that the polarization magnitude distribution actually is consistent with the predicted one, without the high-inclination objects. However, in the case of lobe dominant, steep spectrum radio quasars, the direction is observed to be *parallel* to the radio jets (Stockman, Angel, & Miley 1979). Thus these early papers implicitly assumed that the jets emerged from the sides of the disks. A possible way out is to claim that for these radio-loud objects, the polarization derives from an optical blazar component, but this is observationally untenable because the radio core spectra decline sharply by the millimeter region (e.g., Antonucci et al 1990, Knapp and Patten 1991 and van Bemmel & Bartoldi 2001). Also as noted above, Seyfert 1 galaxies probably show the same effect, and they have very weak radio cores.

In the models of Laor, Netzer and Piran (ca. 1990 vintage), absorption opacity simply dilutes the effects of scattering in terms of magnitudes of polarization, resulting in a strong rise in P with frequency followed by a sharp decline below the Lyman edge. Antonucci *et al.* (1996) and Koratkar *et al.* (1998) tried to test the polarization predictions (low in the optical, higher in the UV longward of the Ly edge). No object seems to have the expected behavior. In fact several showed very high polarization, only below ~ 750Å in the rest frames (Koratkar *et al.* 1995). Regarding the latter, I'm reminded of an apochryphal Eddington quote: never trust an observation until it's confirmed by theory.

It was discovered later (actually rediscovered, e.g., Gnedin and Silant'ev 1978 and probably earlier ones; some modern papers are Matt, Fabian & Ross 1993, Blaes and Agol 1996) that Laor's seemingly reasonable way of accounting for the effects of true absorption is really not correct in most of parameter space. The current state-of-the-art as far as I know is Blaes and Agol 1996. They can get a slight parallel polarization, at least in models that produce too-few ionizing photons, *modulo* Comptonization. But the latest wisdom says the disks will be completely Faraday-depolarized (Agol and Blaes 1996; Blandford, this volume). Perhaps the observed "parallel" polarization is impressed downstream (though inside the BLR).

Getting the AGN geometry right at *any* distance from ground-zero would be valuable, and Chen & Halpern (1990) made some very interesting, very specific predictions for the polarization behavior in double-humped broad H-α profiles — the latter were argued to arise in a thin disk at a larger radius than the BBB. While these "smoking gun" predictions weren't confirmed (Antonucci, Hurt & Agol 1996, Corbett *et al.* 1998) it was later shown that a range of polarization behavior could occur in the disk model, if certain assumptions were relaxed (Chen, Halpern & Titarchuk 1997).

4.2. *Other Diagnostics*

I'll make brief mention of some other tests of the accretion disk paradigm for completeness. My point of view is spelled out in more detail in Antonucci 1999.

4.2.1. *Lyman edge*

Most thermal models predict observable changes in the spectra at the position of the Ly edge. There are certain rather narrow regions of parameter space in which they are not present at a detectable level.

The Shakura-Sunyaev disk generally leads to an edge in absorption, as in stellar atmospheres in which this spectral region is energetically important. Kolykhalov & Sunyaev (1984) considered this in some detail. No such edges are seen, however (Antonucci et al 1989)†. The key test to determine whether or not a candidate edge really comes from a relativistic disk rather than foreground material is a *lack* of an accompanying set of sharp absorption lines: A. Kinney and I once thought we had some candidate disk edges but they all failed this test. *Often people attribute an observed edge to an accretion disk without bothering with or even mentioning this basic test.*

Later authors pointed out that Kolykhalov and Sunyaev were unable to consider surface gravities below that expected for supergiant stars, and that lower gravities and so densities were quite reasonable and result in smaller edges. It's amusing that a surface gravity and so density much *higher* than previously contemplated might also do the trick: (Rozanska *et al.* 1989, but see also Czerny & Pojmanski 1990 and Czerny & Zbyszewska 1991). Also the range of gravitational redshift of the emitting elements and (for high-inclination disks) the range of Doppler shifts would tend to smear the edges out. For certain parameters they would be very hard to detect. Invoking too much inclination would require reconsideration of the polarization distribution, and of consistency with the Unified Models.

It is very interesting to consider now the flux and polarization behavior at the Ba

---

† Small features have been claimed in composites (e.g., Zheng *et al.* 1998, but see Tytler & Davis 1993), but I don't think credibly in individual objects (see Appenzeller *et al.* 1998 for a great observation of 3C273). Since the absorption-line test described above probably wouldn't work well in composites, I'd assume such a feature is unlikely to be due to a disk edge. At least that's the lesson Kinney and I learned from our follow-up observations of disk edge candidates.

edge. In some quasars the BBB polarization is ∼ 1%, and wavelength-independent. More importantly, it arises inside the BLR because neither the broad emission lines nor the "Small Blue Bump" atomic emission in the 2500–3500Å region are polarized. Thus the polarized flux plot is a wonderful way to scrape off the atomic emission and see what the BBB is doing in this region, as mentioned in Sec. 1. So far it looks like there is *no* Ba feature in the BBB (Antonucci 1988; Schmidt & Smith 2000). Explanations of the lack of any Ly edge feature which depend upon relativistic effects, or other effects occuring in the hot innermost annuli, would not work here. In some cases as noted below, the disk continuum in the optical region must match rather red observed spectra (spectral index ≲ −0.5). (The negative slope isn't a results of foreground reddening since we see no downturn in the UV.) *Models must achieve this without producing a Ba edge feature.*

Much stricter constraints on, or detections of, a Ba edge feature would be a *very* worthwhile spectropolarimetric project. (I haven't been able to get telescope time to do it!)

4.2.2. *Spectral energy distribution*

Early accretion disk models predicted positive spectral indices, well longward of an exponential cutoff, whereas almost all quasars and AGN have negative spectral indices (two good studies of the latter are Neugebauer *et al.* 1987 and Francis 1996). The observations weren't fitted to the model optical/UV directly, but only after subtraction of an "infrared power law," extrapolated under the optical (e.g., Malkan 1983). Indeed, Laor (1990) states that such a power law is required. Only trouble is, everyone is now convinced that the IR is dust emission, which must drop like a stone at 1μ and cannot legitimately be extrapolated under the optical (e.g., Barvainis 1992 and references therein). It isn't yet clear whether pure disk models can fit the optical observations (including the lack of Ba edge).

Although the dependence of disk maximum temperature on luminosity (for a given L/L_{Edd}) is only to the 1/4 power in the standard disk, that's enough to predict dramatic differences between the turnover frequencies of high luminosity quasars compared with low luminosity Seyferts. No such difference is found (e.g., Walter & Fink 1993; Mineo *et al.* 2000 show that quasars have a steep rise below 500eV just like Seyferts, naïvely at least suggesting a similar temperature). In fact in the optical range people have typically found that the more luminous AGN are flatter (hotter): see e.g., Fig. 5 from Mushotzky & Wandel 1989.

A more robust prediction is that for a given object, the fitted temperature should vary as the brightness changes. This is seen *qualitatively* in the UV for most objects.

In a particular case, the extremely luminous quasar HS 1700+6416, the spectrum extends to far too high a frequency for the "standard" model (Reimers *et al.* 1989), though hard photons could always come from Comptonization (Siemiginowska & Dobrzycki 1990). I've noticed also lately that it's become socially acceptable to assume that the inner edge of the optically thick disk can be anywhere needed to help fit a model. Previously, attempts were made to fit disks which extend into the last stable circular orbit. This may be reasonable physically, but it conforms to the pattern that a new parameter is adopted for each new observational fact. Certainly the disk model has shown no predictive power.

Finally there is the interesting issue of microlensing variations. Rauch & Blandford (1991) showed that for the Einstein cross *any* optically thick thermal model which can produce the optical-region SED must have a thermodynamic emissivity much greater than 1! Others have disagreed e.g., Czerny, Jaroszynski & Czerny (1994), on the grounds that 1) everyone knows a plain disk doesn't fit the optical slope anyway, so we can invoke another component there from outside the microlensed region, and 2) the observed "caustic crossing" may have been a rare event. I like the Rauch and Blandford color-temperature constraint because a disk could, in fact, produce the observed SEDs given a certain heating of the outer annuli by the inner one, but on the other hand, recent data suggest that the variation analyzed was, in fact, unusual and is legitimately modeled by a 2–3σ event.

4.2.3. *Variability*

As far as I know, Alloin *et al.* (1985) should get the most credit for pointing out explicitly that AGN variations are much, much too fast and too phase-coherent with respect to wavelength for any quasistatic model. Almost equally important in this context is the fine quasar spectral variability study by Cutri *et al.* (1985). Somehow these papers didn't sink in for a decade or so, with claims being made that the problem was first discovered by a much later monitoring campaign on NGC 5548! The latter did provide the best limits on any lag between the long- and short-wavelength variations which was so tight as to require communication between the two relevant annuli at of order light speed (Krolik *et al.* 1991)! This is a profound fact that shouldn't be ignored. In the disk models, this is much shorter than the sound crossing time as well as the viscous time — it really requires something tapping into the basic energy with speeds of order the speed of light! This broadband variability, and the zero spectral index of the variable part of the spectra, are more consistent in principle with hot but optically thin emission (Barvainis 1993).

Excellent recent data on variability can be found in Giveon *et al.* 1999. The rates of flux variations may surprise you.

The rapid in-phase variability has led to the speculation that the energy is actually dissipated in a "corona," thought also to produce the X-rays in the $\sim$ 5keV region. Then photons from the corona could heat the disk, rather than internal dissipation. This is a real "non-starter" since it's energetically untenable for all objects except a few at the lowest luminosities. Remember that the rapid in-phase variability is a generic problem, known at least since 1985 to be applicable to luminous quasars as well as Seyferts. See the sketches in Koratkar and Blaes 1998, reproduced here as Fig. 7 for a clear picture of these SEDs.

Although the energetics is totally damning for heating by the observed X-ray source, I'll also mention that the 5-keV X-ray and optical continua do not vary together as expected in any object, though "second-order" predictions (or postdictions) are somewhat as expected in at least one object, according to Nandra *et al.* (1998, 2000). There sure doesn't seem to be any near-IR vs. X-ray relationship (Done *et al.* 1990). Also a prediction of an unseen Ly edge in emission might be a problem with external illumination of a disk (Sincell & Krolik 1997).

The X-ray Fe K-α profiles look like they could come from Kerr disks and I thought showed at least that there are passive disks present in AGN. However, the line (and "Compton hump") variability is virtually inexplicable in that or any other picture; see Weaver 2000 for a brief review.

I can't think of any accretion-disk predictions that have come true, as far as producing the BBB is concerned. I don't even think the passive-disk predictions for the K-α line count for much until its variability properties can be reconciled with the disk picture. To paraphrase Vince Lombardi, in science prediction isn't everything. It's the only thing.

5. Conclusions

Polarization is a basic property of photons, just like frequency. It is often almost as densely coded with information! Radio astronomers generally make polarization observations automatically. If optical astronomers did the same, we might have some great discoveries. Many years ago the polarimetry optics absorbed a lot of photons, but now the total-flux spectrum accumulates almost as fast with the polarimeter as without it. Perhaps polarimetry should be the default in the optical for some programs. In particular, I don't see why anyone would take total-flux spectra or images of distant AGN or ULIRGs or Scuba sources when they can get the polarization almost for free.

REFERENCES

AGOL, E. & BLAES, O. 1996 Polarization from magnetized accretion discs in active galactic nuclei. *MNRAS* **282**, 965.

ALLOIN, D., PELAT, D., PHILLIPS, M. & WHITTLE, M. 1985 Recent spectral variations in the active nucleus of NGC 1566. *ApJ.* **288**, 205.

ALLOIN, D., BARVAINIS, R., GORDON, M. A. & ANTONUCCI, R. R. J. 1992 CO emission from radio quiet quasars - New detections support a thermal origin for the FIR emission. *AAP* **265**, 429.

ALMAINI, O., LAWRENCE, A. & BOYLE, B. J. 1999 The AGN contribution to deep submillimetre surveys and the far-infrared background. *MBRAS* **305**, L59.

ANTONUCCI, R. R. J. 1982(a) Optical polarization position angle versus radio source axis in radio galaxies. *Natur.* **299**, 605.

ANTONUCCI, R. R. J. 1982(b) Optical flux and polarization spectra compared with radio maps of radio galaxies. *Ph.D. thesis*, 1.

ANTONUCCI, R. R. J. 1983 Optical polarization position angle versus radio structure axis in Seyfert galaxies. *Natur.* **303**, 158.

ANTONUCCI, R. R. J. & ULVESTAD, J. S. 1985 Extended radio emission and the nature of blazars. *ApJ.* **294**, 158.

ANTONUCCI, R. 1988, Polarization of active galactic nuclei and quasars. *Supermassive Black Holes*, 26.

ANTONUCCI, R. R. J., KINNEY, A. L. & FORD, H. C. 1989 The Lyman edge test of the quasar emission mechanism. *ApJ.* **342**, 64.

ANTONUCCI, R. & BARVAINIS, R. 1990 Narrow-line radio galaxies as quasars in the sky plane. *ApJ. Lett.* **363**, L17.

ANTONUCCI, R., BARVAINIS, R. & ALLOIN, D. 1990 The empirical difference between radio-loud and radio-quiet quasars. *ApJ.* **353**, 416.

ANTONUCCI, R. 1993 Unified models for active galactic nuclei and quasars. *ARA&A* **31**, 473.

ANTONUCCI, R., HURT, T. & MILLER, J. 1994 HST ultraviolet spectropolarimetry of NGC 1068. *ApJ.* **430**, 210.

ANTONUCCI, R., GELLER, R., GOODRICH, R. W. & MILLER, J. S. 1996, The Spectropolarimetric Test of the Quasar Emission Mechanism. *ApJ.* **472**, 502.

ANTONUCCI, R., HURT, T. & AGOL, E. 1996 Spectropolarimetric Test of the Relativistic Disk Model for the Broad H alpha Line of ARP 102B. *ApJ. Lett.* **456**, L25.

ANTONUCCI, R. 1999 Constraints on Disks Models of The Big Blue Bump from UV/Optical/IR Observations. *ASP Conf. Ser. 161: High Energy Processes in Accreting Black Holes*, 193.

APPENZELLER, I. *et al.* 1998 ORFEUS II Far-ultraviolet Observations of 3C 273: The Instrinsic Spectrum. *ApJ. Lett.* **500**, L9.

BARTHEL, P. D. 1987 Feeling uncomfortable. *Superluminal Radio Sources*, 148.

BARVAINIS, R. 1992 Do accretion disks exist? IR through radio observations. *Testing the AGN Paradigm*, 129.

BARVAINIS, R. 1993 Free-free emission and the big blue bump in active galactic nuclei. *ApJ.* **412**, 513.

BERRIMAN, G. 1989 The origin of the optical polarizations of Seyfert 1 galaxies. *ApJ.* **345**, 713.

BEST, P. N., LONGAIR, M. S. & ROETTGERING, J. H. A. 1997 HST, radio and infrared observations of 28 3CR radio galaxies at redshift Z of about 1. *MNRAS* **292**, 758.

BLAES, O. & AGOL, E. 1996 Polarization near the Lyman Edge in Accretion Disk Atmosphere Models of Quasars. *ApJ. Lett.* **469**, L41.

BRINDLE, C., HOUGH, J. H., BAILEY, J. A., AXON, D. J., WARD, M. J., SPARKS, W. B. & McLEAN, I. S. 1990 An Optical and Near Infrared Polarization Survey of Seyfert and Broadline Radio Galaxies - Part Two - the Wavelength Dependence of Polarization. *MNRAS*, **244**, 604.

BROWNE, I. W. A., CLARK, R. R., MOORE, P. K., MUXLOW, T. W. B., WILKINSON, P. N., COHEN, M. H. & PORCAS, R. W. 1982 MERLIN observations of superluminal radio sources. *Natur.* **299**, 788.

CAPETTI, A., AXON, D. J., MACCHETTO, F., SPARKS, W. B. & BOKSENBERG, A. 1995 HST Imaging Polarimetry of NGC 1068. *ApJ.* **446**, 155.

CAPETTI, A., MACCHETTO, F., AXON, D. J., SPARKS, W. B. & BOKSENBERG, A. 1995 Hubble Space Telescope Imaging Polarimetry of the Inner Nuclear Region of NGC 1068. *ApJ. Lett.* **452**, L87.

CAPETTI, A. *et al.* 2000 Hubble Space Telescope Infrared Imaging Polarimetry of Centaurus A: Implications for the Unified Scheme and the Existence of a Misdirected BL Lacertae Nucleus. *ApJ.* **544**, 269.

CHEN, K. & HALPERN, J. P. 1990 Spectropolarimetric test of the relativistic disk model for the broad emission lines of active galactic nuclei. *ApJ. Lett.* **354**, L1.

CHEN, K., HALPERN, J. P. & TITARCHUK, L. G. 1997 Polarization of Line Emission from an Accretion Disk and Application to ARP 102B. *ApJ.* **483**, 194.

CHIABERGE, M., CAPETTI, A. & CELOTTI, A. 1999 The HST view of FR I radio galaxies: evidence for non-thermal nuclear sources. *AAP* **349**, 77.

CHIABERGE, M., CAPETTI, A. & CELOTTI, A. 2000 The HST view of the FR I / FR II dichotomy. *AAP* **355**, 873.

COHEN, M. H., OGLE, P. M., TRAN, H. D., GOODRICH, R. W. & MILLER, J. S. 1999 Polarimetry and Unification of Low-Redshift Radio Galaxies. *A.J.* **118**, 1963.

COLLIN-SOUFFRIN, S., DYSON, J. E., McDOWELL, J. C. & PERRY, J. J. 1988 The environment of active galactic nuclei. I - A two-component broad emission line model. *MNRAS*, **232**, 539.

CORBETT, E. A., ROBINSON, A., AXON, D. J., YOUNG, S. & HOUGH, J. H. 1998 The profiles of polarized broad Halpha lines in radio galaxies. *MNRAS* **296**, 721.

CUTRI, R. M., WISNIEWSKI, W. Z., RIEKE, G. H. & LEBOFSKY, M. J. 1985 Variability and the nature of QSO optical-infrared continua. *ApJ.* **296**, 423.

CZERNY, B. & POJMANSKI, G. 1990 Lyman edges in AGN accretion discs. *MNRAS* **245**, 1P.

CZERNY, B. & ZBYSZEWSKA, M. 1991 Comptonization of the Lyman edge in active galactic nuclei. *MNRAS* **249**, 634.

CZERNY, B., JAROSZYNSKI, M. & CZERNY, M. 1994 Constraints on the Size of the Emitting Region in an Active Galactic Nucleus. *MNRAS* **268**, 135.

DEY, A., VAN BREUGEL, W., VACCA, W. D. & ANTONUCCI, R. 1997 Triggered Star Formation in a Massive Galaxy at Z = 3.8: 4C 41.17. *ApJ.* **490**, 698.

DONE, C., WARD, M. J., FABIAN, A. C., KUNIEDA, H., TSURUTA, S., LAWRENCE, A., SMITH, M. G. & WAMSTEKER, W. 1990 Simultaneous multifrequency observations of the Seyfert 1 galaxy NGC 4051 - Constant optical-infrared emission observed during large-amplitude X-ray variability. *MNRAS* **243**, 713.

EKERS, R. D. & SIMKIN, S. M. 1983 Radio structure and optical kinematics of the cD galaxy Hydra A /3C 218/. *ApJ.* **265**, 85.

ELVIUS, A. 1978 Polarization of light in the Seyfert galaxy NGC 1068. *AAP* **65**, 233.

ERACLEUS, M. AND HALPERN, J. 2001 A certified LINER with broad variable emission lines. *Ap. J.* (in press); also astro-ph/0101050.

EVANS, A. S., FRAYER, SURACE, J. A., SANDERS, D. B. 2001 Molecular Gas in Infrared-Excess, Optically-Selected QSOs and the Connection with Infrared Luminous Galaxies. *AJ*, in press - also astro-ph 0101308.

FRANCIS, P. J. 1996 The continuum slopes of optically selected QSOs. *Publications of the Astronomical Society of Australia* **13**, 212.

GELLER, R. M., SAULT, R. J., ANTONUCCI, R., KILLEEN, N. E. B., EKERS, R., DESAI, K. & WHYSONG, D. 2000 Cosmological Halos: A Search for the Ionized Intergalactic Medium. *ApJ.* **539**, 73.

GENZEL, R. *et al.* 1998 What Powers Ultraluminous IRAS Galaxies? *ApJ.*, 498, 579.

GIVEON, U., MAOZ, D., KASPI, S., NETZER, H. & SMITH, P. S. 1999 Long-term optical variability properties of the Palomar-Green quasars. *MNRAS* **306**, 637.

GNEDIN, I. N. & SILANTEV, N. A. 1978 Polarization effects in the emission of a disk of accreting matter. *Soviet Astronomy* **22**, 325.

GONZÁLEZ DELGADO, R. M., HECKMAN, T., LEITHERER, C., MEURER, G., KROLIK, J., WILSON, A. S., KINNEY, A. & KORATKAR, A. 1998 Ultraviolet-Optical Observations of the Seyfert 2 Galaxies NGC 7130, NGC 5135, and IC 3639: Implications for the Starburst-Active Galactic Nucleus Connection. *ApJ.* **505**, 174.

GOODRICH, ROBERT W. 1989 Spectropolarimetry of 'narrow-line' Seyfert 1 galaxies. *ApJ.* **342**, 224.

GOODRICH, R. W. & MILLER, J. S. 1994 Spectropolarimetry of high-polarization Seyfert 1 galaxies: Geometry and kinematics of the scattering regions. *ApJ.* **434**, 82.

GOODRICH, R. W., MILLER, J. S., MARTEL, A., COHEN, M. H., TRAN, H. D., OGLE, P. M. & VERMEULEN, R. C. 1996 FSC 10214+4724: A Gravitationally Lensed, Hidden QSO. *ApJ. Lett.* **456**, L9.

GOPAL-KRISHNA, KULKARNI, V. K. & WIITA, P. J. 1996 The Linear Sizes of Quasars and Radio Galaxies in the Unified Scheme. *ApJL.* **463**, L1.

GOPAL-KRISHNA & BIERMANN, P. L. 1998 Are ultra-luminous infrared galaxies the dominant extragalactic population at high luminosities? *AAP* **330**, L37.

HECKMAN, T. *et al.* 1995 The Nature of the Ultraviolet Continuum in Type 2 Seyfert Galaxies. *ApJ.* **452**, 549.

HEISLER, C. A., LUMSDEN, S. L. & BAILEY, J. A. 1997 Visibility of scattered broad-line emission in Seyfert 2 galaxies. *Natur.* **385**, 700.

HURT, T., ANTONUCCI, R., COHEN, R., KINNEY, A. & KROLIK, J. 1999 Ultraviolet Imaging Polarimetry of Narrow-Line Radio Galaxies. *ApJ.* **514**, 579.

IMPEY, C. D. & NEUGEBAUER, G. 1988 Energy distributions of blazars. *A.J.* **95**, 307.

JAFFE, W., FORD, H., FERRARESE, L., VAN DEN BOSCH, F. & O'CONNELL, R. W. 1996 The Nuclear Disk of NGC 4261: Hubble Space Telescope Images and Ground-based Spectra. *ApJ.* **460**, 214.

KAY, L. E. 1994 Blue spectropolarimetry of Seyfert 2 galaxies. 1: Analysis and basic results. *ApJ.* **430**, 196.

KISHIMOTO, M. 1999 The Location of the Nucleus of NGC 1068 and the Three-dimensional Structure of Its Nuclear Region. *ApJ.* **518**, 676.

KISHIMOTO, M., ANTONUCCI. R., CIMATTI, A., HURT, T., DEY, A., VAN BREUGEL, W., SPINRAD, H. 2000 UV Spectropolarimetry of Narrow-line Radio Galaxies.

KNAPP, G. R. & PATTEN, B. M. 1991 Millimeter and submillimeter observations of nearby radio galaxies. *A.J.* **101**, 1609.

KOLLGAARD, R. I., WARDLE, J. F. C., ROBERTS, D. H., GABUZDA, D. C. 1992 Radio constraints on the nature of BL Lacertae objects and their parent population. *Astronomical Journal* **vol. 104**, p. 1687.

KOLYKHALOV, P. I. & SUNYAEV, R. A. 1984 Radiation of accretion disks in quasars and galactic nuclei. *Advances in Space Research* **3**, 249.

KORATKAR, A., ANTONUCCI, R. R. J., GOODRICH, R. W., BUSHOUSE, H. & KINNEY, A. L. 1995 Quasar Lyman Edge Regions in Polarized Light. *ApJ.* **450**, 501.

KORATKAR, A., ANTONUCCI, R., GOODRICH, R. & STORRS, A. 1998 Below the Lyman Edge: Ultraviolet Polarimetry of Quasars. *ApJ.* **503**, 599.

KORATKAR, A. & BLAES, O. 1999 The Ultraviolet and Optical Continuum Emission in Active Galactic Nuclei: The Status of Accretion Disks. *PASP* **111**, 1.

KROLIK, J. H. & BEGELMAN, M. C. 1986 An X-ray heated wind in NGC 1068. *ApJ. Lett.* **308**, L55.

KROLIK, J. H. & BEGELMAN, M. C. 1988 Molecular tori in Seyfert galaxies - Feeding the monster and hiding it. *ApJ.* **329**, 702.

KROLIK, J. H., HORNE, K., KALLMAN, T. R., MALKAN, M. A., EDELSON, R. A. & KRISS, G. A. 1991 Ultraviolet variability of NGC 5548 - Dynamics of the continuum production region and geometry of the broad-line region. *ApJ.* **371**, 541.

LAING, R. A. 1988 The sidedness of jets and depolarization in powerful extragalactic radio sources. *Natur.* **331**, 149.

LANDAU, R. *et al.* 1986 Active extragalactic sources — Nearly simultaneous observations from 20 centimeters to 1400Å. *ApJ.* **308**, 78.

LAOR, A. 1990 AGN Accretion Discs — Part Three — Comparison with the Observations. *MNRAS* **246**, 369.

LAOR, A., JANNUZI, B. T., GREEN, R. F. & BOROSON, T. A. 1997 The Ultraviolet Properties of the Narrow-Line Quasar I ZW 1. *ApJ.* **489**, 656.

LARA, L., MÁRQUEZ, I., COTTON, W. D., FERETTI, L., GIOVANNINI, G., MARCAIDE, J. M. & VENTURI, T. 1999 The broad-line radio galaxy J2114+820. *New Astronomy Review* **43**, 643.

LEIGHLY, K. M. 2000 STIS Ultraviolet Spectral Evidence for Outflows in Extreme Narrow-Line Seyfert 1 Galaxies. *astro-ph/0012173*

LUTZ, D., VEILLEUX, S. & GENZEL, R. 1999 Mid-Infrared and Optical Spectroscopy of Ultraluminous Infrared Galaxies: A Comparison. *ApJ. Lett.*, **517**, L13.

MADEJSKI, G. M., ŻYCKI, P., DONE, C., VALINIA, A., BLANCO, P., ROTHSCHILD, R. & TUREK, B. 2000 Structure of the Circumnuclear Region of Seyfert 2 Galaxies Revealed by Rossi X-Ray Timing Explorer Hard X-Ray Observations of NGC 4945. *ApJ. Lett.* **535**, 87.

MAIOLINO, R., SALVATI, M., BASSANI, L., DADINA, M., DELLA CECA, R., MATT, G., RISALITI, G. & ZAMORANI, G. 1998. Heavy obscuration in X-ray weak AGNs. *AAP* **338**, 781.

MALKAN, M. A. 1983 The ultraviolet excess of luminous quasars. II - Evidence for massive accretion disks. *ApJ.* **268**, 582.

MARCONI, A., SCHREIER, E. J., KOEKEMOER, A., CAPETTI, A., AXON, D., MACCHETTO, D. & CAON, N. 2000 Unveiling the Active Nucleus of Centaurus A. *ApJ.* **528**, 276.

MARTEL, A. R. 1996 Spectropolarimetry of high-polarization Seyfert 1 galaxies. *Ph.D. Thesis*, 80.

MARTEL, A. 1998 New Hα Spectropolarimetry of NGC 4151: The Broad-Line Region-Host Connection. *ApJ.* **508**, 657.

MARTIN, P. G., THOMPSON, I. B., MAZA, J., & ANGEL, J. R. P. 1983 The polarization of Seyfert galaxies. *ApJ.* **266**, 470.

MATHEWSON, D. S. & FORD, V. L. 1970 Polarization observations of 1800 stars. *MEMRAS* **74**, 139.

MATT, G., FABIAN, A. C. & ROSS, R. R. 1993 X-Ray Photoionized Accretion Discs - Ultraviolet and X-Ray Continuum Spectra and Polarization. *MNRAS* **264**, 839.

MCLEAN, I. S., ASPIN, C., HEATHCOTE, S. R. & MCCAUGHREAN, M. J. 1983 Is the polarization of NGC1068 evidence for a non-thermal source? *Natur.* **304**, 609.

MILLER, J. S. & ANTONUCCI, R. R. J. 1983 Evidence for a highly polarized continuum in the nucleus of NGC 1068. *ApJ. Lett.* **271**, 7.

MILLER, J. S. & GOODRICH, R. W. 1990 Spectropolarimetry of high-polarization Seyfert 2 galaxies and unified Seyfert theories *ApJ.* **355**, 456.

MILLER, J. S., GOODRICH, R. W. & MATHEWS, W. G. 1991 Multidirectional views of the active nucleus of NGC 1068. *ApJ.* **378**, 47.

MINEO, T. *et al.* 2000 BeppoSAX broad-band observations of low-redshift quasars: spectral curvature and iron Kalpha lines. *AAP* **359**, 471.

MIYAJI, T., WILSON, A. S. & PEREZ-FOURNON, I. 1992 The radio source and bipolar nebulosity in the Seyfert galaxy NGC 3516. *ApJ.* **385**, 137.

MUSHOTZKY, R. F. & WANDEL, A. 1989 On the ratio of the infrared-to-ultraviolet continuum to the X-rays in quasars and active galaxies. *ApJ.* **339**, 674.

NANDRA, K., CLAVEL, J., EDELSON, R. A., GEORGE, I. M., MALKAN, M. A., MUSHOTZKY, R. F., PETERSON, B. M. & TURNER, T. J. 1998 New Constraints on the Continuum Emission Mechanism of Active Galactic Nuclei: Intensive Monitoring of NGC 7469 in the X-Ray and Ultraviolet. *ApJ.* **505**, 594.

NANDRA, K., LE, T., GEORGE, I. M., EDELSON, R. A., MUSHOTZKY, R. F., PETERSON, B. M. & TURNER, T. J. 2000 Origin of the X-Ray and Ultraviolet Emission in NGC 7469. *ApJ.* **544**, 734.

NEUGEBAUER, G., GREEN, R. F., MATTHEWS, K., SCHMIDT, M., SOIFER, B. T. & BENNETT, J. 1987 Continuum energy distributions of quasars in the Palomar-Green Survey. *ApJ. Supp.* **63**, 615.

OWEN, F. N. & LEDLOW, M. J. 1994 The First Stromlo Symposium: The Physics of Active Galaxies. ASP Conference Series, Vol. 54, 1994, G.V. Bicknell, M.A. Dopita, and P.J. Quinn, Eds., p. 319.

OWEN, F. N., EILEK, J. A. & KASSIM, N. E. 2000 M87 at 90 Centimeters: A Different Picture. *ApJ.* **543**, 611.

PIER, E. A. & KROLIK, J. H. 1993 Infrared Spectra of Obscuring Dust Tori around Active Galactic Nuclei. II. Comparison with Observations. *ApJ.* **418**, 673.

RAUCH, K. P. & BLANDFORD, R. D. 1991 Microlensing and the structure of active galactic nucleus accretion disks. *ApJ. Lett.* **381**, L39.

REIMERS, D., CLAVEL, J., GROOTE, D., ENGELS, D., HAGEN, H. J., NAYLOR, T., WAMSTEKER, W. & HOPP, U. 1989 The luminous quasar HS1700+6416 and the shape of the 'big bump' below 500 A. *AAP* **218**, 71.

REYNOLDS, C. S., DI MATTEO, T., FABIAN, A. C., HWANG, U. & CANIZARES, C. R. 1996 The 'quiescent' black hole in M87. *MNRAS* **283**, L111.

RODRIGUEZ-PASCUAL, P. M., MAS-HESSE, J. M. & SANTOS-LLEO, M. 1997 The broad line region of narrow-line Seyfert 1 galaxies. *AAP* **327**, 72.

RYDBECK, G., WIKLIND, T., CAMERON, M., WILD, W., ECKART, A., GENZEL, R. & ROTHERMEL, H. 1993 High resolution (C-12)O(2-1) observations of the molecular gas in Centaurus A. *AAP* **270**, L13.

SALVATI, M. & MAIOLINO, R. 2000 Where are the Type 2 AGNs? *Large Scale Structure in the X-ray Universe*, Proceedings of the 20–22 September 1999 Workshop, Santorini, Greece, eds. Plionis, M. & Georgantopoulos, I., Atlantisciences, Paris, France, p. 277

SAMBRUNA, R. M., CHARTAS, G., ERACLEOUS, M., MUSHOTZKY, R. F. & NOUSEK, J. A. 2000 Chandra Uncovers a Hidden Low-Luminosity Active Galactic Nucleus in the Radio Galaxy Hydra A (3C 218). *ApJL.* **532**, L91.

SANDERS, D. B., PHINNEY, E. S., NEUGEBAUER, G., SOIFER, B. T. & MATTHEWS, K. 1989 Continuum Energy Distributions of Quasars: Shapes and Origins. *ApJ.* **347**, 29.

SARAZIN, C. L., KOEKEMOER, A. M., BAUM, S. A., O'DEA, C. P., OWEN, F. N. & WISE, M. W. 1999 X-Ray Properties of B2 1028+313: A Quasar at the Center of the Abell Cluster A1030. *ApJ.* **510**, 90.

SCHMIDT, G. D. & MILLER, J. S. 1980 The spectrum and polarization of the nucleus of NGC 4151. *ApJ.*, **240**, 759.

SCHMIDT, GARY D. & SMITH, PAUL S. 2000 Evidence for Polarized Synchrotron Components in Radio-Optical Aligned Quasars. *ApJ.* in press, or astro-ph0008168.

SIEMIGINOWSKA, A. & DOBRZYCKI, A. 1990 Accretion disk in the high-redshift quasar HS 1700 + 6416. *AAP* **231**, L1.

SINCELL, M. W. & KROLIK, J. H. 1997, The Vertical Structure and Ultraviolet Spectrum of X-Ray–irradiated Accretion Disks in Active Galactic Nuclei. *ApJ.* **476**, 605.

SINGAL, A. K. 1993 Evidence against the unified scheme for powerful radio galaxies and quasars. *MNRAS* **262**, L27.

SINGH, K. P. & WESTERGAARD, N. J. 1992 Radio structure of the Seyfert galaxy Markarian 509. *AAP* **264**, 489.

STOCKMAN, H. S., ANGEL, J. R. P., & MILEY, G. K. 1979 Alignment of the optical polarization with the radio structure of QSOs. *ApJ. Lett.* **227**, 55.

SURACE, J. A., SANDERS, D. B. & EVANS, A. S. 2000 High-Resolution Optical/Near-Infrared Imaging of Cool Ultraluminous Infrared Galaxies. *ApJ.* **529**, 170.

THOMPSON, I. B. & MARTIN, P. G. 1988 Optical polarization of Seyfert galaxies. *ApJ.* **330**, 121.

TRAN, H. D., MILLER, J. S. & KAY, L. E. 1992 Detection of obscured broad-line regions in four Seyfert 2 galaxies. *ApJ.* **397**, 452.

TRAN, H. D. 1995(a) The Nature of Seyfert 2 Galaxies with Obscured Broad-Line Regions. II. Individual Objects. *ApJ.* **440**, 578.

TRAN, H. D. 1995(b) The Nature of Seyfert 2 Galaxies with Obscured Broad-Line Regions. III. Interpretation. *ApJ.* **440**, 597.

TRAN, H. D., COHEN, M. H. & GOODRICH, R. W. 1995 Keck Spectropolarimetry of the Radio Galaxy 3C 234. *A.J.* **110**, 2597.

TYTLER, D. & DAVIS, C. 1993 On the Lack of Emission or Absorption in the Lyman Continua of Qsos: what is the Source of the Lyman Continuum Radiation, and where are the Broad Line Clouds? *American Astronomical Society Meeting* **183**, 9106.

ULVESTAD, J. S., ANTONUCCI, R. R. J. & GOODRICH, R. W. 1995 Radio properties of narrow-lined Seyfert 1 galaxies. *A.J.* **109**, 81.

VAN BEMMEL, I. AND BERTOLDI, F. preprint 2001 Millimeter observations of radio-loud active galaxies. *astro-ph 0101137*

VERMEULEN, R. C., OGLE, P. M., TRAN, H. D., BROWNE, I. W. A., COHEN, M. H., READHEAD, A. C. S., TAYLOR, G. B. & GOODRICH, R. W. 1995 When Is BL Lac Not a BL Lac? *ApJ. Lett.* **452**, L5.

VIGNATI, P. *et al.* 1999 BeppoSAX unveils the nuclear component in NGC 6240. *AAP* **349**, L57.

WALTER, R. & FINK, H. H. 1993 The Ultraviolet to Soft X-Ray Bump of SEYFERT-1 Type Active Galactic Nuclei. *AAP* **274**, 105.

WEAVER, K. A. 2000 Probing Dense Matter in the cores of AGN: Observations with RXTE and ASCA To appear in "Proceedings of X-ray Astronomy '99 - Stellar Endpoints, AGN and the Diffuse Background," 2000. G. Malaguti, G. Palumbo & N. White (eds), Gordon & Breach (Singapore)

WOLFE, A. M. 1978 *Pittsburgh Conference on BL Lac Objects*, Pittsburgh, Pa., University of Pittsburgh, 1978. 439 p.

WROBEL, J. M. & LIND, K. R. 1990 The double-lobed blazar 3C 371. *ApJ.* **348**, 135.

YOUNG, S., HOUGH, J. H., EFSTATHIOU, A., WILLS, B. J., AXON, D. J., BAILEY, J. A. & WARD, M. J. 1996 Scattered broad optical lines in the polarized flux spectrum of the FR II galaxy 3C 321. *MNRAS* **279**, L72.

ZHENG, W., KRISS, G. A., TELFER, R. C., GRIMES, J. P. & DAVIDSEN, A. F. 1998 A Composite HST Spectrum of Quasars. *ApJ.* **492**, 855.

Compact Objects and Accretion Disks

By ROGER BLANDFORD[1], ERIC AGOL[1],
AVERY BRODERICK[1], JEREMY HEYL[2],
LEON KOOPMANS[1], HEE-WON LEE[3]

[1]Theoretical Astrophysics, Caltech, Pasadena, CA 91125, USA
[2]Center for Astrophysics, 60 Garden St., Cambridge, MA 02173, USA
[3]Yonsei University, Seoul, Korea

Recent developments in the spectropolarimetric study of compact objects, specifically black holes (stellar and massive) and neutron stars are reviewed. The lectures are organized around five topics: disks, jets, outflows, neutron stars and black holes. They emphasize physical mechanisms and are intended to bridge the gap between the fundamentals of polarimetry and the phenomenology of observed cosmic sources of polarized radiation, as covered by the other lecturers. There has been considerable recent progress in spectropolarimetry from radio through optical frequencies and this is producing some unique diagnostics of the physical conditions around compact objects. It is argued that there is a great need to develop a correspondingly sensitive polarimetric capability at ultraviolet through γ-ray energies.

Spectropolarimetric observations, particularly those at radio and optical wavelengths, have played an important role in high energy astrophysics. From the discovery of synchrotron radiation to the first good evidence for AGN unification, from the polarization patterns in the coherent emission of radio pulsars to the discovery of variable, linear polarization in the absorption troughs of broad absorption line quasars, polarization studies often provide the best and sometimes the only clue we have as to the geometric disposition of the emitting elements in these diverse sources when we cannot resolve them directly.

These notes summarize lectures delivered by Roger Blandford at the XII Canary Islands Winter School on Astrophysical Spectropolarimetry. They are written up with the assistance of Eric Agol (Disks), Leon Koopmans (Jets), Hee-Won Lee (Outflows), Jeremy Heyl (Neutron Stars) and Avery Broderick (Black Holes) The lectures were intended to provide a bridge between the general physical foundations of polarimetry and its practical description presented at the school by Drs. Landi Degl'Innocenti and Keller and the observationally oriented lectures of Drs. Antonucci and Hildebrand. They also make some important connections to solar, stellar (especially white dwarf) and maser polarimetry as described by Drs. Stenflo, Mathys and Elitzur, respectively. They are organised around five generic astrophysical sources: disks, jets, winds, neutron stars, and black holes. In each case a cursory motivation is provided by summarizing some relevant observations and presenting some of the key issues that polarimetry can help to resolve. This is followed by an heuristic discussion of some relevant physical mechanisms in a manner which, it is hoped, will allow them to be applied elsewhere followed by a brief account of how they have been used so far and some suggestions of possible future investigations.

1. Disks

1.1. *Motivation*

Accretion disks are commonly found when gas, with angular momentum, is gravitationally attracted towards a central massive body (Frank, King, & Raine 1992). First described in the context of Laplace's nebular hypothesis and first seriously analyzed by Lüst (1952) they have been observed around black holes, neutron star and white dwarf binaries (Shapiro & Teukolsky 1983), around protostars, and especially within active galactic nuclei, including quasars (Blandford, Netzer, & Woltjer 1990, Krolik 1999). It is this last type of disk that provides us with much of our most detailed observations. Although it is not part of my task to discuss them, observations of young stellar objects and cataclysmic variables are turning out to be particularly instructive and much of what follows is informed by the results of these studies.

Disks are planar structures and if their opacity is predominantly scattering, by either free electrons or dust grains, then the direction and strength of the polarization tells us about the orientation and inclination of the disk as well as the location of the continuum source. As we shall discuss, (cf also Antonucci, Hildebrand, these proceedings), most astrophysical disks are associated with jets or bipolar outflows and when these can be resolved, they may represent the projected rotation axis of the inner disk. (As we shall also see, disks can be warped and this axis can change with radius and, in the case of AGN, it may be quite different from the axis of the host galaxy.)

We wish to use polarization observations to determine what disks are really like. Unfortunately, the current observational capability is limited. Polarimetry in the radio, the near infrared and the optical regions of the spectrum is really quite good by astronomical, (although not solar), standards. Optically, spectropolarimetry has been performed at the 0.001 level down to $R \sim 18$. Measurement in the mid and far infrared is more of a challenge, though one that has been met in the far infrared, (Hildebrand, these proceedings).

However, to understand the inner disk we need ultraviolet and X-ray polarimetry. The former was carried out for a while on bright quasars using the HST Faint Object Camera, as we shall describe in section 3 below, although this has proved to be a little controversial. X-ray polarimetry has really only been accomplished successfully on a few bright sources (Mészáros *et al.*1988). There are plans to fly a more sensitive polarimeter on Spectrum-X. It will become clear that there is a very strong scientific case to be made to develop X-ray polarimetry. There is also a strong incentive to develop a γ-ray polarimetric capability though, here, the technical challenges are even greater. In principle, Compton telescopes operating at $\sim$ MeV energies, record polarimetric information though, in practice, it has proven to be almost impossible to extract this signal from observations taken to date.

1.2. *Observation*

A particularly good example of an AGN accretion disk can be found in NGC 4258 (cf Elitzur, these proceedings). Here water maser observations reveal a resolved, disk orbiting a 43 million solar mass black hole (beyond all reasonable doubt). We have believed for a long while that gas moves radially inwards through this disk as a result of a hydromagnetic torque that transports angular momentum outward. The binding energy that is released by the infalling gas can be radiated away and this process accounts for the most luminous of quasars and binary X-ray sources. It can also be responsible for driving powerful outflows, as we shall see. Evidence that disks can extend all the way down to the central compact object has been provided by ASCA X-ray observations of relativistically-broadened fluorescent iron emission lines from low luminosity Seyfert and LINER galaxies. (This interpretation has been somewhat controversial; observations, with superior sensitivity and spectral resolution, from XMM-Newton are therefore eagerly awaited.)

Not all disks are thin. There are good phenomenological reasons to suspect that the disks contained in many Seyfert and LINER galaxies thicken over some radii to form

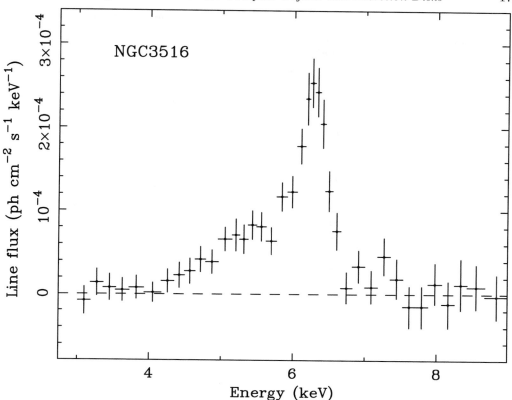

FIGURE 1. Broad Fe $K\alpha$ line from NGC 3516 observed with ASCA by Nandra *et al.*(1999).

dense, obscuring torii. Theoretically, it has been proposed that the inner regions of disks that are supplied with gas at a rate that is either much higher or much lower than the Eddington rate (given by $\dot{M}_{\text{Edd}} = L_{\text{Edd}}/c^2 = 4\pi GM/c\kappa$) will thicken because the gas will be unable to cool and the inflow may even become quasi-spherical. Similarly, the disk in NGC 4258 is clearly warped and this is thought to be a common occurrence. It has even been proposed that radiative torques acting on disks can turn them over (Pringle 1997). Polarization observations offer the opportunity to probe the complex geometry of these flows.

Another example of a subtle arrangement of the emitting elements is provided by X-ray observations of the thin disks in Seyfert galaxies. The current working model is that only a fraction of the binding energy released by the accreting gas is radiated from the disk photosphere (with an effective temperature in the keV range). The remainder is dissipated in a hot corona, presumably as a consequence of magnetic flaring followed by reconnection and hydromagnetic wave damping. The heated electrons can then scatter the escaping soft photons and, as the corona is probably Thomson thick, this will lead to a power-law tail in the hard X-ray spectrum. However, roughly half of these photons will strike the disk where they can be absorbed if they have low energy and suffer inelastic Compton recoil loss at high energy. The reflected spectrum will therefore be convex and be imprinted with line features, most famously, the 6.4 keV $K\alpha$ line of Fe (Fabian *et al.*2000 and references therein, Fig. (1)). On this basis, it has been argued that the widths of lines in Seyfert and LINER galaxies like MCG 6-30-15 imply that the central black hole is rapidly spinning. However, the details depend upon the relative location of the region where the Comptonisation is taking place and the region of the disk responsible for most of the reflection. There are two ways to test these models using more detailed observations. The first is to use a technique called reverberation mapping (e.g. Young & Reynolds 2000). This requires monitoring the variation of the line and the continuum simultaneously and deriving the lag in the variation of the former in response to the latter. This tells us about the geometry and size of the line-emitting region. The second technique is to use polarization observations which will test the geometry.

1.3. *Physical Processes*

We now summarize some relevant physical processes. A good general reference is Rybicki & Lightman (1979).

1.3.1. *Thomson Scattering*

Classical Thomson scattering is strongly polarizing. A free electron can be considered as a Larmor dipole driven by the electric field of the incident wave (with polarization vector $\vec{e}$). The scattered power into polarization is $\vec{e'}$ is given by

$$\frac{d\sigma}{d\Omega'} = r_e^2 (\vec{e} \cdot \vec{e'})^2 \,, \tag{1.1}$$

where $r_e = e^2/m_e c^2 = 2.82 \times 10^{-13}$ cm is the classical electron radius. Note that when the scattering angle is $\phi = 90°$, the radiation is 100 percent polarized.

Averaging over incident polarization and summing over final polarization states gives the familiar differential cross section

$$\frac{d\sigma}{d\Omega'} = \frac{1}{2} r_e^2 (1 + \cos^2 \phi) \,. \tag{1.2}$$

Integrating over solid angle gives the total Thomson cross section

$$\sigma_T = \frac{8\pi}{3} r_e^2 = 6.65 \times 10^{-25} \text{cm}^2 \,. \tag{1.3}$$

1.3.2. *Compton scattering*

At X-ray and γ-ray energy, we must take account of the electron recoil. Conserving energy and linear momentum, we obtain an expression for the scattered energy ϵ' in terms of the incident energy ϵ

$$\epsilon' = \frac{\epsilon}{1 + \frac{\epsilon}{m_e c^2}(1 - \cos\phi)} \,. \tag{1.4}$$

Averaging over ϕ for small values of ϵ, we obtain the mean energy shift

$$<\Delta\epsilon> = -\frac{\epsilon^2}{m_e c^2} \,. \tag{1.5}$$

The Thomson cross section must be replaced by the Klein-Nishina cross section

$$\frac{d\sigma}{d\Omega'} = \frac{1}{2} r_e^2 \left(\frac{\epsilon'}{\epsilon}\right)^2 \left[\frac{\epsilon'}{\epsilon} + \frac{\epsilon}{\epsilon'} - \sin^2\phi\right] \,, \tag{1.6}$$

which emphasizes forward over backward scattering. High energy scattering is nearly as strongly polarizing as Thomson scattering. Averaging over ϕ again for small angle, we obtain

$$\sigma = \sigma_T \left(1 - \frac{2\epsilon}{m_e c^2} + \dots\right) \,. \tag{1.7}$$

For a large photon energy, $\epsilon \gg m_e c^2$, we have

$$\sigma \sim \frac{3\sigma_T}{8\epsilon}\left[\log\left(\frac{2\epsilon}{m_e c^2}\right) + \frac{1}{2}\right]. \tag{1.8}$$

The rate at which the photons heat the electrons through Compton recoil is therefore given for $\epsilon \ll m_e c^2$ by

$$W_+ = n_e \sigma_T c \int d\epsilon N(\epsilon)\frac{\epsilon^2}{m_e c^2} = n_e \sigma_T c U \frac{<\epsilon>}{m_e c^2}, \tag{1.9}$$

where $N(\epsilon)$ is the photon number density per unit energy, $U = \int d\epsilon \epsilon N$ is the photon energy density and $<>$ should be interpreted as an energy density-weighted photon energy.

These expressions describe the energy loss in the initial rest frame of the scattering electron. However when the plasma is hot the electron will be moving and this will cause the photon to experience a Doppler shift. To $O(v/c)$, blue shifts balance redshifts and there is no net energy change when the electron distribution is isotropic. However, it is apparent what there will be a net energy gain to $O(v/c)^2 = O(kT/m_e c^2)$, because the rate of approaching collisions will exceed the rate of receding collisions. We can therefore express the net rate of energy loss by the electrons in terms due to the Doppler shift as

$$W_- = n_e \sigma_T c U \frac{xkT}{m_e c^2}, \tag{1.10}$$

where x is a number that we can fix by observing that there should be energy balance, $W_+ = W_-$, when we use a dilute black body of temperature T

$$N(\epsilon) \propto \epsilon^2 \exp[-\epsilon/kT]. \tag{1.11}$$

(It is necessary to use a dilute black body to avoid having to consider nonlinear, induced Compton scattering.) As $<\epsilon>= 4kT$, we deduce that $x = 4$.

If this were the dominant physical process, then the equilibrium electron temperature in a given radiation field would be

$$T_c = \frac{<\epsilon>}{4k}. \tag{1.12}$$

However, accretion disk corona are probably heated through reconnection and hydromagnetic turbulence, and the temperature is probably quite non-uniform and hard to predict in detail. An additional complication is that the thermalization timescales are actually quite long compared with the disk dynamical timescale and so the plasma is likely to have a significant suprathermal component (Gierlinski *et al.*1999).

More generally, we can deduce the form of the kinetic equation for the photon distribution. As the individual photon energy shifts are small, this will have the form of a modified diffusion equation in energy space. However, as the scattering angles are typically large, we cannot regard this as a diffusion in momentum space. We therefore just consider an isotropic radiation field to bring out some principles (although this approximation is inappropriate for computing polarization, since an isotropic radiation field creates zero net polarization upon scattering). As Compton scattering conserves the number density of photons the equation must have the form

$$\frac{\partial N}{\partial t} = -\frac{\partial F}{\partial \epsilon}, \tag{1.13}$$

where F is the flux of photons in energy space. Now for a dilute black body, Eq. (1.11), F will be linear in N, as long as we can ignore induced scattering, and, as it represents a diffusion, F will contain the first derivative of N with respect to energy. As F must vanish for a dilute black body it must have the form

$$F(\epsilon) = -g(\epsilon)\left(\frac{\partial(N/\epsilon^2)}{\partial\epsilon} + \frac{N}{\epsilon^2 kT}\right), \tag{1.14}$$

for some function $g(\epsilon)$. We next multiply this equation by ϵ and integrate over energy and use either Eq. (1.9) or Eq. (1.10) to identify the function $g(\epsilon) = n_e\sigma_T c\epsilon^4 kT/m_e c^2$. The resulting (Fokker-Planck) equation is,

$$\frac{\partial N}{\partial t} = \frac{n_e\sigma_T}{m_e c}\frac{\partial}{\partial\epsilon}\epsilon^2\left[\epsilon^2 kT\left(\frac{\partial(N/\epsilon^2)}{\partial\epsilon}\right) + N\right]. \tag{1.15}$$

This formalism describes the behavior of electrons interacting with a dilute gas of photons inside a box with reflecting walls. If we need to take account of induced scattering processes, then Eq. (1.14) must be modified so $F = 0$ when $N(\epsilon)$ has the Planck form,

$$N(\epsilon) = \frac{8\pi\epsilon^2}{h^3 c^3}\left[\exp(\epsilon/kT) - 1\right]^{-1}. \tag{1.16}$$

The result is that $N \to N(1 + N)$, in the second term in brackets in Eq. (1.15) which is then known as the Kompaneets equation and is central to discussions of the transfer of radiation through hot plasmas.

Clearly the radiation as described by this isotropic formalism will be unpolarized. In order to describe the polarization of a cosmic source, we must tackle the radiative transfer. There are three approaches that have commonly been followed.

(*a*) *Escape Probability Formalism* This is the simplest approach. We add a term to the right hand side of Eq. (1.15)

$$-\frac{Nc}{R(1 + \tau)}. \tag{1.17}$$

The extra factor $1 + \tau$ takes into account the impeded photon escape when the Thomson depth τ exceeds unity. This approach, which is most commonly used, although instructive as far as the spectrum goes, is not much help when it comes to polarization.

(*b*) *Intensity Formalism* Provided that we restrict our attention to simple shapes – slabs, spheres etc. – we can incorporate the energy space transport within the equation of radiative transfer using a scattering kernel. This can then be solved by taking moments and imposing a closure relation in the standard manner. It is possible to include polarization though this leads to quite involved equations.

(*c*) *Monte Carlo Formalism* In many ways the most versatile method is the same one used in nuclear reactors for the transport of neutrons. This is to follow individual photons, within the scattering region starting with energies, locations and directions that are selected according to a prescribed distribution using random numbers. Polarization is relatively easy to handle, and most polarization is created in the last few scatterings, reducing the computational burden which occurs at large optical depths. All of this is quite straightforward in principle, though, in practice, Monte Carlo simulations are quite an art as a variety of ingenious tricks have to be used to reduce the variance with a finite amount of computer time.

1.3.3. Dust Scattering

Dust scattering is more complex than Thomson scattering as the cross section depends on wavelength, grain size, and grain composition (cf Hildebrand, these proceedings). In the limit when the wavelength is much greater than the grain size, Rayleigh scattering applies, which has the same angular cross section as Thomson scattering, but scales as λ^{-4}, a fact which has been used to distinguish electron scattering from dust scattering in some Seyfert 2 galaxies. For a range of different sizes and compositions, it is impossible to express the dust scattering cross section in a simple formula, but extensive numerical calculations have been carried out by, e.g., Draine & Lee (1984), Zubko & Laor (2000).

1.3.4. *Faraday Rotation*

The next relevant physical process is Faraday rotation. When electromagnetic wave modes propagate through a plasma, their phase velocities will be changed from c. To lowest order, the eigenmodes are circularly polarized with phase velocity difference

$$\Delta V_\phi = 2c \frac{\omega_p^2 \omega_G}{\omega^3} \cos \alpha \,, \tag{1.18}$$

where $\omega_p = (4\pi n_e e^2/m_e)^{1/2}$ is the electron plasma frequency, ω_G is the electron gyro frequency and α is the angle between the ray and the magnetostatic field. If we decompose a linear polarized wave into two circularly polarized modes that propagate through the medium and then recombine the modes after they leave the medium, then there will be a net rotation of the plane of polarization through an angle

$$\frac{d\Phi}{d\tau_T} = \frac{\omega}{2} \int ds \frac{\Delta V_\phi}{c^2} = 0.1 \left(\frac{\lambda}{500\,\mathrm{nm}}\right)^2 \left(\frac{B_\parallel}{1\,\mathrm{G}}\right) \,. \tag{1.19}$$

Polarization observations can tell us as much about the intervening medium as about the source.

1.3.5. *Relativistic Radiative Transfer*

Another interesting complication is that special and general relativistic effects will affect the transfer of radiation from the disk to us. This is particularly interesting for line radiation. The non-relativistic Doppler shift will broaden the profile of a line formed at the photosphere of a rotating disk, with the blue wing coming from the approaching limb and the red wing from the receding limb. The gravitational shift (which is not separated from the Doppler shift in a general relativistic calculation) will accentuate the red wing whose extent depends upon how close the inner edge of the disk gets to the hole. A further effect is that rays will be deflected by the gravitational pull of the central black hole so that an image of an accretion disk observed from near the equatorial plane would exhibit the back side of the disk. (There are ambitious proposals to deploy an X-ray interferometer in space which could exhibit these and other effects.)

Of more direct relevance to this school, is the behavior of the polarization. There are two main effects. Firstly, aberration changes the emission angle, and consequently the emitted polarization, in the rest frame of the orbiting gas. Secondly, the plane of polarization is rotated as the ray propagates near the black hole (Fig. (2)). As we discuss in more detail in §5, the polarization direction is parallel-propagated along null geodesics. All of this is straightforward, if somewhat tedious, to compute. Specific models for continuum emission have been computed by Laor *et al.*(1990) and Connors *et al.*(1980), while relativistic effects on line polarization were computed by Chen & Eardley (1991). It may be possible to use observations of the rotation of the polarization direction with wavelength, in a spectral line, or in the continuum if the wavelength is a measure of the effective radius of the disk, to measure the spin of the hole.

1.4. *Interpretation*

Having outlined some of the relevant physical mechanisms, let us return to the problem with which we began this section. "How much polarization do we expect from an accretion disk and what do we observe?" We can split the problem into two parts – the escape

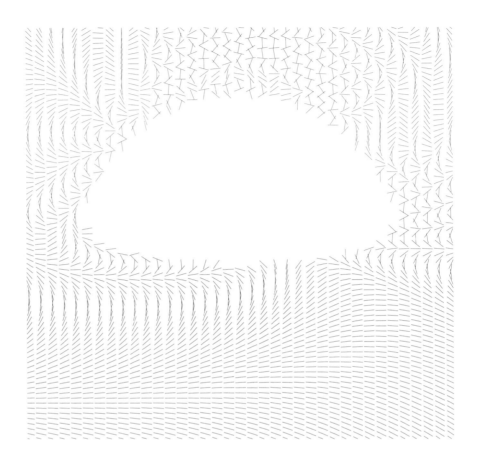

FIGURE 2. Polarization angle as a function of position for an electron-scattering dominated thin accretion disk around a Kerr black hole (a=0.998) viewed at infinity from an angle of 75°. The figure is $20GM/c^2$ in size.

of photons emitted in the disk and the behavior of the scattered photons. For a given atmosphere, the problem is linear and we can superpose the two components. However, if we try to solve self-consistently for the ionization and thermal state of the atmosphere, the problem becomes nonlinear.

1.4.1. *Electron Scattering*

Radiative transfer in a pure scattering, plane parallel atmosphere is a classical problem that was solved for a Thomson scattering kernel, initially analytically, by Chandrasekhar (1960) and then in greater generality by Angel (1969) using a Monte Carlo approach. The answer is that the emergent polarization varies with inclination, having a value $p = 0.12$ when the atmosphere is viewed horizontally and $p = 0.02$ when viewed at the most probable inclination of 60° and, of course, $p = 0$ when viewed normally. Real disk atmospheres also have an absorptive opacity and this will reduce (or increase) the emergent polarization significantly (Hubeny *et al.*2000). In addition, we now believe that

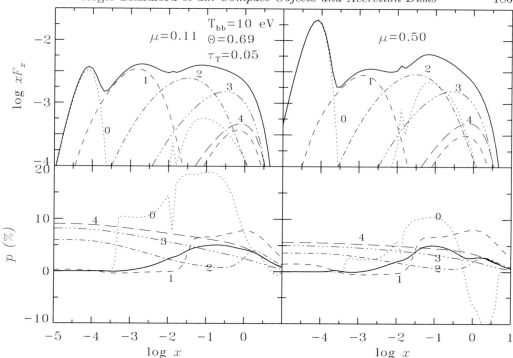

FIGURE 3. Spectrum and polarization versus frequency ($x = h\nu/m_e c^2$) of hot, plane-parallel corona ($T_e = 0.11\, m_e c^2$, $\tau_T = 0.05$) above a cold disk viewed at two inclination angles, $\mu = \cos i = 0.11$ and 0.5. The numbers label various scattering orders for reflection, while the solid lines show the total spectrum and polarization (from Poutanen & Svensson 1996). No relativistic propagation effects have been included.

accretion disks are strongly magnetized. The rationale for this is that ionized accretion disks are known to be unstable to developing strong internal magnetic fields with interior magnetic pressures estimated to be $\sim 1 - 10$ percent of the gas pressure. This magnetic field will surely be carried out beyond the photosphere and into the coronae (discussed above) by buoyancy forces. Furthermore, magnetic pressure is likely to dominate gas pressure in an accretion disk corona, just like in the solar corona.

When we consider the specific parameters appropriate to observed disks, we find that thermal emission should be unpolarized based on the following argument. Consider a given disk annulus, the radiation pressure at the photosphere, $aT^4/3$, is smaller than the gas pressure within the disk. Because observations at a given thermal wavelength peak near $\lambda \sim hc/kT$, we can derive a lower limit on the magnetic field strength $B \gtrsim 10^2 (\lambda/500\,\text{nm})^{-2}\,\text{G}$. Since photons traverse differing optical depths after their last scattering and the magnetic field will likely have significant inhomogeneities, any polarization will be erased by the Faraday rotation $\langle d\Phi/d\tau_T \rangle \gtrsim 10$, *independent* of the wavelength observed or the physical size of the disk (Agol & Blaes 1996).

Thus, it should come as no surprise that AGN disks are generally only polarized by a small amount (Antonucci, this volume). Furthermore, the polarization that is observed may be imprinted extrinsically. Purported rises in polarization below 912 Å in a few quasars observed with HST contradict old theoretical predictions and stand as a challenge to disk theory (Koratkar & Blaes 1999).

Turning to X-ray wavelengths, where non-thermal emission means that we might ignore

Faraday rotation, but cannot ignore reflection, predictions of the emergent polarization under a variety of models have been presented by Matt, Fabian, & Ross (1993) and Poutanen & Svensson (1996) and references therein, as shown in Fig. (3). The rather flat X-ray spectrum is created by Compton scattering of thermal emission from the accretion disk, which is partly due to absorbed X-rays. The electron-scattering reflection feature is suppressed at low energies, $\lesssim 8$ keV, by X-ray bound-free absorption opacity, and at high energies, $\gtrsim 100$ keV, reduced by electron recoil; consequently, the largest X-ray polarizations should lie between these energies. The magnitude of the polarization may depend upon the exact geometry and placement of the coronal emission regions, an unexplored problem.

1.4.2. *Dusty Disks*

The outer parts of accretion disks may be cool enough ($T < 1800$ K) to be inhabited by dust grains. If the dusty disk drives a wind, or is inflated or warped, then dust will scatter the light from the inner disk, imprinting a polarization signature from the infrared to ultraviolet. It may in practice be quite difficult to distinguish between a dusty disk, torus, or outflow using polarization, as the level of polarization expected is quite small, $\sim 1\%$, and depends on the details of the dust model, e.g. Königl & Kartje (1994). In addition, dust extinction can create polarization if the grains, charged by collisions with ions, are aligned by magnetic fields, inducing polarization at the percent level as well.

1.5. *Summary*

- Disks are commonly found in accreting systems.
- Model accretion disks can create strong polarization both in transmission and in reflection, throughout the electromagnetic spectrum.
- Measurement of the variation of linear polarization with wavelength can, in principle, reveal a lot about the disk structure and the location of coronal emission sites.
- However, the situation is, in practice, more complex, particularly at optical wavelengths, where external illumination, warping, and especially Faraday rotation are likely to be very important.
- Even when polarization cannot be measured, its effects are so large under conditions of strong electron scattering that radiative transfer calculations should include polarization.
- Monte Carlo techniques are well-suited for computing polarization in a given model.
- There is a very strong case for developing X-ray polarimetry.

2. Jets

2.1. *Motivation*

Jets, or more generally bipolar outflows, are also surprisingly common. They have been studied in association with active galactic nuclei (AGN), binary X-ray sources, young stellar objects, novae and so on. Jet speeds are typically a few times the escape velocity from the central object; in the case of black holes, bulk Lorentz factors of $\gamma \sim 10$ are inferred. (Gamma ray bursts may also produce jets with Lorentz factors $\gamma \sim 300$.) Jets are so common that it has been speculated that they may be an essential concomitant of accretion flow – the channels through which the liberated angular momentum and perhaps also much of the energy leave the system. The challenge to the astrophysicist is to explain how jets are powered and collimated. However, even after decades of work, major theoretical and observational questions about their origin, collimation and even their constituents still remain.

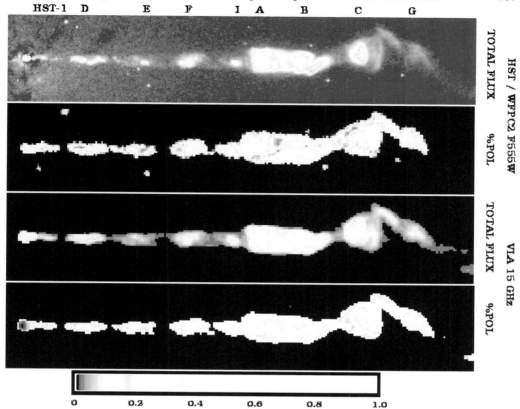

FIGURE 4. False-color representations of the total intensity and polarization of the M87 Jet in the optical (HST F555W, top two panels) and radio (VLA 14.5 GHz, bottom two panels). The HST observations were carried out in May 1995, while the VLA observations were done in February 1994. All maps were rotated so that the jet is along the x-axis, and are convolved to 0.23″ resolution (from Perlman et al. 1999).

There are, generically, two proposed origins for the jet power: the central object (black hole, neutron star or protostar) and the accretion disk (e.g. Blandford *et al.*1990). In both cases, the energy derives from differential rotation. For example in the case of a Keplerian disk that extends down to the surface of a non-rotating, unmagnetized star, as much energy is released in the boundary layer as in the disk. An extreme case is presented by the Crab (Weisskopf *et al.*2000) and Vela (Helfand, Gotthelf & Halpern 2001) pulsars which exhibit prominent jets without there being any accretion disk, presumably.

One of the best observational approaches to investigate the mechanisms which produce jets is to determine the jet composition at radii where they can be observed directly. In the case of black-hole jets, the plasma is likely to be electron-ion if the jet originates from a disk, or electron-positron if it derives from the black hole. Polarization observations have been prominent in attempts to distinguish between these two possibilities. Most contemporary explanations of the collimation invoke strong magnetic fields, though in most cases, the argument for magnetic collimation is a theoretical one, based upon eliminating the alternatives. One exception to this is the bipolar outflow associated with young stellar objects (YSOs) where polarization observations strongly support the notion that the magnetic field is dynamically important (Akeson & Carlstrom 1999).

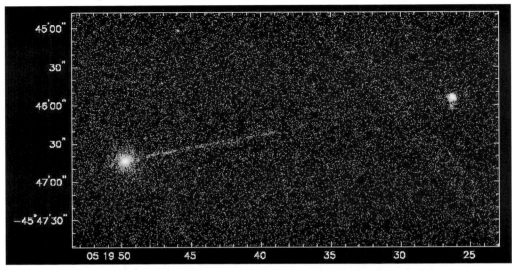

FIGURE 5. Gray-scale representation of the full-resolution Chandra image of the nucleus, jet and western hot-spot of Pictor A (from Wilson *et al.*2001).

2.2. *Observations*

The first jet observed was that in the nucleus of the elliptical galaxy M87 in the Virgo cluster. The modern representation is given in Fig. (4) (Perlman *et al.*1999). In this case, the jet emerges from no more than 100 times the gravitational radius of the central hole ($M = 3 \times 10^9$ M$_\odot$; $GM/c^2 = 4 \times 10^{14}$ cm) and propagates radially outward for a distance $\sim 3 \times 10^{22}$ cm, seven orders of magnitude larger. The jet, however, is neither homogeneous nor smooth. Superimposed upon an overall decrease in surface brightness as the jets expand away from the central hole are strong side to side variations and bright features. These may reflect a time-dependence at the jet origin or independent, local instabilities. At small radii, the M87 jet is strongly one-sided, and this is generally attributed to Doppler beaming and adduced as evidence for relativistic outflow.

Another good example of an extragalactic jet is Pictor A, which has recently been observed at X-ray energies using Chandra (Fig. (5)); Wilson, Young & Shopbell, 2001). It shows strikingly efficient collimation and powerful emission from the western hot spot, which is also prominent at optical and radio wavelengths. The Galactic source GRS 1915+105 shows mild superluminal motion which is interpreted in terms of moving features, probably internal shocks, with space velocities $\sim 0.9c$ (e.g. Mirabel & Rodriguez 1994). This is mild compared with most extragalactic compact jets where the speeds are more typically $\sim 0.99\,c$, corresponding to bulk Lorentz factors $\gamma \sim 7$. Indeed, there are observational indications that much larger bulk Lorentz factors are produced in extragalactic jets. Independent evidence that these jets are relativistic comes from γ-ray observations, principally done with the EGRET detector on Compton Gamma Ray Observatory. These showed that those jets that are beamed towards us, and which are collectively known as blazars, are often powerful γ-ray sources. The total electromagnetic spectrum of blazars (and similar sources) comprises a broad band synchrotron radiation spectrum extending from low radio frequencies to an upper frequency between optical and X-ray wavelengths. The same electrons are responsible for an inverse Compton component that can extend up to TeV energies (with variability times as short as ~ 30 min). (We can only observe TeV emission from relatively local sources because TeV photons from cosmologically distant sources will be absorbed through pair production on the intergalactic infrared background.) The fastest jets may well be associated with γ-ray bursts, if they are indeed beamed, for which speeds of $\sim 0.999995c$ have been inferred.

2.3. *Physical Processes*

2.3.1. *Synchrotron Radiation*

As described in Dr. Landi Degli'Innocenti's contribution (for more details see also Rybicki & Lightman 1979), the synchrotron power radiated by an ultra-relativistic electron with energy $\gamma m_e c^2$ in a field of strength B is given by

$$P = \frac{4}{3}\gamma^2 \sigma_T c U_B \,, \tag{2.20}$$

where $U_B = B^2/8\pi$ is the magnetic energy density and we have averaged over pitch angle. The corresponding radiative cooling time is

$$t_S = \frac{5 \times 10^8}{B^2 \gamma} \,\mathrm{s} \,, \tag{2.21}$$

where B is measured in Gauss. The characteristic frequency radiated is

$$\nu_c = \gamma^2 B \,\mathrm{MHz} \,. \tag{2.22}$$

The polarization of single particle emission varies from 2/3 for $\nu \ll \nu_c$ to 1 for $\nu \gg \nu_c$. The electric vector is perpendicular to the projected magnetic field direction. Averaging over a power law distribution of relativistic electrons, $dN/d\gamma = K\gamma^{-s}$, it can be easily shown that the observed intensity is

$$I_\nu \propto K B^{1+\alpha} \nu^{-\alpha} \,, \tag{2.23}$$

where the spectral index $\alpha = (s-1)/2$. The net degree of linear polarization is

$$\frac{Q}{I} = \frac{s+1}{s+7/3} \,. \tag{2.24}$$

Synchrotron radiation is also naturally circular polarized to an extent

$$\frac{V}{I} \sim 3/\gamma \,, \tag{2.25}$$

dependent upon the detailed angular distribution function and the viewing angle.

An important issue for what follows is the viability of a synchrotron maser. This can be shown to be impossible for ultra-relativistic emission in vacuo. Essentially, in order for a maser to operate, it is necessary that there be a population inversion and that the emissivity at a given frequency decreases sufficiently rapidly with increasing energy and this does not happen with regular synchrotron emission. (It can however arise when relativistic electrons emit synchrotron radiation in a plasma, although in practice the conditions for this to occur are rather restrictive. This is known as the Razin effect.)

Of more relevance is what happens when the electrons are no more than mildly relativistic. The emission is then confined to a series of harmonics of the fundamental gyro frequency ω_G/γ. As the central frequency of a harmonic decreases with increasing energy, there are frequencies and directions where maser action is possible, given a population inversion. Cyclotron masers are likely to be highly polarized. The polarization from an electron-ion plasma will be elliptical; that from an equal pair plasma will be purely linear.

When the brightness temperature $T_B = I_\nu c^2/2k\nu^2$ of a synchrotron source approaches the kinetic temperature of the emitting electrons, $T_k \sim \gamma m_e c^2/3k \propto (\nu/B)^{1/2}$, the radiation will be absorbed. The optically thick radiation from a source will have a brightness temperature that is limited to this value and so the optically thick equivalent of Eq. (2.23) is

$$I_\nu \propto \nu^{5/2} B^{-1/2} .$$

(2.26)

When we consider the linear polarization of a self-absorbed source, we observe that the electrons emitting in the field-perpendicular polarization at a given frequency will have slightly lower energies than those emitting in the field-parallel polarization. Therefore the brightness temperature of the field parallel emission will be slightly larger than that of the field perpendicular emission. The degree of linear polarization from a power-law relativistic electron distribution function can be computed to be

$$\frac{Q}{I} = \frac{-3}{6\,s + 13} .$$

(2.27)

An example is the supermassive black hole candidate in the Galactic center, Sagittarius A*, which has no linear polarization up to frequencies of 86 GHz (cf Hildebrand, these proceedings), but shows surprisingly strong circular polarization (e.g. Bower 2000). The degree of CP increases sharply with frequency. Whereas both advection-dominated-accretion-flow (ADAF) models and accretion-disk-powered-jet models can account for the spectrum from centimeter wavelengths to X-rays of Sagittarius A*, they have distinct polarization characteristics which might in the near future be able to distinguish between the two models.

2.3.2. *Inverse Compton Scattering*

Inverse Compton (or more properly Thomson) scattering, in which a highly energetic particle transfers momentum to a low-energy photon, is very similar to synchrotron radiation. If the radiation field is isotropic, then the power is given by Eq. (2.20) with the magnetic energy density U_B replaced by the radiation energy density, $U_{\rm rad}$. As photons are conserved in Thomson scattering, the mean photon frequency is boosted by an average factor

$$\frac{\nu'}{\nu} = \frac{4}{3}\gamma^2 ,$$

(2.28)

where ν' is the scattered frequency and ν is the incident frequency. One power of γ arises from the Lorentz transformation into the electron rest frame; the second comes from the scattering back into the original frame. The polarization observed will be generically be $\sim 1/\gamma$ unless the incident radiation field is both highly anisotropic and polarized, in which case a strongly polarized scattered spectrum can be emitted. This can arise when, for example, radiation is scattered into a beam by a warped disk. Similarly, circular polarization in the incident radiation will be partly retained in the scattered radiation.

2.3.3. *Inverse Compton Limit*

The comparison of synchrotron radiation and inverse Compton scattering leads to what, for historical reasons, is called the inverse Compton limit (Kellermann & Pauliny-Toth 1969). The way the argument is traditionally expressed is that the ratio of the Compton power radiated by an electron to the synchrotron power can be written as

$$\frac{L_{C^{-1}}}{L_S} = \frac{U_S}{U_B} \propto \frac{\nu_S^3 T_B}{B^2} \propto T^5 \nu_S ,$$

(2.29)

in obvious notation, and where we have assumed that the source is self-absorbed at the observing frequency. If we set $\nu_S \sim 1 - 10$ GHz, then the brightness temperature is limited to $T_B \sim 2 \times 10^{12}$ K if this ratio is not to exceed unity.

The original concern was that if the ratio did exceed unity then the second order

Compton scattering would be even greater than the first order scattering and so on. Of course this can't go on for too many orders because the Klein-Nishina limit will limit the scattering. Nowadays we think we can identify the synchrotron and the inverse Compton components and so we know their ratio and can deduce the source brightness temperature which is quite insensitive to its value.

2.3.4. *Kinematics of Bulk Relativistic Motion*

In the case of a relativistic jet, it is often easier to compute the radiation spectrum in the comoving (primed) frame of the emitting plasma and then perform a Lorentz boost into the (unprimed) frame of the observer. The frequency will be boosted by the Doppler factor (e.g. Blandford & König 1979),

$$\delta = \frac{\nu'}{\nu} = \frac{1}{\gamma(1 - \beta\cos\theta)}, \qquad (2.30)$$

where θ is the scattering angle in the observer frame and $\beta = v/c$ is the bulk velocity of the plasma in the jet. Note that for a jet beamed toward us with $\theta < \gamma^{-1}$, the Doppler factor is $\delta \sim \gamma$. Note also that the rate of change of observer time $t_{\rm obs}$ to proper time τ satisfies

$$\frac{d\tau}{dt_{\rm obs}} = \delta. \qquad (2.31)$$

As a consequence, the observed transverse speed of a feature moving with the jet speed is given by

$$\beta_{\rm obs} = \gamma\beta\sin\theta\frac{d\tau}{dt_{\rm obs}} = \frac{\beta\sin\theta}{1 - \beta\cos\theta}. \qquad (2.32)$$

This has a maximum value $\gamma\beta$ for $\theta = \cos^{-1}\beta$ and consequently the expansion can be "superluminal" when $\beta > 0.71$. The kinematics of real jets is undoubtedly more complex and the space motion of shock features must be distinguished from the speed of the emitting plasma. Frequently, observers make the approximate identification $\beta_{\rm obs} \sim \theta^{-1} \sim \delta \sim (dt/dt_{\rm obs}) \sim \gamma$ when interpreting measurements of compact extragalactic radio sources.

The behavior of polarization under a Lorentz boost is straightforward. The k-vector swings forward along the direction of motion making an angle θ' with the direction of motion in the plasma rest frame and an angle θ in the jet frame, where $\sin\theta' = \delta\sin\theta$. If we imagine the k-vector as being rotated in this manner, then the electric and magnetic fields associated with individual photons will be similarly rotated about a direction $\vec{k} \times \vec{B}$ so that $\vec{k}$, $\vec{E}$ and $\vec{B}$ continue to form an orthogonal triad (Fig. (6)).

Because the Planck distribution function is Lorentz scalar, and the brightness temperature enters only in the ratio ν/T, it is clear that the brightness temperature must transform in the same manner as frequency. This implies that if the inverse Compton limit is applied in the observer frame, the brightness temperature measured by a radio astronomer can be as high as $\sim 2 \times 10^{12}\,\delta$ K. This is particularly germane at this time because radio astronomers are able to estimate these brightness temperatures, both directly using ground and orbiting VLBI, and indirectly by carefully analyzing refractive interstellar scintillation. Using the limits on the source size derived from these observations and its relation to the source flux-density (i.e. the Rayleigh-Jeans equation), one can derive an apparent surface brightness temperature and therefore the value of δ, assuming the intrinsic brightness temperature cannot exceed the inverse Compton limit significantly. The values that are found require bulk Lorentz factors $\gamma \sim \delta \sim 30$ and may be even higher.

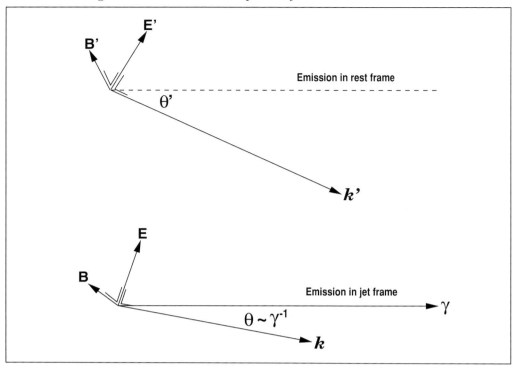

FIGURE 6. The Lorentz boost of polarization.

2.3.5. *Faraday Conversion*

We discussed Faraday rotation in a cold, non-relativistic plasma in the last section. We must now consider what happens in an ultra-relativistic pair plasma. On symmetry grounds, the eigenmodes must be linearly polarized and are usually labeled ordinary, where the electric vector along the direction $\vec{k} \cdot \vec{B}$, and extraordinary, where it is not. If a linearly polarized wave, obliquely polarized with respect to the magnetic field is incident upon the plasma, then it can be decomposed into ordinary and extraordinary modes that will propagate with slightly different phase velocities. In this manner, circular polarization will be created. In other words, there is a conversion of U to V. The sense of circular polarization will be given by the sign of $(\vec{e} \cdot \vec{B})(\vec{e} \cdot \vec{k} \times \vec{B})$ and so in order to have a measurable circular polarization from a cosmic source, it is necessary to have a preferred field orientation. This phenomenon is known as Faraday conversion (e.g. Jones & O'Dell 1977).

Because Faraday conversion is caused by the lowest energy relativistic electrons, it can serve as a probe of the low energy end of the electron energy distribution. Faraday conversion is furthermore proportional to $e^2 B^2/m^2$, i.e. independent from the sign of the particle's charge. An equal mixture of electrons and positrons can therefore produce Faraday conversion, but not rotation. A comparison of linear and circular polarization might therefore probe the constituents of the jet (e.g. electron-positron versus electron-ion pairs).

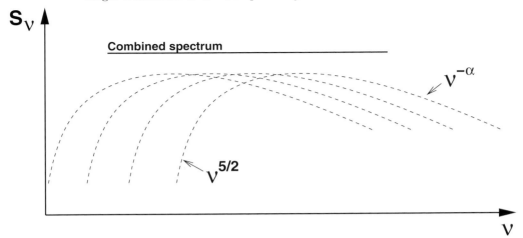

FIGURE 7. A superposition of self-absorbed sources peaking at different frequencies can create a combined 'flat' spectrum.

2.4. *Interpretation*

2.4.1. *Shocks and the Integrated Spectrum*

The emitting element in powerful synchrotron jets is thought to be a relativistic shock wave and a typical source will comprise the emission from several of them. In the limit, a jet can be thought of as accelerating relativistic electrons over its length with a field strength that diminishes with radius. The total flux density observed at a given frequency is dominated by the emission from the radio photosphere, where the optical depth is roughly unity. Blazars and similar sources generally have "flat" spectra, i.e. $-0.5 < \alpha < 0.5$ and can be interpreted as the superposition of a series of self-absorbed sources at successive radii, each peaking at successively lower frequencies (Fig. (7)).

2.4.2. *Jet Composition*

On the basis of the fraction of the jet energy per relativistic electron responsible for the synchrotron radio emission that we observe to be radiated on an outflow timescale, it has been tentatively deduced that jets cannot carry protonic "baggage" and so must comprise electron-positron pairs with a low energy cutoff in the distribution function (e.g. Reynolds *et al.*1996).

Wardle *et al.*(1998) measure a large degree of circular polarization in 3C279, a source where the linear polarization is also quite high, and so there cannot be too much normal Faraday rotation. In order to explain the circular polarization, they have to invoke a large population of mildly relativistic electrons and positrons. On this basis they conclude that relativistic jets comprise pair plasmas at least at the radii where they are directly observed.

2.4.3. *Coherent Emission Mechanisms*

All of this calls into question the fundamental synchrotron hypothesis. If the deduced jet powers are unreasonably large then we should certainly be prepared to consider the possibility that the radio emission, or at least its compact and variable part, may be due to a coherent emission mechanism. This is not unreasonable. After all, the sun and Jupiter support high brightness coherent emission under far more docile conditions. Furthermore, a shock front is a very natural environment in which strong and unstable currents are likely to be induced and these are commonly observed to radiate coherently in plasmas. Probably the most likely possibility is coherent cyclotron emission emerging from very much more compact regions than under the jet hypothesis. This requires the magnetic field strength to be hundreds or even thousands of Gauss and so the pressures must be much larger than in the jets. Coherent cyclotron emission is likely to be strongly circularly polarized, unless there are equal numbers of electrons and positrons.

2.4.4. *Microlensing and Refractive Scintillation*

Both microlensing by compact objects in the line-of-sight and refractive scintillation by density fluctuations in the Galactic ionized medium can introduce non-intrinsic variability of the compact structures in jets of extra-galactic radio sources (e.g. knots or shock fronts). For significant variability to occur, these structures must have an angular size of the order the Fresnel scale for scintillation or Einstein radius in the case of microlensing (e.g. Koopmans & de Bruyn 2000). Both scales are typically several micro-arcseconds. The expected time-scale of variability is determined by the transverse velocity of the source compared to the scattering medium (i.e. the compact objects or the Galactic ionized ISM) and is typically hours to weeks in the case of refractive scintillation or weeks to months in the case of microlensing.

Whereas polarization is typically little affected by either scintillation or microlensing (which retains polarization angle), in both cases only the most compact source structures vary significantly. If these structures have different polarization degrees or angles compared with the flux-density weighted average over the source, the net results will be polarization variability that strongly correlates with the non-intrinsic source variations. A correlation between changes in polarization and non-intrinsic flux variations of extra-galactic radio sources (for example in intra-day variables (IDVs), which strongly scintillate) could therefore provide information on the polarization properties of the most compact micro-arcsecond scale jet structures, which are impossible to observe directly in any other way.

There are several other ways that the Galactic ionized ISM can introduce non-intrinsic variations in the polarization of radio sources (including radio jets), or even induce circular polarization. The simplest case is that of extreme scattering events (ESEs), where large localized overdensities in the Galactic ionized medium move into the line-of-sight to the radio source. The enhanced electron column density increases the Faraday rotation and could result in an observable change in the polarization angle of the radio source, as well as in a change of its flux-density due to refractive lensing. Besides Faraday conversion, which converts linear polarization to circular polarization, if a strong gradient in the rotation measure over the source exist, the scintillation patterns of left and right-hand polarized wavefronts will be slightly displaced. This will introduce a time-variable circular polarization that is strongly correlated with the scintillation-induced flux-density variations (Macquart & Melrose 2000), but independent of the degree of linear polarization (which is not the case for Faraday conversion).

2.4.5. *Inverse Compton Scattering*

A major concern in interpreting the inverse Compton X-ray and γ-ray observations of blazars is the source of the incident photons. In the lower power objects that can be observed at TeV energies, these are thought to be synchrotron photons emitted locally within the jet. In the higher power quasars where there is a powerful photoionizing continuum as well as a relativistic jet coming towards us, most of the incident photons are thought to originate from the disk and to be scattered into the jet. However, this is not certain and X-ray (or γ-ray) polarimetry could validate this, because these scattered X-rays of external origin would be highly polarized.

2.5. *Summary*

- Jets appear to be a common and perhaps even a necessary feature of accreting or possibly even just simply rotating systems.
- Despite much observational progress, the fundamental questions concerning the origin, composition and collimation of jets remain unanswered.
- Black hole jets are formed ultra-relativistically as required to account for their superluminal motion, high radio brightness temperature and prodigious γ-ray emission.
- Radio polarimetry of relativistic jets is helping us to deduce their composition although coherent emission mechanisms cannot be ruled out.
- X-ray and γ-ray polarimetry could help refine models of relativistic jets by probing jets close to their origins.

3. Outflows

3.1. *Motivation*

Somewhat paradoxically, accreting systems commonly exhibit outflows. The reason for this behavior is simple. As gas accretes onto a compact object it must release its gravitational binding energy. When this is possible, it will do so by radiating. However, this may not be possible when the gas accretes much faster than the Eddington rate, $\dot{M}_{\rm Edd} = 4\pi GM/\kappa_T c$, the photons will be trapped by electron scattering and the energy can only be carried off by a bulk outflow. Even if the radiation is not trapped, then it can still drive an outflow if it encounters gas with an opacity $\kappa \gg \kappa_T$, for example with dust grains or resonance lines. Either, or more likely both, of these processes are believed to drive the outflows associated with Broad Absorption Line Quasars (BALQs) and this lecture will be primarily about these objects, although the principles involved are more generally applicable. The broad absorption lines by which these quasars are distinguished are associated with the ultraviolet resonance lines of the common ions. (Recall that quasars are mostly at high redshift and so these lines are observed in the visible.) They show absorption troughs, extending to the blue of the regular, broad emission lines by which quasars are identified spectroscopically. This is just what is also seen in the star P Cygni, though the BALQ relative velocities, $\sim 0.1c$, which are what one might escape for gas escaping from the vicinity of a black hole, are much larger than those encountered in P Cygni (Weymann *et al.*1991).

However, we do not understand the flow of gas around the black hole and the location of the emission line and the absorption line gas is still quite controversial. Models of broad line clouds have been constructed in which the gas flows "in, out, round or about" and there may be elements of each of these four kinematic classes in real sources. It is probably safe to conclude that the flow is not very simple; otherwise we would have already understood it using a technique called "reverberation mapping". In this technique it is supposed that the lines vary in direct response to the photoionizing continuum. By monitoring them both, it is possible to construct a Greens function response of the emission line gas, This can then be compared with the predictions of simple kinematic models. It has been possible to use this technique to locate the gas in several instances. However, the details of the velocity field remain controversial.

Despite this, most authors have assumed that the outflow of the absorbing gas is roughly equatorial and radial, originating from close to the hole. This implies that the quasar will only be classified as a BALQ when the observer direction is also close to the

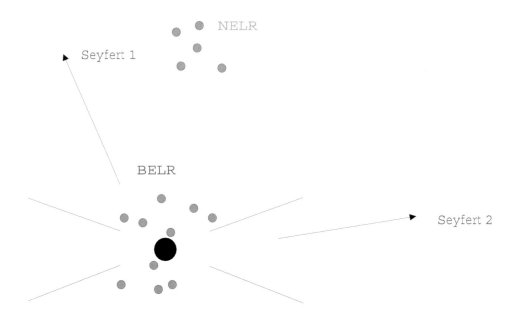

FIGURE 8. Orientation model for Type 1 and Type 2 Sefert galaxies. In the simplest scheme, only the narrow line region can be seen from a Setfert 2 galaxy; the broad line region lies behind a thick equatorial ring of obscuring gas and dust.

equatorial plane; otherwise she will see a regular radio-quiet quasar (Fig. (8)). The goal is to use polarization observations to see if this is truly the case (Antonucci 1993).

3.2. *Observation*

As we have remarked, a typical BALQ spectrum shows broad emission lines like CIV $\lambda1548$, accompanied by broad troughs extending to the short wavelength end of the spectrum with widths up to $\sim 30,000$ km s^{-1}. The emission lines are generally unpolarized, though both the semi-forbidden line CIII]$\lambda1909$ and Lyα can be polarized, as we shall discuss. The troughs which are caused by approaching gas, represent photons that are removed from the radiation field and scattered sideways. They also remind us that momentum is taken out of the radiation field so that the gas is accelerated (Arav *et al.*1995). (Of course this may not be the only accelerating force involved, though it is simplest to assume that it is.)

The absorption troughs themselves are not black which presumably means that gas moving along different directions is scattering radiation into our line of sight. In addition, although the troughs are relatively smooth, they do show velocity structure and a variable degree of polarization that can be as large as ~ 0.2 a characteristic feature of resonance scattering. However, resonance scattering is not the only means of producing linear polarization; electron scattering can do the job just as well and some observers have preferred this explanation.

In what follows, we shall confine our attention to resonance scattering for two reasons. Firstly, we observe resonance scattering directly and its contribution to the opacity is three to four orders of magnitude larger than that of electron scattering. Secondly, the physics is much more interesting than that of electron scattering! Furthermore we shall confine our attention to the polarization and the implications that it has for the kinematics, as opposed to the dynamics of the flow.

However, we must mention some additional observational clues as to the nature of BALQs. Firstly, BALQs are both radio- and X-ray-quiet. In some models this implies that there is a highly ionized region that can absorb the X-rays and transmit the ultraviolet radiation. In other interpretations, the X-rays are never emitted in the first place. Secondly the BAL phenomenon is pretty much confined to quasars; the lower power Seyfert galaxies do not exhibit these broad troughs. Thirdly, the continuum (after subtracting the galactic contribution) is fairly uniformly polarized with $p \sim 0.01 - 0.05$, suggesting electron scattering in a disk corona is responsible. This is presumably located inside the absorption line gas. As mentioned above, it is also possible that this same scattering occurs outside the absorption line gas in which case it would probably be responsible for the polarization of the absorption line troughs. Finally, a large polarization, increasing towards shorter wavelength, has been reported at wavelengths longward of the Ly edge in a few, regular quasars. (It should be emphasized, though, that these HST observations were very difficult to make and the results have been controversial.)

3.3. *Physical Processes*

3.3.1. *Resonance Transitions*

Let us first consider the levels of intermediate Z ions adopting the Russell-Saunders approximation. We can distinguish the Li-like ions, such as CIV, NV, OVI, the Be-like ions, such as CIII, NIV, OV and the B-like ions such as CII, NIII, OIV. The electronic state is determined by the principal quantum numbers for the valence electrons (n, ℓ), the total orbital and spin angular momenta (L, S) for the term and the total angular momentum J for the level which, in turn is divided into $2J+1$ sublevels. This information is encoded in the quantum mechanical designation of the level. For example, the ground state of OIV is $2s^2 2p\ ^2P^o_{1/2}$ which means that one of the three valence electrons is unpaired in a $2p$ level with total orbital angular momentum $L = 1$ (i.e. a P state) and total spin $S = 1/2$, hence the superscript $2S + 1 = 2$. There are two possible choices for J and the lower energy one, by Hund's rule, has $J = 1/2$ as designated in the final subscript. (The final superscript, o, indicates an odd parity, which must change under a permitted (electric dipole) transition.) The electric dipole selection rules are that $\Delta \ell = \pm 1$; $\Delta S = 0$; $\Delta L = 0, \pm 1$; $\Delta J = 0, \pm 1$, (except that an $J = 0 \rightarrow 0$ transition is forbidden); and $\Delta M = 0, \pm 1$, (except that a $M = 0 \rightarrow 0$ transition is forbidden).

3.3.2. *Singlets*

Now consider a singlet transition such as the Be-like CIIIλ977. The ground state has $J = 0$ and only one sublevel with $M = 0$; the excited state has $J = 1$ and only permitted transitions to $M = \pm 1$ need be considered. The radiation pattern depends only upon the angular parts of that wavefunction, through the Wigner-Eckart theorem. When the scattering angle $\theta = \pi/2$, the degree of linear polarization can be computed to be $p(\pi/2) = 1$. For a general scattering angle,

$$p(\theta) = \frac{p(\pi/2) \sin^2 \theta}{1 + p(\pi/2) \cos^2 \theta}, \tag{3.33}$$

just as for Thomson scattering. This formula is generally true for electric dipole transitions and so all we need compute is $p(\pi/2)$.

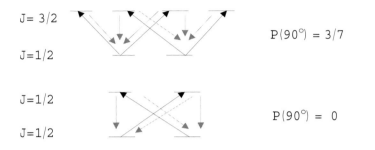

DOUBLET J=1/2 to 1/2, 3/2; eg CIV 1558, Ly α1216

$P(90°) = 3/7$

$P(90°) = 0$

FIGURE 9. Atomic transitions that contribute to the overall polarization for a typical double transition. If the doublet is resolved then the individual components have perpendicular polarizations of 3/7 and 0. If it is not, then the average polarization is 3/11

3.3.3. *Doublets*

The next most complicated case is the Li-like doublet transition, eg CIVλ1550. Here the ground level has $J = 1/2$ and there are two possible excited levels of which the lower energy level is $J = 1/2$. All transitions between sublevels are permitted and the net polarization is $p(\pi/2) = 0$. The higher energy excited level, associated with the shorter wavelength transition, has $J = 3/2$ and averaging over all of the permitted transitions gives $p(\pi/2) = 3/7$. If we average over both transitions according to their statistical weights, then we end up with $p(\pi/2) = 3/11$ Fig. (9).

$Ly\alpha$, has a similar type of transition. Here the the wavelength separation of the two transitions is so small that the doublet will not be resolved and averaging over the two excited levels makes sense. However, in the case of CIV, the energy difference is larger than the likely thermal width, so that one could, for example, imagine continuum photons propagating out of an expanding flow, encountering the polarizing $J = 1/2 \rightarrow 3/2$ transition first and then encountering the $J = 1/2 \rightarrow 1/2$ transition which erases all of this polarization. It is clear that the polarization is sensitive to the nature of BALQ outflows.

3.3.4. *Triplets*

The next simplest case is the B-like triplet transitions such as CII λ1335. In this case, there are two choices for the ground state $J = 1/2, 3/2$ and two for the excited state, $J = 3/2, 5/2$; the selection rules forbid direct transitions with $J = 1/2 \rightarrow 5/2$. The energy difference between the two ground levels is small enough that they should be equally populated by collisions in the absence of radiative transitions.

It is helpful to introduce the ionization parameter U which is the ratio of the number density of hydrogen ionizing photons to the electron density, designated n. When the radiative excitation rate, $\sim 10^4 U n_{10}$ s^{-1}, exceeds the collisional excitation rate $\sim 600 n_{10}$ s^{-1}, the ground state sublevels will be populated in an unequal fashion that must be computed by solving for all the transitions. (Note that both of these rates are likely to be much less than the spontaneous, de-excitation rate and so collisional de-excitation of the excited states is generally thought not to be an issue for permitted lines under AGN conditions.) Under these conditions, when the ground sublevels are radiatively mixed, the resulting polarizations will differ from the values computed assuming as statistical population of the ground sub-levels due to collisions. In this case, the polarization is increased from $p(\pi/2) = 0.21$ to $p(\pi/2) = 0.38$. This increase in predicted polarization is typical.

3.3.5. *Supermultiplets*

The next level of complication arises when an ion in a single ground state can be excited into several different excited states under radiative mixing conditions. In order to solve for the population of the different sublevels and the polarization, we must consider these distinct multiplets together.

3.3.6. *Magnetic Mixing*

A final variation, which is quite likely to be relevant in an AGN, arises when the magnetic field is strong enough that the cyclotron frequency, $\omega_G = 1.8 \times 10^7 (B/1\mathrm{G})$ rad s^{-1} exceeds the radiative excitation rate. In this case, the relevant eigenstates are referred to the magnetic field direction rather than the normal to the scattering plane and density matrices have to be used to attack the problem in general. This is also known as the Hanle effect and is discussed at greater length here in the solar context by Dr. Stenflo. An important consideration is the degree of Faraday polarization. The rotation of the plane of polarization is given by

$$\Delta\Phi \sim 4 \times 10^{-5} \left(\frac{N}{10^{20}\,\mathrm{cm}^{-2}} \right) \left(\frac{B}{1\,\mathrm{G}} \right) \left(\frac{\lambda}{1000\,\text{Å}} \right)^2 . \tag{3.34}$$

This is unlikely to be a factor at ultraviolet wavelengths, but could be significant in the optical.

3.4. *Interpretation*

3.4.1. *Emission Line Clouds*

There is a standard model of the emission line gas based upon the notion of an emission line cloud, a stratified slab of gas of size $\sim 10^{13}$ cm, located at a radius $R \sim 0.3$ pc from the continuum source with a density $\sim 10^{10}$ cm^{-3}. The ionization state of the gas is determined by the relative importance of photoionization and recombination which is, in turn, controlled by the ionization parameter, U. Typically this is $U \sim 0.1$. These clouds have a photoionization temperature $T \sim 10,000$ K and an equivalent sound speed ~ 10 km s^{-1}. However they are moving with speed $\sim 10,000$ km s^{-1} and Mach number $M = 1000$ through a hotter and more tenuous confining medium. This is patently absurd! Nonetheless, this model does provide a good representation of the ratios of the observed line strengths. What is clearly required is a convincing dynamical model that retains the successful features of the atomic astrophysics. For the moment, we just consider the polarization in the context of the cloud model.

If we consider a CIV $\lambda 1550$ photon propagating out of a "standard" cloud, the optical depth for solar abundance of carbon, mostly in a triply ionized state, is $\sim 10^5$. Under these conditions, photons do not diffuse spatially out of the cloud as might, at first, be guessed. Instead, they undergo a random walk in frequency and escape when they migrate into the wings of the line where the cloud becomes transparent. The polarization really has to be computed using a Monte Carlo simulation and, under these conditions of high optical depth, it is not surprising that it is too small to be measured, even when there is velocity shear giving an anisotropic escape probability. The one conspicuous exception is the semi-forbidden (intercombination line violating the selection rule $\Delta S = 0$) CIII] line $\lambda 1909$. Here the optical depth is closer to 10 and substantial linear polarization $p \sim 0.01$, was predicted and indeed has been reported in this line (Lee 1994, Cohen *et al.*1995, but see Ogle *et al.*1999).

There is a rather different story when external photons are scattered by the emission line clouds. Here we expect a high albedo and roughly half the photons will undergo just one scattering (Korista & Ferland 1998). If the distribution of scatterers is anisotropic, then we might also expect to detect a linear polarization signal in permitted lines. This is not usually seen, which suggests that a particular line of sight contains at most one cloud at a given wavelength. This is a strong constraint upon models of the velocity distribution.

3.4.2. *Rayleigh Scattering by Lyα*

Hydrogen is the most abundant element so scattering of Lyα photons is likely to occur far into the wings of the line. The fine structure level splitting between $2P_{3/2}$ and $2P_{1/2}$ is quite small for this transition and off-resonance scattering is characterized by the (classical) Rayleigh scattering phase function (e.g. Stenflo 1980). This process is called Rayleigh scattering when the initial and final states of the atom or ion are identical.(This distinguishes it from Raman scattering which arises when they are different.) Rayleigh scattering has been clearly observed in symbiotic stars (Nussbaumer, Schmid, & Vogel 1989). For large velocity shifts ΔV, the scattering optical depth exceeds unity for

$$\frac{\Delta V}{10^4 \text{ km s}^{-1}} \simeq \left[\frac{N_{HI}}{3 \times 10^{22} \text{ cm}^{-2}} \right]^{1/2}, \qquad (3.35)$$

(Lee & Blandford 2000).

Of particular interest is the case of Lyα, because in the damping wings the scattering phase function becomes that of the classical Rayleigh function, which enhances polarization. The fundamental reason why this is the case is that near the line center, where the optical depth is large, most photons migrate in frequency space faster than they do in real space. They are therefore comparatively insensitive to the cloud shape and large scale velocity shear. However, in the wings of the line, the optical depth is much smaller and a large scale pattern in the cloud shapes, for example, translates into a measurable linear polarization. (Effects like this have been reported in high column density supershells associated with starburst galaxies (Lee & Ahn 1998).)

If the accretion disk of a quasar is warped so that some part of it is shadowed from the direct exposure to the central engine or has a thickness that increases slower than linearly with radius, a large column density N_{HI} may exist in the shaded region and Rayleigh reflection of Lyα is expected. In this case we may expect up to 10 percent polarization in the Lyα wings, which is consistent with ~ 7 percent polarization reported from the radio-quiet quasar PG 1630+377 (Koratkar *et al.*1995)

3.4.3. *Absorption Line Clouds*

A somewhat analogous situation is found for the absorbing clouds observed directly in the BALQs. These are believed to be located outside the emission line region where the ionization parameter $U \sim 1$ and the size is estimated to be even smaller than the size of the emission line clouds $\sim (s/V)^2 R \sim 10^{11}$ cm. (In a quite different type of model, it has been proposed that the emitting and absorbing gas originates from very much smaller radii and forms part of a space-filling flow, Murray *et al.*1995. Many of the following considerations apply to this model as well.)

The actual kinematics of line formation can be quite complicated. This is generally handled under the Sobolev approximation (Rybicki & Hummer 1978). An incident continuum photon is scattered when it is resonant with a permitted transition taking place in the rest frame of the outflowing gas. The scattered line is redshifted in frequency by $V_{\parallel}/\lambda$. The surface occupied by gas resonant with a fixed observer frequency is called a Sobolev surface and can have a fairly convoluted shape. Some photons may encounter several Sobolev surfaces before escaping for good. The optical depth to absorption by ion X through a single Sobolev surface where the parallel velocity varies monotonically is

$$\tau = \int ds\, n_X \sigma = \left|\frac{dV_{\parallel}}{ds}\right|^{-1} \int dV_{\parallel} n_X \sigma = \left|\frac{dV_{\parallel}}{ds}\right|^{-1} \frac{c}{\nu} n_X \int d\nu\, \sigma(\nu)\,. \tag{3.36}$$

Substituting numerical values,

$$\tau = 0.3 X_{-4} f \frac{\lambda}{1000\text{\AA}} \frac{dN_{20}}{dV_9}\,, \tag{3.37}$$

where $X = 10^{-4} X_{-4}$ is the abundance of the ion, f is the oscillator strength and dN_{20}/dV_9 is the hydrogen column density (in units of 10^{20} cm^{-2}) per unit velocity (in units of 10,000 km s^{-1}). Resonance line scattering by the common ions occurs at column densities about three orders of magnitude smaller than those required for electron scattering (Lee & Blandford 1997).

We can compute the polarization expected from a particular model using Monte Carlo simulations. It turns out to be possible to give factor 2 estimates for the polarization, by multiplying expressions which describe the most important factors in producing the integrated polarization,

$$p \sim p(\pi/2) D(\tau) A G\,, \tag{3.38}$$

In this equation, $D(\tau)$ is a depolarizing factor that takes into account multiple scattering. It is typically roughly fit by an exponential $D = \exp(-\tau/b)$. A is a factor that takes into account the anisotropy in the escape probability. If the flow is uniformly expanding, it will be very difficult for the photons to escape in the radial direction and far easier for them to escape tangentially. This roughly doubles the polarization. Conversely, if the outflow is in the form of a jet, then small scattering angles will be favored with lower net polarization. Finally G is a geometrical factor which is supposed to account for the observed gas distribution. If the outflow is confined to an equatorial fan subtending a solid angle $\Delta\Omega$, then $G \sim \Delta\Omega/3$ and the electric vector will lie parallel to the projected symmetry axis. For a jet, the polarization will be perpendicular to the axis (which should be coincident with the radio axis). More extensive observations than have been possible so far will be needed to test the hypothesis that the lines are due to resonance scattering, through the dependence on atomic type (through $p(\pi/2)$), and to decide upon the flow geometry. On this basis, the strongly polarized lines are expected to be HeIλ584, OVλ630, Ne Iλ736, NIVλ765, CIIIλ977, SiIIIλ1206, AlIIIλ1671 and MgIλ2852. The modestly polarized lines include CIIλ687, OIVλ789 and NIIIλ991 and weakly polarized lines include OVIλ1034,

HIλ1216, NVλ1240, SiIVλ1396, CIVλ1549 and MgIIλ2798. CIIλ858, NIIIλ764, and OIVλ609 are expected to be unpolarized and can be used to measure the amount of electron scattering.

3.4.4. *Polarization beyond the Lyman Edge*

Observations of the rest ultraviolet continuum from a few high redshift quasars have shown a strong polarization increasing irregularly to shorter wavelengths shortward of the Lyman continuum (Koratkar *et al.*1998). This may be as large as $p \sim 0.2$, although the observations were extremely difficult and are consequently a bit uncertain. One possible explanation is that there are several highly polarizing singlet lines in this region, like HeIλ584, OVλ630, NeIλ736, NIVλ765, or CIIIλ977. If the outflow speeds associated with these relatively high ionization lines are large, $V \sim 0.1c$, then it is possible that the lines could overlap enough to give an apparent continuum polarization. It would be good to have the capability to repeat these observations.

3.5. *Summary*

• Spectropolarimetry provides a powerful diagnostic of the disposition of the broad emission and absorption line gas in quasars.

• Resonance scattering should be variably polarized with the degree and direction dictated by fundamental considerations of atomic astrophysics and the flow geometry. The radiation observed in the troughs may be the scattered photons removed from other lines of sight.

• By contrast, the optical continuum exhibits a fairly constant polarization, suggestive of electron scattering.

• Emission lines are generally unpolarized, excepting the semi-forbidden line CIII]λ1909. This is consistent with the standard cloud model of emission line formation. BAL troughs are variably polarized.

• Large polarization rising with frequency has been reported to the blue of the Lyman continuum. This may be due to blends of strongly polarized, prominent singlet transitions

4. Neutron Stars

4.1. *Motivation*

Although physicists and astronomers (most famously Baade and Zwicky) were quick to appreciate the possibility that $\sim 10^{57}$ neutrons could assemble to form a self-gravitating neutron star, it was not until the discovery of radio pulsars in 1967, that there was compelling evidence that they really existed. To date we have cataloged over a thousand radio pulsars, know of hundreds of accretion-powered neutron stars in X-ray binaries, and are starting to find isolated neutron stars accreting from the interstellar medium. In addition, five radio, or rotation-powered, pulsars are observed to pulse at optical wavelengths (Chakrabarty & Kaspi 1998), at least seven as γ-ray pulsars (Thompson 2000), and ~ 40 are detectable at X-ray energies (Becker, 2000).

For the astronomer neutron stars are the most common result of evolution of a massive star. However, far from being an endpoint, they represent a rebirth often in a more luminous state than the progenitor star. For the physicist, neutron stars provide a magnificent cosmic laboratory, allowing us to witness the behavior of cold nuclear matter at supranuclear densities, the indirect effects of extremely high T_C conductivity and superfluidity, and, as we shall see, the consequences of magnetic field strengths perhaps nine orders of magnitude greater than we can sustain on earth.

Neutron stars are also of special interest to the polarimetrist as they have already furnished the strongest and most rewardingly variable signals of any cosmic sources. It is possible to follow the change in the polarization (sometimes nearly completely polarized) through individual pulses from bright radio pulsars. It is also possible to study the average polarization properties of large samples of pulsars, viewed from a range of vantage points and, thereby build up a picture of the magnetic field geometry and try to determine the site of and the mechanism for their high brightness emission. Unfortunately, there is still no polarimetric capability at X-ray wavelengths where accreting neutron stars in binary systems emit most of their radiation. However, very strong linear polarization is anticipated and the details should be no less prescriptive of the emission.

A third class of object, in addition to the accretion- and rotation-powered pulsars that is of particular interest at the moment is the magnetar. It appears that a minority of neutron stars are formed with super-strong magnetic field $\sim 10^{14} - 10^{15}$ G. As predicted by Thomson and Duncan (1995), these magnetars decelerate very quickly but still have a larger reservoir of magnetic energy that can be tapped to power γ-ray bursts. These field strengths are well in excess of the quantum electrodynamical critical field $B_c \equiv m_e^2 c^3 / e\hbar = 4.4 \times 10^{13}$ G, where the cyclotron energy of an electron equals its rest mass. This, in principle, allows us to test the theory in a regime that is qualitatively quite different from that in which impressively high precision tests have already been made. (There is no real anxiety that the theory is suspect above the critical field but, as is the case with general relativity, there is a strong interest in performing the check.) Another class of X-ray source is the Anomalous X-ray Pulsars. These are possibly a late evolutionary phase of magnetars.

In this section, we shall discuss the expected polarimetric properties of all three types of sources, although there are only observations of rotation-powered pulsars.

4.2. *Observation*

4.2.1. *Rotation-Powered Pulsars*

Radio pulsars are spinning, magnetised neutron stars (e.g. Lyne & Smith 1998). The majority have spin period between 0.1s and 3s and surface magnetic field strengths $\sim 10^{12}$ G, estimated from the rate at which they appear to slow down. The neutron stars themselves appear to be mostly formed with masses quite close to the Chandrasekhar mass ~ 1.4 M$_\odot$. Their poorly measured radii are ~ 10 km, consistent with there having central densities a few times nuclear as the best models of the nuclear equation of state imply. (We really do not know the interior composition at all well. It could be mostly neutrons or contain a large fraction of protons, hyperons, pions or even free quarks. Accurate measurements of the radius along with the rate of cooling will provide important constraints on the equation of state of nuclear matter.) Other, impressive vital statistics of neutron stars include escape velocities $\sim 0.3c$, surface gravities $\sim 10^{14}$ cm s^{-2} and maximum spin frequencies (that are nearly attained in observed objects) ~ 1 kHz. The radio emission from pulsars has extremely high brightness temperatures which can, by some estimates, exceed $\sim 10^{30}$ K.

The integrated pulse profiles of radio pulsars frequently show one, two or three pulses. This, and other observational evidence, has been interpreted in terms of an emission model where there is a strong "core" beam of emission close to the magnetic axis surrounded by a weaker "cone" beam. When the observer latitude is similar to that of the magnetic axis, a single, dominant core component is seen. Increasing (or decreasing) the observer latitude leads to a three peaked, cone-core-cone pattern. When the observer is more inclined to the magnetic axis, only the two cone components will be seen. These pulse profiles exhibit strong, broad band linear polarization varying through the pulse. Values $p \sim 1$ are consistently measured at certain pulse phases in certain pulsars. The position angle swings regularly through the main pulse with a total swing that can be as high as $\sim 180°$. Strong circular polarization is also commonly measured near the center of the pulse, with the handedness often changing sign.

Individual pulses, whose polarimetric properties can be measured in the strongest pulsars, are no less interesting. They show individual emission units, known as subpulses, with durations typically a few degrees of pulsational phase. These can "drift" through the pulses appearing at progressively earlier or later phases in successive pulses. These subpulses often exist in one of two orthogonal polarization states. Even shorter timescale features known as microstructure (or now even nanostructure) has been well documented and these too can exhibit high, though complex, polarization properties.

The optical pulses from the Crab pulsar in the Crab Nebula have been particularly well studied. The pulse profile is cusp-like and the plane of polarization swings smoothly through $\sim 70°$, while the degree varies between $p = 0.1$ and 0.5. Another famous optical pulsar is associated with the Vela supernova remnant, shows similar strong, variable linear polarization.

4.3. *Physical Processes*

4.3.1. *Curvature Radiation*

The large measured brightness temperatures imply that the radio emission is produced by a coherent process. One widely discussed possibility is that the emission is some variant on coherent curvature emission whereby bunches of charged particles stream outward along the curving, roughly dipolar magnetic field lines from the star with ultrarelativistic speed (Lorentz factors γ of several hundred) and radiate like giant electrons. The emission properties are like those already summarized for synchrotron radiation, with the important difference that the radius of curvature of the orbit, R, is energy-independent. The characteristic emission frequency is $\omega \sim \gamma^3 c/R$, lying in the radio band for $R \sim 10 - 100$ km. The radiation from an individual bunch is beamed within an angle $\sim \gamma^{-1}$ to the direction of motion.

At a particular pulse phase, the observer sees emission from a curve though the magnetosphere where the line of sight is tangent to the magnetic field. There is thought to be a radius-to-frequency mapping so that the emission at a given frequency is concentrated over an interval of radius along this curve, and that this radius decreases with increasing frequency. The polarization from a tangent point will be quite strongly linearly polarized with electric vector parallel to the projected curvature vector on the sky. As the pulsar spins, this projected curvature vector will rotate on the sky and a characteristic swing of the plane of polarization will be produced. This is known as the rotating vector model. If we view an individual bunch from one side of its orbital plane then the other, we will see one sense of circular polarization followed by the opposite sense. This mechanism clearly has the ingredients to explain the radio polarization observations. However, it is a bit puzzling how a totally polarized pulse can be formed in this manner. One possible explanation is that the bunches are quite strongly flattened and they radiate most strongly perpendicular to their flattening plane, where the polarization will be most strongly linear.

4.3.2. *Maser Processes*

There has recently been a resurgence of interest in maser emission models. It is relatively easy to imagine that the necessary population inversion will develop in the outflowing plasma. A typical pulsar can develop an EMF of $\sim 10^{13} - 10^{16}$ V and a small fraction of this potential difference developing in a transient "gap" will create a fast stream of electrons and/or positrons that can stream through more slowly moving (though still ultrarelativistic) plasma. What is a bit harder is to find suitable wave-particle interactions that can lead to an overall negative absorption coefficient. As an illustration of the difficulty, note that synchrotron radiation *in vacuo* has an absorption coefficient

$$\kappa = \frac{1}{8\pi\nu^2} \int dN \frac{1}{\gamma^2} \frac{\partial}{\partial\gamma} \gamma^2 p_\nu(\nu, \gamma) . \tag{4.39}$$

The single particle emissivity, $p_\nu(\nu, \gamma)$ is proportional to $\nu^{1/3}\gamma^{-2/3}$ at low frequency and increases with energy at high frequency. Hence, synchrotron absorption is necessarily positive and maser action is precluded, independent of the particle distribution function. Similar conclusions have been drawn for other emission mechanisms in vacuum.

However this conclusion does not necessarily follow if there is a plasma present. One particularly interesting case is the so-called anomalous cyclotron (otherwise known as cyclotron-Cerenkov) resonance between a wave with angular frequency ω and wave vector $\vec{k}$ interacting with an electron (or positron) moving with velocity $\vec{v}$, in a field where the non-relativistic gyro frequency is ω_G. This occurs if

$$\omega - k_\parallel v_\parallel = -\omega_G/\gamma . \tag{4.40}$$

where $\parallel$ refers to the component along the magnetostatic field.

This equation needs some interpretation (Lyutikov, Blandford & Machabeli 1999). Consider, for simplicity, a circular polarized wave propagating along the magnetic field. A resonance satisfying the condition Eq. (4.40) clearly requires that the phase velocity of the wave be less than c. If we transform into the guiding-center frame of the electron, then the wave angular frequency changes sign indicating that it is propagating in the opposite direction along the magnetic field, with the same sense of circular polarization. This means that it resonates with particles gyrating in the opposite sense around the field (i.e. with opposite charge) than is the case in a regular cyclotron resonance. Put another way the particle outruns the wave so that, in the frame where the wave is at rest, the particle follows the electric vector as it spirals around the magnetic field. A consequence is that, in the electron guiding-center frame, a quantum of wave energy has negative energy. Therefore in exciting the electron to a higher state of gyration it emits a quantum and *vice versa*. (In the rest frame, population inversion now requires having more particles in the lower gyrational state!) Not surprisingly, this arrangement can lead to maser action in the outer magnetosphere and this has been proposed as a mechanism to produce the core emission. The modes are naturally circular polarized, although the handedness depends upon the details of the electron and positron distribution functions.

This is not the only possible way to have a maser process. There is a second resonance associated with the curvature drift motion v_{drift} of the gyrating electron as it moves along the curving magnetic field

$$\omega - k_\parallel v_\parallel = k_\perp v_{\text{drift}} . \tag{4.41}$$

This will be perpendicular to the curvature plane and will consequently produce emission with polarization orthogonal to that predicted by the rotating vector model. This mechanism has been invoked to account for the cone emission (Fig. (10)).

4.3.3. *Propagation Effects*

There is unfortunately a complication (Arons & Barnard 1986, Lyutikov *et al.*1999, Hirano & Gwinn 2001). The emitted radiation must propagate through the outer magnetosphere. This can lead to genuine absorption at the normal cyclotron resonance by more slowly moving electrons. Landau damping is also a possibility. Furthermore, there are refractive effects that may imprint additional polarization on the emergent radiation in much the same way as occurs in the ionosphere. There can be mode conversion, for

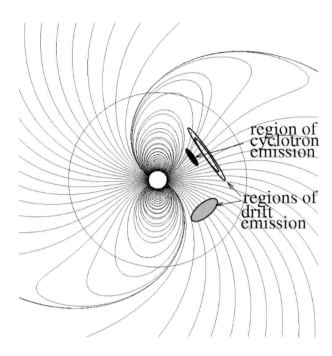

FIGURE 10. Location of core and conal emission regions in a maser model of pulsar emission (after Lyutikov *et al.*1999).

example from a subluminal ordinary mode to a propagating electromagnetic wave. Finally, and perhaps most interestingly from a physics perspective, there can be non-linear scattering effects (Lyutikov, 1998). At the high brightness temperature (or, equivalently, large occupancies of individual quantum mechanical states) found in pulsar radiation there will be a large amount of scattering between different radiation beams. These can be mediated by individual electrons, in which case the interaction is known as induced Compton scattering, or by collective wave modes of the plasma (known as Raman scattering when the scatterer is an electrostatic wave). There is not space to discuss these rather complex processes further, save to remark that, the associated matrix elements have a strong sensitivity to polarization and frequency and that this can be used to identify them. The theory is starting to match the observations in its richness!

Some of these propagation effects can also be relevant in the interstellar medium.

4.3.4. *Thomson Scattering in a Strong Magnetic Field*

The cross section for Thomson scattering must be changed if there is a strong magnetic field present (e.g. Mészáros 1992). In the limit when the wave angular frequency $\omega <<$ ω_G, or $E << 44(B/10^{12}\text{G})$ keV, the electrons are constrained to move along the magnetic field like beads on a wire. The dominant cross section is between polarization states with $\vec{k}, \vec{E}, \vec{B}$ coplanar. It is clearly given by

$$\frac{d\sigma}{d\Omega} = r_e^2 \sin^2\theta \sin^2\theta' \, , \tag{4.42}$$

where θ, θ' are the angles made by the incident and scattered wave vectors with the magnetostatic field.

4.3.5. *Inverse Compton Scattering in a Strong Magnetic Field*

This anisotropy in the cross section introduces an additional complication to inverse Compton radiation. This is because the incident photon propagates in a direction making an angle $\sim \gamma^{-1}$ to the magnetic field and so the total scattering cross section as given by Eq. 4.42 is reduced by a factor $\sim \gamma^{-2}$ from the Thomson value assuming that the frequency in the electron rest frame (ω') is less than ω_G. However, under these circumstances, we must also consider the effect of the "$\vec{E} \times \vec{B}$" drift of the electron perpendicular to the magnetic field. This will produce an oscillatory electron motion that is larger than the field parallel motion by a factor $\gamma \omega'/\omega_G$. The azimuth and incident polarization-averaged cross section will be given by

$$\frac{d\sigma}{d\Omega} = \frac{r_e^2}{4} \left(\frac{\omega}{\omega_G} \right)^2 (1 + \cos^2 \theta)(1 + \cos^2 \theta') , \qquad (4.43)$$

for incident frequency ω satisfying $\omega_G/\gamma\omega < \omega_G$. In practice, this leads to a rather complex polarization pattern.

4.3.6. *Quantum Electrodynamical Effects*

The virtual electron-positron plasma that comprises the QED vacuum affects the propagation of radiation through it. These are clearly likely to be of importance when the magnetic field strength approaches the critical field strength, B_c.

An external magnetic field causes photons of different polarizations to travel at slightly different speeds. The CP-invariance of electrodynamics tells us that the two modes must be linearly polarized. Specifically, the indices of refraction of both modes differ from unity and are given by (Heyl & Hernquist 1997a):

$$n_\perp = 1 + \frac{\alpha}{4\pi} \frac{8}{45} \left(\frac{B_\perp}{B_c} \right)^2 + \cdots , \qquad (4.44)$$

$$n_\parallel = 1 + \frac{\alpha}{4\pi} \frac{14}{45} \left(\frac{B_\perp}{B_c} \right)^2 + \cdots , \qquad (4.45)$$

for $\hbar\omega \ll m_e c^2$ and $B \ll B_c$. $B_\perp$ is the component of the magnetic field perpendicular to the propagation direction of the photon. A photon in the perpendicular polarization has its electric field vector perpendicular to the projection of the magnetic field into the transverse plane, and similarly for the parallel polarization.

For fields stronger than B_c, the index of refraction for the mode with the electric field perpendicular to the external magnetic field saturates at

$$n_\perp = 1 + \frac{\alpha}{4\pi} \frac{2}{3} \sin^2 \theta + \cdots , \qquad (4.46)$$

while the index for the other mode increases without limit

$$n_\parallel = 1 + \frac{\alpha}{4\pi} \frac{2}{3} \sin^2 \theta \frac{B}{B_c} + \cdots . \qquad (4.47)$$

Therefore, there are two natural limits to the behavior. In the weak field regime the vacuum polarization may be sufficiently strong to decouple the polarization states as they propagate through the magnetosphere (Heyl & Shaviv 2000, 2001). Normal pulsars as well as strongly magnetized white dwarfs fall in this regime. In the strong-field regime magnification and distortion of the image of the neutron star surface may become important in addition to the decoupling of the propagating modes (Shaviv, Heyl & Lithwick 1999).

The processes of photon splitting and one-photon pair production are forbidden in field-free regions, the first by Furry's theorem and the second by four-momentum conservation. However, in the strong magnetic fields surrounding a neutron star, both processes may be important. Photon splitting most strongly affects photons in the $\perp$-mode which may split into two photons in the $\parallel$-mode (Adler 1971, the mode-naming convention used here is opposite to that used by Adler); this both distorts and polarizes the photon spectrum.

One-photon pair production has an energy threshold of $\hbar\omega \sin\theta > 2m_e c^2$ for photons in the $\parallel$-mode. The threshold for photons in the $\perp$-mode is slightly larger $\hbar\omega \sin\theta > m_e c^2(1 + \sqrt{1 + 2B/B_c})$. Near the thresholds, in strong magnetic field especially, the cross-section for this process is complicated by the formation of the pair in discrete Landau levels or a positronium bound state (Daugherty & Harding 1983; Usov & Melrose 1996).

4.4. Interpretation

4.4.1. Rotation-powered Pulsars

Partly because it provides such a straightforward interpretation of the polarization data, the curvature radiation model is probably the favorite explanation for pulsar radio emission. However, it does have some drawbacks. One of these is that it is difficult to maintain a compact bunch for very long as the electrons travel along trajectories with different radii of curvature and are subject to radiation reaction. Another problem is that it has proven hard to find a suitable plasma instability which will allow charge particle bunches to grow. A third and currently controversial observational claim is that the emitting area is much larger than expected if the bunches form in the inner magnetosphere as the rotating vector model requires. Specifically, Gwinn, *et al.*(2000) find that the size of the Vela pulsar emission region is roughly ten per cent of its light cylinder radius, $\sim c/\Omega$, suggesting that the emission originates in the outer magnetosphere. Conversely, Cordes (2001) finds that the source is unresolved.

The maser explanation, by contrast, only works in the outer magnetosphere and the average direction of polarization should be orthogonal to the projection of the pulsar spin axis on the sky. This can be tested using the Crab and Vela pulsars where X-ray jets are observed which, although their formation is not understood, are presumed to be along the projected spin axis. In the case of the Crab pulsar, the situation is ambiguous because it is unclear if the two pulses come from one or two magnetic poles. However, in the case of the Vela pulsar, the electric vector is unambiguously perpendicular to the projected spin axis, consistent with the maser model. It could also be consistent with curvature radiation if propagation effects are important. It should be possible to discriminate between these two models observationally.

The pulsed optical radiation and X-ray radiation seen from several radio pulsars is generally thought to be incoherent synchrotron and inverse Compton radiation, respectively. However, the location of the emission region, and, in the case of the X-rays, the source of the incident photons (the surface of the star or coherent radio emission from the magnetosphere) is currently undecided. Suffice it to remark it here that polarization arguments figure prominently in these debates.

4.4.2. Accretion-powered Pulsars

In addition to being strongly polarizing, the opacity, Eq. 4.42, is highly anisotropic. So, even though we cannot measure the X-ray polarization directly, at present, it does have a strong effect on what we observe. In particular, the mass that accretes onto a spinning neutron star with a surface field $\sim 10^{12}$ G is likely to be channeled toward the magnetic poles and, if the accretion rate is large enough, there will be a significant Thomson opacity at the poles. However, it will be much easier for the radiation to escape along the direction of the magnetic field than in the transverse direction. For this reason, X-rays are thought to emerge in two broad pencil beams about the magnetic axis as is observed.

4.4.3. *Surface Emission from Isolated Neutron Stars and Magnetars*

The atmospheres of neutron stars are thought to emit strongly polarized radiation (Pavlov & Shibanov 1978). The opacities in the two polarization modes of the atmospheric plasma may differ by several orders of magnitude (Lodenqual *et al.*1974). The opacity in the extraordinary mode (i.e. $E \perp B$) is generally a factor of $(\omega/\omega_G)^2$ smaller than in the ordinary mode. Since the atmospheres are typically at a temperature of several million degrees, the natural place to study this emission is in the X-rays. Furthermore, as we shall see, the vacuum significantly affects the propagation of radiation passing through it at X-ray and higher energies.

Although the emission at the surface may nearly be fully polarized, one observes radiation from regions with various magnetic field directions. In this vein, Pavlov & Zavlin (2000) argue that the net polarization in the X-rays is on the order of ten percent and decreases for more compact stars. However, this treatment ignores the fact that QED renders the vacuum birefringent. The field strength varies sufficiently gradually, that is

$$\frac{\nabla |\Delta k|}{|\Delta k|} \ll |\Delta k|, \tag{4.48}$$

where

$$|\Delta k| = \frac{\alpha}{4\pi} \frac{2}{15} \frac{\omega}{c} \left(\frac{B_\perp}{B_k}\right)^2, \tag{4.49}$$

in the weak field regime, that the two polarization modes are decoupled. Radiation produced at the surface with its polarization direction perpendicular to the local magnetic field direction will keep its polarization perpendicular to the field even as it passed through regions where the field direction changes. The observed polarization reflects the direction of the field at a distance

$$r \approx 1.2 \times 10^7 \left(\frac{\mu}{10^{30} \text{ Gcm}^3}\right)^{2/5} \left(\frac{\nu}{10^{17} \text{ Hz}}\right)^{1/5} (\sin \beta)^{2/5} \text{ cm}, \tag{4.50}$$

from the center of the star (Heyl & Shaviv 2000, 2001). Here μ is the magnetic dipole moment of the neutron star, and β is the angle between the dipole axis and the line of sight. QED ensures that the strongly polarized radiation at the surface of the neutron star remains strongly polarized until it is detected; therefore, the simple detection of strongly polarized X-rays from the atmosphere of a neutron star will verify a thus far untested prediction of QED. Cheng & Ruderman (1979) used a similar argument to account for the strong polarization of radio emission from pulsars.

Although, the QED process of one-photon pair production plays a crucial role in radio pulsars by fueling the plasma that produces the emission (*e.g.* Daugherty & Harding, 1982), the threshold for the reaction is much higher than the typical energies from the surface emission. The cross-section for photon splitting increases dramatically with increasing photon energy, $\propto E^6$ (Heyl & Hernquist 1997b) and is only important above 10 keV even in the strongest magnetized sources.

4.5. *Soft-Gamma Repeaters*

Thompson and Duncan first argued that the soft-gamma repeaters are neutron stars fueled by a dynamic magnetic field whose strength greatly exceeds B_c. In their quiescent state, these objects emit thermal radiation from their surfaces and the discussion of the previous subsection applies. If their surface fields are sufficiently strong (the surface field is expected to exceed the value estimated by spin down of $\sim 10^{15}$ G), magnetic lensing may be important for photons whose polarization is parallel to the magnetic field (*i.e.* the ordinary mode); however, thermal emission in this mode appears to be strongly suppressed.

However, what makes the soft-gamma repeaters unique is that they burst. In fact, the soft-gamma repeater, SGR 1900+14, is the only object beyond our solar system to have had contemporary geophysical consequences (it ionized the nightside upper atmosphere nearly to daytime levels). This soft gamma-ray emission is generally well below the threshold for one-photon pair production, but photon splitting should degrade the energies of the photons by at least a factor of two and polarize them by converting photons in the extraordinary mode to the ordinary mode (Baring & Harding 1997). Observing this tracer of photon splitting would require gamma-ray polarimetry.

4.6. *Summary*

• Magnetized neutron stars provide cosmic laboratories where we can observe unique polarization effects in action and use them to identify the emission mechanism.

• Radio pulsars offer the richest polarization data set outside the solar system. They are strongly diagnostic of the emission mechanism and the effects of propagation.

• Accretion-powered pulsars introduce new effects associated with strong field anisotropic emission and scattering in the $\sim 10^{12}$ G surface fields. Even though the X-ray polarization is not yet measured, it is important in determining the total spectrum and pulse profile.

• X-ray pulsars allow us to address important physics questions, like the composition and compressibility of cold matter at supra-nuclear density.

• The simple detection of strongly polarized X-rays from the atmosphere of a neutron star (or optical radiation from neutron stars with $B > 10^{13}$ G) will verify the prediction that QED renders the vacuum birefringent and provide an estimate of the radius of the star itself.

• The convincing case that magnetars exist with surface fields well in excess of the critical field, $(4.4 \times 10^{13}$ G), offers the equally exciting, (though observationally very challenging) prospect of testing quantum electrodynamics in a regime far removed from terrestrial investigation.

5. Black Holes

5.1. *Motivation*

There is now very good evidence for the existence of black holes in the universe. They appear to be a common endpoint of the evolution of massive stars in our Galaxy and nearby galaxies and we know of roughly ten good cases where the dynamically determined mass significantly exceeds the Oppenheimer-Volkoff limit (for neutron stars) ~ 2.5 M$_\odot$ and the Chandrasekhar limit (for white dwarfs) ~ 1.4 M$_\odot$. In addition, there are many more cases of transient X-ray sources where the circumstantial evidence, in the absence of dynamics, is pretty convincing. A significant fraction of massive stars must end their life this way.

Similarly, dynamical studies of the nuclei of nearby galaxies reveal the presence of "massive dark objects" which, if they were, for example, clusters of compact objects, would be very short-lived. Identifying them with massive black holes is by far the most conservative conclusion to draw. It appears that the nuclei of most normal galaxies, including our own, contain black holes with masses in the range $\sim 3 \times 10^6 - 3 \times 10^9$ M$_\odot$. (There are speculative suggestions that there may be a large population of intermediate mass black holes, perhaps relics of the first generation of stars.)

The existence of black holes is, arguably, the most far-reaching implication of the general theory of relativity. The theory has been probed in the weak field regime and passed all quantitative tests with an accuracy that can be as small as $\sim 3 \times 10^{-4}$. It is in the nature of the theory that, if we understand the laws of physics under these circumstances, it is simply a question of geometry to describe strong field environments, when the equivalent Newtonian potential approaches c^2. If the theory in its essential simplicity is correct, then the metric of an asymptotically flat black hole spacetime (excluding some mathematical niceties) is essentially given. Indeed, in one of the greatest successes of mathematical physics, we have a closed form version of the metric of a spinning black hole, known as the Kerr metric, and essentially all classical physics that can be discussed in a flat spacetime can also be discussed around a black hole; there are no difficulties of principle. There are, however, considerable difficulties in execution (which have mostly been overcome using numerical calculations.) Although we know of more general spacetimes that include a gravitationally significant charge or orbiting mass, we believe that these are irrelevant to observed black holes and that astronomers need only be concerned with the Kerr metric.

However, it is logically possible that the theory of general relativity could be wrong or incomplete and that, as a consequence, black holes are fundamentally and observably different from their general relativistic description. For this reason, it is vitally important that we try to find ways to probe the spacetime around black holes, now that we know where to find them. In this regard, observing black holes provides a far more telling test of relativity theory than cosmological observations. This is because cosmological observations are seriously compromised by our deep ignorance of the nature of dark matter and energy as well as the effects of evolution.

We already know that gravitational waves exist. Binary pulsars are observed to lose orbital energy at rates that agree with theory to a fraction of a percent. However, this mostly tests linearized theory even in the sources. The direct detection of gravitational waves, which we hope will happen one day, is also is only a linear test. The ultimate test of general relativity is to make detailed observations of gravitational waves from coalescing black holes; an observation that I suspect will be not be technologically feasible for some time. In addition to testing strong field relativity, this can also provide much useful astrophysical information on galaxy merger rates, AGN evolution and so on. Computing the wave forms in necessary generality is a major challenge to computational science. What is relevant in the present context is that gravitational waves have natural polarization states, just like electromagnetic waves, and much of the information from these coalescences will be encoded in the polarization details. (There are other strong field sources of gravitational radiation that have been considered, notably topological defects like cosmic strings. Unlike the case with black holes, there is no observational evidence yet for their existence. However, if they are ever discovered, then it may well be their gravitational radiation polarization that is their distinctive signature. Computing this polarization is a good project which appears to have been mostly ignored.)

There is a second and quite different reason for being interested in black holes. This is that we do not understand properly how they work. We have already introduced jets, disks, and outflows; and discussed how polarization observations can teach us about their properties. We already know, from direct observation, that all three of these continue down to relatively close (in logarithmic terms) to the black hole. This is where most of the energy is released. However, we do not understand how this all happens and how the flow around the black hole depends upon the mass and the spin of the hole as well as the accretion rate and the immediate environment. It appears that the answers to these questions will only be found by exploring the black hole itself.

A third contextual aspect of this study is that we are beginning to suspect that black holes have a much larger and more active role in galactic and extragalactic astronomy than used to be the case. It is increasingly likely that gamma ray bursts are associated with the formation or augmentation of black holes and that these have major environmental impacts on their surroundings and could soon become useful cosmological probes. Black hole transients provide the dominant hard X-ray emission of galaxies like our own and create powerful outflows. The discovery of dormant, or near-dormant black holes in the nuclei of normal galaxies has affirmed the long-standing black hole model of active galactic nuclei, including quasars and giant, double radio sources. However, the implications have much broader implications than the properties of AGN *per se*. The quasars themselves provide the best cosmologically distant beacons that we have and they allow us to study the intergalactic medium, matter and cosmography. Furthermore, it is becoming increasingly apparent that they have an much more active role in the very formation of galaxies, both in the initiation and perhaps in the cessation of the process.

The long term observational goal, then, is to verify that black holes are described by the Kerr metric and to measure their masses and spins in such a way as to elucidate their role in stellar and galactic evolution. In this section, we will try to show how polarization observations can contribute to meeting this objective.

5.2. *Observation*

We have already described most of the relevant observations of black holes including the fairly strong dynamical measurements of their masses and the Fe K line emission which provides the strongest evidence to date that black holes spin relatively rapidly. The most relevant, existing observation for the purpose of this lecture are the measurements of linear and circular polarization.

5.2.1. *Sgr A**

The center of our Galaxy appears to be identified with the radio source Sgr A*. There is now excellent dynamical evidence that is a "dark, compact object" with a mass 2.6×10^6 M$_\odot$ and a black hole is by far the most conservative interpretation. The source is nearly at rest and stars can be tracked moving (and accelerating) around it (Ghez *et al.*2000). The source has a spectrum which peaks at ~ 300 GHz and the image is broadened at radio wavelengths by interstellar scattering. At 43 GHz, the average scatter-broadened size is reported to be $\sim 3 \times 10^{13}$ cm, (Lo *et al.*1999) and to scale roughly $\propto \lambda^2$. Sgr A* is a weak and soft X-ray source. Interestingly, a 106 d periodicity in the radio emission has also been reported (Zhao, Bower & Goss 2001).

Sgr A* has long been known to have negligible linear polarization at radio and mm wavelengths. This is not a surprise because the Faraday rotation is expected to be quite high so that the differential (in both angle and frequency) rotation is also large enough to depolarize all measurements. What is a surprise (as discussed here by Hildebrand) is that Aitken *et al.*(2000) measure ~ 10 percent linear polarization at 150 GHz, using the SCUBA instrument on the James Clerk Maxwell Telescope, although there is an upper limit of ~ 1 percent at 86 GHz (Bower *et al.*2000). Clearly there is a need to confirm the SCUBA measurement.

More recently, it has been discovered that Sgr A* exhibits quite strong circular polarization, (Bower, Falcke & Backer 2000). The 5 GHz degree of circular polarization appears to have been stable at a value of ~ -0.003 for nearly twenty years. At higher frequencies, up to ~ 43 GHz, the degree of circular polarization appears to increase up to a few percent and become increasingly variable, doubling in a few days.

5.3. *Physical Processes*

5.3.1. *Spinning Black Holes*

The spacetime around a spinning black hole is described by the Kerr metric expressed in Boyer-Lindquist coordinates with $G = c = 1$

$$ds^2 = -(1 - 2mr/\rho^2)dt^2 - (4amr\sin^2\theta/\rho^2)dtd\phi + (\rho^2/\Delta)dr^2 + \rho^2d\theta^2$$
$$+ (r^2 + a^2 + 2mra^2\sin^2\theta/\rho^2)\sin^2\theta d\phi^2 \,, \tag{5.51}$$

where

$$\rho^2 = r^2 + a^2\cos^2\theta \,, \tag{5.52}$$
$$\Delta = r^2 - 2mr + a^2 \,, \tag{5.53}$$

and m is the mass such as would be measured by the orbit of a distant satellite, and $a < m$ is the specific angular momentum of the hole, as could be measured operationally by the precession rate of a gyroscope.

There is an event horizon, $\mathcal{H}$, which is located where $\Delta = 0$ i.e. where the radial coordinate $r = r_+ = m + (m^2 - a^2)^{\frac{1}{2}}$. Particles on timelike or null geodesics must be inwardly moving within r_+ which leads to the interpretation that $\mathcal{H}$ represents a surface of no return. The four velocity, $\vec{u} = \{dt/d\tau, dr/d\tau, d\theta/d\tau, d\phi/d\tau\}$ of a material particle satisfies

$$g_{\alpha\beta}u^\alpha u^\beta = -1 \,. \tag{5.54}$$

The equation of a photon, following a null geodesic is given by $g_{\alpha,\beta}dx^\alpha dx^\beta = 0$, supplemented with equations representing the conservation of energy and angular momentum as well as an additional integral of the motion.

The angular velocity $\Omega = d\phi/dt$ of a particle, orbiting with fixed r, θ, therefore satisfies

$$u^{02}[g_{00} + 2\Omega g_{0\phi} + \Omega^2 g_{\phi\phi}] = -1 \,. \tag{5.55}$$

This implies that $\Omega_{\min} < \Omega < \Omega_{\max}$ where $\Omega_{\min} > 0$ when $r_+ < r < r_e \equiv m + (m^2 - a^2\cos^2\theta)^{1/2}$. The radius r_e is known as the static limit and the region between it and the horizon, where inertial frames are dragged by the spin of the hole, is known as the ergosphere. A particular significance of the ergosphere is that orbits of negative energy (including rest mass) exist within it. As $r \to r_+$ at the event horizon,

$$\Omega_{\min} \to \Omega_{\max} \to \Omega_H \equiv a/(r_+^2 + a^2)^{1/2} \,, \tag{5.56}$$

the angular velocity of the hole.

A remarkable theorem due to Hawking states that the area of the horizon, which can be computed from the metric to be $A = \int_{\mathcal{H}}(g_{\theta\theta}g_{\phi\phi})^{1/2}d\theta d\phi = 4\pi(r_+^2 + a^2)$ cannot decrease. We can use this to define a so-called irreducible radius r_0 and irreducible mass m_0 through

$$r_0 = 2m_0 = (A/4\pi)^{1/2} \,. \tag{5.57}$$

This immediately implies that $a = r_0^2\Omega_H$. It turns out that the area is proportional to the thermodynamic entropy. Now imagine that we exchange some mass and some angular momentum with the hole, reversibly (and therefore at constant area) from just outside the horizon. These must be added according to

$$dm = \Omega d(am) \,. \tag{5.58}$$

This equation can be integrated to give

$$m = \frac{m_0}{[1 - (\Omega r_0)^2]^{1/2}} \,. \tag{5.59}$$

Imposing the condition $a < m$, we find that there is a mass $m - m_0 < 0.29m$ that can, in principle, be extracted from the hole. The main way that this is thought to occur naturally is through the agency of large scale magnetic field that threads the event horizon of the black hole. This magnetic field can exert a torque on the hole, similar to the magnetic torques acting upon the sun and neutron stars, for example. Black hole spin provides a plausible power source for high energy phenomena like ultrarelativistic jets and gamma ray bursts and this is one reason why black holes are commonly thought to be spinning rapidly. (Even if the spin is not a significant power source, then the specific angular momentum of the gas that accretes onto a black hole is generally so large that it is very hard to imagine slowly spinning holes ever being formed.)

For present purposes, though, what is most important is that, in a rapidly spinning hole, the accreting matter can form a disk extending quite close to the horizon. Specifically, if we consider circular Keplerian orbits around a hole then these are stable down to a radius of marginal stability which is located at $6m$ for a non-rotating (Schwarzschild) hole and approaches the horizon as $a \to m$. In addition, it is possible for strong pressure gradients within the disk to support matter in non-Keplerian orbits inside $6m$. This means that gas may survive quite a long while in and around the ergosphere before crossing the horizon or being ejected. The dominant emission may come from this region, and as the radiation escapes, its trajectory and the propagation of its polarization can be significantly influenced by the curvature of the spacetime. This provides us with a potential probe of the Kerr metric.

5.3.2. *Geometrical Optics of Plasma Waves in Flat Space*

We are interested in the propagation of plasma waves in the curved spacetime around a black hole. However, for the moment, let us consider the propagation of waves in flat space under geometrical optics. This is appropriate because the wavelengths that we are considering, at least for electromagnetic radiation, are always much smaller than the horizon radius. It is convenient to exploit the analogy with Hamiltonian particle dynamics. Under the eikonal approximation, we can define a phase ϕ such that $\nabla\phi = \vec{k}$ and $\partial\phi/\partial t = -\omega$. The phase velocity is, as usual, defined by $\vec{V}_\phi = \omega/\vec{k}$. We assume the existence of a dispersion relation

$$\omega = \Omega(\vec{k}, \vec{x}, t) \,. \tag{5.60}$$

Equivalently, there is a Hamilton-Jacobi equation

$$\frac{\partial\phi}{\partial t} + \Omega(\nabla\phi, \vec{x}, t) = 0 \,, \tag{5.61}$$

which must be satisfied. The three Hamilton equations are

$$\frac{d\vec{k}}{dt} = -\nabla\Omega \,, \tag{5.62}$$

$$\frac{d\vec{x}}{dt} = \frac{\partial\omega}{\partial\vec{k}} \equiv \vec{V}_g \,, \tag{5.63}$$

$$\frac{d\omega}{dt} = \frac{\partial \Omega}{\partial t},\tag{5.64}$$

where

$$\frac{d}{dt} \equiv \frac{\partial}{\partial t} + \vec{V}_g \cdot \nabla,\tag{5.65}$$

and V_g is recognized as the group velocity. These three equations govern the propagation of plasma modes in a spatially inhomogeneous and temporally varying medium, under the short wavelength approximation. In a cold, unmagnetized plasma the dispersion relation is $\Omega = (\omega_P^2 + c^2 k^2)^{1/2}$ and $V_g = c^2/V_\phi$. We can think of wave quanta – plasmons – and the energy they carry, as moving along a path $\vec{x}(t)$ with the group velocity. These plasmons are conserved; i.e. the wave energy density U can be shown to obey a conservation equation of the form

$$\frac{\partial}{\partial t}\left(\frac{U}{\omega}\right) + \nabla \cdot \left(\frac{U\vec{V}_g}{\omega}\right) = 0.\tag{5.66}$$

Consider, for example, shear Alfvén waves. The dispersion relation is $\omega = \vec{k} \cdot \vec{B}/(4\pi\rho)^{1/2}$ and the group velocity is $\vec{V}_g = \vec{B}/(4\pi\rho)^{1/2}$. The wave packets propagate along the magnetic field along with the energy although $\vec{k}$ can be directed at a large angle to $\vec{B}$.

The propagation of the polarization can be most simply approached by decomposing the given wave into its normal modes, propagating each along the direction of the group velocity, and compute the relative change in phase (to lowest order in the eikonal approximation this is simply $\int dx \cdot \vec{k}$ along the path.) Of course the character (i.e. polarization, local phase velocity etc.) of these modes will change, but provided we are in the WKB limit, the modes are distinguished, non-degenerate (see below), and there is no mode crossing (which can occur and has to be handled more carefully), these changes will change adiabatically. As a result, they can be tracked and the total phase difference along a path can be computed.

The total flux can be computed by using the conservation of intensity along the path. Equivalently, we say that the phase space density of individual quanta of wave excitation, in individual modes is conserved along paths. If there is emission or absorption along the path then it is straightforward to write down the equation of radiative transfer and use the local emission and absorption coefficients to evolve the intensity (e.g. Rybicki & Lightman 1979; Bekefi 1966).

In practice, of course, all of this can easily become quite involved. However it is important to understand the principles because these alert us to the sort of effects we might expect to observe.

5.3.3. *Magnetized Accretion Disk*

As a more pertinent illustration of some of these ideas, let us consider electromagnetic wave modes propagating through an accretion disk containing a strong, though disordered, magnetic field.

Consider a magnetoactive plasma with $X = \omega_P^2/\omega^2, Y = \omega_G/\omega < 1$ where ω_P is the plasma frequency and ω_G is the electron gyro frequency. Under so-called "quasi-longitudinal" conditions - essentially when $\cos\theta < Y$, where θ is the angle between $\vec{k}$ and $\vec{B}$, the electromagnetic eigenmodes are elliptically polarized with axis ratio $r = 1 \pm Y\sin\theta\tan\theta$ and phase velocity difference $\Delta V = cXY\cos\theta$.

Now suppose that synchrotron (or cyclotron) radiation is emitted within an accretion disk of thickness H. The major axis of the polarization ellipse will be Faraday rotated at a rate $\Delta\Phi/ds = \Delta V\omega/2c^2$. Now if, as we expect, the magnetic field direction reverses often along a ray and if, as also anticipated, $X|Y|\omega H/c \gg 1$, then we expect that the emergent linear polarization will be vanishingly small (cf §1.3.4). (The limiting polarization along an individual ray is likely to be determined by the decrease in the density rather than the magnetic field strength.)

The circular polarization is a bit more problematic. If we suppose that there is a net magnetic field normal to the disk, as is true of some models, then there should be a preferred sense of the circular polarization that is emitted in cyclotron or low energy synchrotron radiation. This will be largely preserved in propagating out of the disk.

If the emitted radiation is effectively unpolarized, we can analyze the production of circular polarization due to Faraday conversion by decomposing each wave into the eigenmodes. Thus consider a single eigenmode propagating out of the disk through a spatially varying magnetic field. We suppose that the variation happens relatively slowly on the scale of the wavelength so that the polarization ellipse adjusts adiabatically (with no mode crossings). Next, suppose that the two eigenmodes are launched with equal amplitude and that the field is uniform. The beating between the two eigenmodes will result in a circular polarization of amplitude $Y \sin\theta \tan\theta$ that changes sign as the plane of linear polarization rotates. If either the Faraday depth is large, or the sign of the magnetic field is as likely to be negative as positive, then the limiting circular polarization that emerges from the disk is equally likely to have either sign and so there will be no net circular polarization.

Now let the magnetic field direction vary along a ray. If the angle θ varies, then there will be a corresponding change in the axis ratio of the polarization ellipse, but still no preference for one sign over the other. However, if the azimuthal angle ϕ relating $\vec{B}$ to $\vec{k}$ changes in a systematic fashion along all rays then a net phase difference between the two modes will develop. In the same way that Faraday conversion creates circular polarization, such a phase difference will also create circular polarization. The main difference is that while Faraday conversion depends upon the direction of the magnetic field, and hence will not lead to a net polarization for randomized fields, the new mechanism depends upon the rate of shearing and only the strength of the field. Thus, it is possible to conceive of situations in which the rate of shearing and the typical length scales over which the magnetic field reverses are related in such a manner that a net circular polarization is produced without a commensurate linear polarization.

This situation is precisely what might be anticipated in a magnetized accretion disk. In the disk interior, the typical field direction will trail to reflect the differential rotation in the disk. However, the field will be swept back by progressively smaller angles as the ray approaches the disk surface, corresponding to a net rotation of the average azimuthal angle ϕ. There will only be a preferred sense of limiting circular polarization, if the magnetostatic field is still changing in this systematic manner over the last radian of Faraday rotation. The net circular polarization will be $\sim c \ln Y / \omega H X$.

These are some of the subtle effects that could be present in an accretion disk and which could, under some circumstances, create measurable polarization, even in the absence of general relativity.

5.3.4. *Geometrical Optics of Vacuum Waves in a Curved Spacetime*

We first consider the propagation of photons in a vacuum surrounding a black hole. These follow orbits called null geodesics, just as material particles follow timelike geodesics. The equation of motion can be expressed in a general coordinate system though, in our case, Boyer-Lindquist coordinates, by saying that the total derivative of the wave vector along the ray vanishes. In index notation, this becomes

$$k^\beta k^\alpha_{;\beta} = 0 \,, \tag{5.67}$$

cf Eq. (5.62).

There are essentially three constants of the motion that describe these orbits, an energy, $-k_0$ an angular momentum, k_ϕ, and a third quantity known as the Carter constant, Q (e.g. Misner, Thorne & Wheeler 1973.) (In fact, we only need the ratios $k_\phi/k_0, Q/k_0$ to define the rays.) Close to the black hole the rays are strongly curved with the consequence that a distant observer, able to resolve a black hole, would be able to see a distorted image of the disk behind the hole, apparently hovering above the hole (cf Fig. (2).). The ray trajectories are given, in general, by the solution of a set of coupled ordinary differential equations that can be partly integrated in terms of elliptic functions (Rauch & Blandford, 1994). Given a model of the emission, for example of the surface emissivity of a thin accretion disk, it is a straightforward, though quite lengthy, exercise to compute the total emergent flux and, indeed, the form of the image that would be resolved if the black hole could be resolved.

Now, turn to the propagation of the polarization of vacuum modes in a curved space-time, specifically outside the horizon of a Kerr hole (Laor, Netzer & Piran 1990). As we have already emphasised, both non-thermal emission (e.g. synchrotron radiation) and electron scattering are likely to create polarized sources of radiation. If we just consider linear polarization for the moment, and the generalization to circular polarization is straightforward, then the question that we must answer is "How do we propagate the plane of polarization from one point to the next along a curving ray?". The answer is that the electric vector is "parallel-transported"(eg Misner, Thorne & Wheeler 1973). What this means is that the unit vector in the direction of the electric field $\hat{e}^\alpha$ changes along the ray such that its magnitude and projection onto the direction of the ray remain constant. An additional constraint arising from Maxwell's equations in vacuum is that the electric vector must be perpendicular to the wave vector. Hence, in index notation, we have

$$k^\beta \hat{e}^\alpha_{;\beta}; \quad \hat{e}^\alpha k_\alpha = 0 . \tag{5.68}$$

These can be solved consistently to propagate the electric vector along a ray.

What is actually done is somewhat different. It turns out that there is another conserved quantity, called the Walker-Penrose (1970) tensor, associated with the photon spinors (the familiar geometrical object that in this case are associated with light-like geodesics, the path taken by photons in vacuum.) This actually involves the electric vector and can be used to relate the polarization at the point of emission to that at the point of observation directly without having to integrate a differential equation (Connors, Stark & Piran 1980). It is then possible to define a transfer function for the polarization and to compute the polarized flux given a specific emission model using the propagated intensity. (Note that, in propagating the intensity, we must correct for the Doppler and gravitational shifts. There is a natural way to do this in general relativity.)

5.3.5. *Geometrical Phase*

The next level of complication is to introduce the plasma into the curved space time. Let us do this in two stages. The first stage is to ignore the magnetic field so that the local dispersion relation takes the form $\Omega = (\omega_p^2(\vec{x}) + c^2 k^2)^{1/2}$. In this case, the refractive index is locally isotropic. This means that the two eigenmodes at a point are degenerate and that we have to formulate a rule to connect the polarization from one point along a path to the next.

The wave packets, which travel at the local group velocity, are no longer moving along null geodesics, but timelike geodesics instead. It turns out that the Hamiltonian equations of motion can be generalized in a covariant manner, provided that one has knowledge of the local linearized dispersion relation at every relevant point in space time. Therefore, there is a prescription for computing the paths. These can be described by a four velocity u^α and an acceleration $a^\alpha = du^\alpha/d\tau$ with respect to a freely-falling frame, where $d\tau$ is an interval of proper time. The standard relativistic way to handle this is to generalize the notion of parallel-transport to Fermi-Walker transport (e.g. Misner, Thorne & Wheeler 1973), which corrects for the non-null motion of the wave packets. The propagation equation becomes

$$\frac{d\hat{e}^\alpha}{d\tau} = \hat{e}^\beta a_\beta u^\alpha - u_\beta \hat{e}^\beta a^\alpha . \tag{5.69}$$

Eq. (5.69) reduces to Eq. (5.68) when Eq. (5.67) is satisfied. This provides a natural basis in which to discuss polarization propagation and phase changes.

It is instructive to consider a wave propagating along a twisting optical fiber, with $\vec{k}$ parallel to the local tangent to the fiber. Here again we have gradients in an isotropic refractive index. In this case, the unit electric vector, along $\hat{\vec{e}}$, must remain perpendicular to the unit wave vector $\hat{\vec{k}}$. When the fiber bends, the change in $\hat{\vec{e}}$ must be along $\hat{\vec{k}}$; there is no other vector to be involved as the medium is isotropic. Therefore we can write down the equation of propagation for the electric vector from first principles.

$$\frac{d\hat{\vec{e}}}{ds} = -\hat{\vec{k}} \left(\hat{\vec{e}} \cdot \frac{d\hat{\vec{k}}}{ds} \right) . \tag{5.70}$$

This is a limiting case of Eq. (5.69).

A good way to visualize what is happening (Berry 1990) is to allow the tangent to the fiber to trace out a path on the unit sphere. $\hat{\vec{e}}$ is tangent to the sphere and it is straightforward to see that rotation angle of $\hat{\vec{e}}$ after traversing a complete circuit equals the solid angle enclosed by that circuit. If we propagate a linearly polarized wave along a twisting fiber, the polarization direction will, in general, be rotated between two points where the fiber is parallel. This experiment has been performed successfully (Chiao *et al.*1989). (Actually this was under conditions when physical as opposed to geometrical optics applies, though the results should be identical.)

This rotation – essentially a phase change between the two circularly polarized eigenmodes – is known as the geometric phase. Geometric phase is a quite general phenomenon in physics and analogs are expected to be relevant to wave propagation in a curved spacetime. The Foucault pendulum provides another example of this general phenomenon. As is well known, a Foucault pendulum at latitude ℓ will only rotate through an angle in inertial space of $2\pi(1 - \sin \ell)$, the solid angle traced out by the radial vector on the unit sphere, as it is carried around a complete circuit in one day by the spinning Earth. Now to see where general relativity may come in, it is helpful to consider a Foucault pendulum at the North pole. According to the above discussion, there is no rotation of the plane of oscillation according to Newtonian dynamics. However, the tiny dragging of inertial frames effect associated with the spin of the earth leads to an equally tiny rotation of the plane of polarization and there have been proposals to measure it. Effects such as these would be much larger near a spinning black hole and could also influence the propagation of electromagnetic waves.

5.3.6. *General Relativistic Magnetoionic Theory*

The second stage is to reinstate the magnetic field which, on general grounds, is surely present. This breaks the degeneracy between the two wave modes. This is akin to changing the pivot of a Foucault pendulum from a point attachment to an axle. As far as is known, there is no generalization of the Carter constant and the Walker-Penrose tensor, to non-null wave modes. The equations for the paths can be integrated and the intensity and the polarization can be propagated using the relativistic generalization of the approach outlined above (Broderick & Blandford, in preparation.)

When a linearly-polarized vacuum wave crosses the ergosphere of a rapidly spinning hole, there is a contribution to the emergent polarization position angle of order unity due to the dragging of inertial frames. We can think of this as a phase difference in of order unity in the two circular polarized modes into which the linear mode can be decomposed. Now, if we introduce a magnetoactive plasma into the path, then there is likely to be a large Faraday rotation per unit length. The total rotation will differ by much more than $O(1)$ along different paths that any emitted linear polarization is likely to be erased. However, as discussed above, the eigenmodes are not completely circular and have an ellipticity $O(XY)$. What this means is that a systematic phase difference $O(1)$ will be introduced between the two modes and that, if the original modes are in phase so that there is no circular polarization, a degree of polarization $O(XY)$ will emerge, independent of the reversals of the magnetic field and variations in the total Faraday rotation along different lines of sight.

These matters deserve further attention.

5.4. *Interpretation*

5.4.1. *Sgr A* and other Low Luminosity AGN*

X-ray observations of Sgr A^* (Baganoff *et al.*2001) have shown the the luminosity is very small ($\sim 4 \times 10^{33}$ erg s^{-1}) and the spectrum is quite soft. Variability on an hour timescale may also have been seen. (As has been argued elsewhere, Blandford & Begelman 1999, this is generally to be expected if most of the mass supplied to the hole is blown away in a wind.) This suggests that the density of gas is very low close to the hole and opens up the possibility that we may be seeing radio or, more likely, mm emission from the ergosphere. Under these conditions, polarization observations could be quite diagnostic of the physical conditions.

The variable circular polarization discussed above, increasing in degree with frequency, has at least three explanations. Firstly, there could be a radius to frequency mapping so that the radio photosphere shrinks with frequency and the field gets stronger so that the energy of the emitting relativistic electrons also decreases. This leads to an increase in the emitted degree of circular polarization (cf §1). Variability studies at high frequency should be quite diagnostic. Secondly, the polarization could be due to a flat space propagation effect along the lines discussed above. It will be particularly interesting to see if the sign of the circular polarization really does not change, as the observations to date may suggest. This could, in principle, be related to the angular velocity of the disk as outlined above, though quantitatively this seems improbable.

The third possibility may be the most unlikely, yet it is the most exciting. This is that the circular polarization reflect directly the geometry of the ergosphere and be due to a general relativistic, propagation effect as outlined above. In the case of Sgr A^*, we expect the gas at high latitude in the ergosphere to be moving with speed $\sim c$. We deduce that the expected degree of circular polarization is:

$$ C \sim XY \sim 10^{-2} \left(\frac{\dot{M}}{10^{22}\,\mathrm{g\,s^{-1}}} \right) \left(\frac{B}{100\,\mathrm{G}} \right) \left(\frac{\lambda}{1\,\mathrm{cm}} \right)^3 . \tag{5.71} $$

It is not impossible that this effect is observable These matter deserve more attention, both observational and theoretical, in Sgr A^* and other, nearby galaxies.

5.4.2. *Imaging the Ergosphere*

Interesting and timely as these ideas may be, radio observations are probably not likely to contribute to a confirmation of the essential features of the Kerr metric until we can actually image the ergosphere. At present, as we have remarked, the best resolution has been achieved in M87 ($\sim 100m$). Probably the best prospects lie with sub mm VLBI observations of Sgr A*, where interstellar scattering precludes resolving the ergosphere at radio wavelengths (Falcke, Melia & Agol 1999). There are also quite futuristic plans to develop X-ray interferometry to achieve analogous goals (Cash *et al.*2000).

5.5. *Summary*

- Black holes are common features of the evolution of massive stars.
- Massive black holes are commonly found in the nuclei of normal galaxies. Presumably they powered active galactic nuclei including quasars and radio sources in the past.
- We have good grounds to be confident in the general theory of relativity and, specifically, the Kerr metric which describes the curved spacetime around a spinning black hole. However, this does not absolve us from the responsibility of testing the theory.
- We also want to understand how black holes accrete and how they form jets as well as their impact on Galactic and extragalactic astronomy.
- Recent observations suggest that we may be observing radio and mm emission from very close to the black hole in Sgr A*. There is consequently interest in developing the magnetoionic theory and radiative transfer in a general relativistic environment.

RB thanks Javier Trujillo Bueno and the Director of the Instituto de Astrofisica de Canarias for their hospitality, his fellow lecturers for their instruction and the students for their attention and questions. The NSF and NASA are acknowleged for support under grants AST 99-00866 and 5-2837 respectively.

REFERENCES

DISKS

AGOL, E. & BLAES, O. M. 1996, MNRAS, 282, 965

ANGEL, J. R. P., 1969, ApJ 158, 219

CHANDRASEKHAR, S., 1960, *Radiative Transfer*, New York, Dover

CHEN, K. & EARDLEY, D. M. 1991, ApJ, 382, 125

CONNORS, P. A., STARK, R. F., & PIRAN, T. 1980, ApJ, 235, 224

DRAINE, B. T. & LEE, H. M. 1984 ApJ, 285, 89

FABIAN, A. C., IWASAWA, K., REYNOLDS, C. S., & YOUNG, A. J. 2000, PASP, 112, 1145

FRANK, J., KING, A., & RAINE, D. 1992, *Accretion Power in Astrophysics*, Cambridge University Press, Cambridge

GIERLINSKI, M., ZDZIARSKI, A. A., POUTANEN, J., COPPI, P. S., EBISAWA, K., & JOHNSON, W. N. 1999, MNRAS, 309, 496

HUBENY, I., AGOL, E., BLAES, O., & KROLIK, J. H., 2000, ApJ, 533, 710

KÖNIGL, A. & KARTJE, J. F. 1994, ApJ, 434, 446

KORATKAR, A. & BLAES, O. M. 1999, PASP, 111, 1

KROLIK, J., 1999, *Active Galactic Nuclei: from the Central Black Hole to the Galactic Environment*, Princeton University Press, Princeton

LAOR, A., NETZER, H., & PIRAN, T. 1990, MNRAS, 242, 560

MATT, G., FABIAN, A. C., & ROSS, R. R., 1993, MNRAS, 264, 839

MESZAROS, P., NOVICK, R., SZENTGYORGYI, A., CHANAN, G. A., WEISSKOPF, M. C., 1988, ApJ, 324, 1056

NANDRA, K., GEORGE, I. M., MUSHOTZKY, R. F., TURNER, T. J., & YAQOOB, T., 1999, ApJ, 523, L17

POUTANEN, J. & SVENSSON, R., 1996, ApJ, 470, 249

PRINGLE, J., 1997, MNRAS, 292, 136

SHAPIRO, S. L. & TEUKOLSKY, S. A., 1983, *Black Holes, White Dwarfs, and Neutron Stars*, Wiley-Interscience, New York

YOUNG, A. J. & REYNOLDS, C. S., 2000, ApJ, 529, 101

ZUBKO, V. G. & LAOR, A., 2000, ApJS, 128, 245

JETS

AKESON, R. & CARLSTROM, J., 1999, ApJ, 491, 254

BLANDFORD, R. D. & KÖNIGL, A., 1979, ApJ 232, 34

BLANDFORD, R. D., NETZER, H., WOLTJER, L., COURVOISIER, T. J. -. & MAYOR, M. 1990, Saas-Fee Advanced Course 20. Lecture Notes 1990. Swiss Society for Astrophysics and Astronomy, XII, Springer-Verlag Berlin Heidelberg New York

BOWER, G., 2000, GCNEWS - Galactic Center Newsletter, vol. 11, p. 4-6 (eds. A. Cotera, H. Falcke, & S. Markoff)

HELFAND, D. J., GOTTHELF, E. V. & HALPERN, J. P., 2001, ApJ, in press

JONES, T. W. & O'DELL, S. L., 1977, ApJ 214, 522

KELLERMANN, K. I. & PAULINY-TOTH, I. I. K., 1969, ApJL, 155, L71

KOOPMANS, L. V. E. & DE BRUYN, A. G., 2000, A&A 358, 793

MACQUART, J.-P. & MELROSE, D. B., 2000, ApJ accepted, (astro-ph/0007429)

MIRABEL, I. F. & RODRIGUEZ, L. F., 1994, Nature 371, 46

PERLMAN, E. S., BIRETTA, J. A., ZHOU, F., SPARKS, W. B. & MACCHETTO, F. D., 1999, AJ 117, 2185

REYNOLDS, C. S., FABIAN, A. C., CELOTTI, A. & REES, M. J., 1996, MNRAS

RYBICKI, G. B. & LIGHTMAN, A. P., 1979, New York, Wiley-Interscience

WARDLE, J. F. C., HOMAN, D. C., OJHA, R. & ROBERTS, D. H., 1998, Nature 395, 457

WEISSKOPF, M. *et al.*, 2000, ApJ, 536, L81

WILSON, A. S., YOUNG, A. J. & SHOPBELL, P. L., 2001, ApJ 546, in press, (astro-ph/0008467)

OUTFLOWS

ANTONUCCI, R., 1993, ARAA, 31, 473

ARAV, N., KORISTA, K. T., BARLOW, T. A. & BEGELMAN, M. C., 1995, Nature, 376, 576

Cohen, M. H. *et al.*, 1995, ApJ, 448, L77

KORATKAR, A., ANTONUCCI, R. R. J., GOODRICH, R. W., BUSHOUSE, H. & KINNEY, A. L., 1995, ApJ, 450, 501

KORATKAR, A., ANTONUCCI, R. R. J., GOODRICH, R. W., AND STORRS, A., 1998, ApJ, 503, 599

KORISTA, K. & FERLAND, G., 1998, ApJ, 495, 672

LEE, H. -W., 1994, MNRAS, 268, 49

LEE, H. -W. & BLANDFORD, R. D., 1997, MNRAS, 288, 19

MURRAY, N., CHIANG, J., GROSSMAN, S. & VOIT, M., 1995, ApJ, 451, 498

OGLE, P. M., COHEN, M. H., MILLER, J. S., TRAN, H. D., GOODRICH, R. W. & MARTEL, A. R.. 1999, ApJS, 125, 1

RYBICKI, G. B. & HUMMER, D. G., 1978, ApJ, 219, 654

WEYMANN, R. J., MORRIS, S. L., FOLTZ, C. B., & HEWETT, P. C., 1991, ApJ, 373, 23

KORATKAR, A., ANTONUCCI, R. R. J., GOODRICH, R. W., BUSHOUSE, H., & KINNEY, A. L.,

1995, ApJ, 450, 501

Lee, H. -W. & Ahn, S. -H., 1998, ApJ, 504, L61

Lee, H. -W. & Blandford, R. D., 2000, ApJ, submitted

Nussbaumer H., Schmid, H. M., & Vogel, M., 1989, A& Ap, 211, L27

Stenflo, J. O., 1980, A& Ap, 84, 68

NEUTRON STARS

Adler, S. L., 1971, Ann. Phys., 67, 599.

Arons, J. & Barnard, J. J., 1986, ApJ, 302, 120

Baring, M. G. & Harding, A. K. 1997, ApJ, 482, 372.

Becker, W., 2000, Adv. Sp. Res., 25, 647

Chakrabarty, D. & Kaspi, V., 1998, ApJ, 498, L37

Cheng, A. F. & Ruderman, M. A. 1979, ApJ, 229, 348.

Cordes, J. M., 2001, ApJ, inpress

Daugherty, J. K. & Harding, A. K. 1982, ApJ, 252, 337.

Daugherty, J. K. & Harding, A. K. 1983, ApJ, 273, 761.

Gwinn, C. R., 2000, ApJ, 531, 902

Heyl, J. S. & Hernquist, L., 1997a, Journ. Phys. A, 30, 6485.

Heyl, J. S. & Hernquist, L., 1997b, Phys. Rev. D, 55, 2449.

Heyl, J. S. & Shaviv, N., 2000, Mon. Not. Royal Astr. Soc., 311, 555.

Heyl, J. S. &, Shaviv, N., 2001, Nature, submitted.

Hirano, C. & Gwinn, C. R., 2001, ApJ, in press

Lodenqual, J., Canuto, V., Ruderman, M. & Tsuruta, S., 1974, ApJ 190, 141.

Lyne, A. G. & Smith, F. G., 1998, Pulsar Astronomy, Cambridge: Cambridge University Press

Lyutikov, M., 1998, MNRAS, 298, 1198

Lyutikov, M., Blandford, R. & Machabeli, G., 1999, MNRAS, 305, 338

Mész'aros, P., 1992. High Energy Radiation from Magnetized Neutron Stars, Chicago: University of Chicago Press

Pavlov, G. G & Shibanov, I. A., 1978, AZh, 55, 373.

Pavlov, G. G & Zavlin, V. E., 2000, ApJ, 529, 1011.

Shaviv, N. J., Heyl, J. S. & Lithwick, Y., 1999, Mon. Not. Royal Astr. Soc., 306, 333.

Thompson, A. C. & Duncan, R., 1995, MNRAS, 275, 255

Thompson, D., 2000, Proc. International Symposium on High Energy Astrophysics, Heidelberg, (in press) astro-ph/0101039

Usov, V. V. & Melrose, D. B., 1996, ApJ, 464, 306.

BLACK HOLES

Baganoff, F. *et al.*, 2001, ApJ, submitted

Bekefi, G., 1966, Radiation Processes in Plasmas, New York: Wiley

Berry, M. V., 1990, Phys. Today, 12, 34

Blandford, R. D. & Begelman, M. C., 1999, MNRAS, 303, L1

Bower, G. C., Falcke, H. & Backer, D. C., 2000, ApJ, 523, L29

Bower, G. C., Wright, M. C. H., Backer, D. C. & Falcke, H., 2000, ApJ, 527, 851

Cash, W., Shipley, A., Osterman, S. & Joy, M., 2000, Nature, 407, 160

Chiao, R. Y., Tomita, A. & Wu, Y-S, 1989, in Geometric Phases in Physics, ed. A. Shapere & F. Wilczek, Singapore: World Scientific

Falcke, H, Melia, F., & Agol, E., 1999, ApJ 528, L13

Ghez, A., Morris, M., Becklin, E. E., Kremenek, T. & Tanner, A., 2000, Nature, 407,349

LAOR, A., NETZER, H. & PIRAN, T., 1990, MNRAS, 242, 560

LO, K.-Y., SHEN, Z.-Q., ZHAO, J.-H. & HO, P. T. P., 1999, The Central Parsecs, ed. Falcke *et al.*, Berkeley:ASP

MISNER, C, THORNE, K. S. & WHEELER, J. A., 1973, Gravitation, San Francisco: Freeman

RAUCH, K. P. & BLANDFORD, R. D., 1994, ApJ, 421, 46

WALKER, M. & PENROSE, R., 1970, Comm. Math. Phys., 18, 265

ZHAO, J.-H., BOWER, G. C. & GOSS, W. M., 2001, ApJ, in press

Astronomical Masers and their Polarization

By MOSHE ELITZUR

Department of Physics and Astronomy, University of Kentucky, Lexington, KY 40502, USA

Maser radiation occurs naturally in interstellar space. Masers provide extremely bright beacons that trace small scale structure in the host environments, which range from comets all the way to external galaxies. The radiation is sometimes, though not always, polarized. When it is, the polarization can reach much higher levels than in thermal sources—the radiation from some masers is fully polarized. This chapter provides an overview of astronomical masers, discusses the differences between maser and non-maser radiation and covers the fundamental theory of maser radiation and its polarization.

In 1963, the first radio emission from an interstellar molecule, the hydroxyl radical OH, was discovered. The ground state of this molecule produces four radio lines at wavelengths of approximately 18 cm (figure 1). When an atomic or molecular system radiates in several lines, certain ratios are expected between the line intensities. However, almost from the start the emission patterns displayed by the four OH lines in most astronomical sources were peculiar, deviating considerably from expectations. In 1965 these peculiarities culminated with the discovery of radio line emission with such exceptional properties, the emitting substance was dubbed 'mysterium' for lack of an obvious explanation. The exceptional properties included extremely high brightness, line widths much narrower than in all previous cases and substantial polarization. It did not take too long, though, to realize that 'mysterium' radiation was simply maser emission from interstellar OH.

The name MASER is an acronym for Microwave Amplification by Stimulated Emission of Radiation. Lasers, which have become ubiquitous in everyday life, operate on the same principles as masers, only involving visible Light instead of Microwave radiation—hence the replacement of the first letter in the name. Masers were actually the first devices built in the laboratory; it took more than five years to extend the effect to lasers. While the construction of a laser or maser device requires special effort on Earth, the effect occurs naturally in interstellar space, thanks to collision rates which are too low to enforce equilibrium populations. By now, maser emission has been detected from many different interstellar molecules including H_2O (water), SiO (silicon monoxide), CH_3OH (methyl alcohol, also known as methanol), NH_3 (ammonia), CH, HCN and H_2CO (formaldehyde). Maser wavelengths range from $\sim$ 1 mm all the way to $\sim$ 30 cm. Masers occur in many environments, including comets, molecular clouds, star-forming regions, evolved stars, supernova remnants and external galaxies with z up to 0.265. Strong maser emission has been detected also from H recombination lines in the young star MWC 34.

These lectures concentrate on the unique physical aspects of masers in astronomical environments. Since polarization is one of the more intricate properties of astronomical masers, the fundamentals of maser theory are described first. Additional reading is available in a textbook (Elitzur 1992a), a number of comprehensive reviews (Cohen 1989; Reid & Moran 1991; Elitzur 1992b; Moran, Greenhill, & Herrnstein 1999) and proceedings of topical meetings (Clegg & Nedoluha 1992; Andersen 1998).

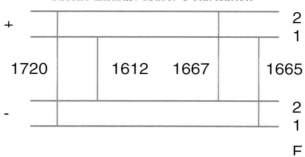

FIGURE 1. Energy levels of the OH ground state, marked by their parity and overall angular momentum (F). Transition frequencies are in MHz

1. Astronomical Masers—Overview

1.1. *Evidence for Maser Action*

How do we know that the radiation from a given source involves maser amplification? After all, the photons do not arrive tagged by the emission mechanism which produced them in the first place. In general, maser radiation distinguishes itself with narrow linewidths, occasional high polarization and, above all, extreme brightness.

When is the brightness considered extreme? It is convenient to express intensity in terms of equivalent brightness temperature T_b, defined as the temperature that a black body should have in order to produce the intensity I_ν observed at frequency ν:

$$I_\nu = B_\nu(T_b). \tag{1.1}$$

In general, the brightness temperature obviously varies with frequency. It becomes frequency independent only when the radiation frequency distribution follows the Planck function.

What sets the scale of brightness temperature in the case of line emission? This process involves two levels separated by an energy gap $\Delta E = h\nu_0$. Denote the statistical weight of each level (the number of degenerate magnetic sub-levels) by g_i (i = 1, 2), the level population by N_i and the population per sub-level by n_i ($= N_i/g_i$). The line emission coefficient is then

$$\epsilon_\nu = N_2 A_{21} h\nu_0 \phi(\nu)/4\pi, \tag{1.2}$$

where $\phi(\nu)$ is the frequency profile arising from the particle velocity distribution and A_{21} is the Einstein A-coefficient of the transition. The line absorption coefficient, taking into account both the absorption and stimulated emission processes, is similarly

$$\kappa_\nu = (N_1 B_{12} - N_2 B_{21})h\nu_0\phi(\nu)/4\pi$$
$$= (n_1 - n_2)g_2 B_{21} h\nu_0 \phi(\nu)/4\pi \tag{1.3}$$

Einstein's relation for the B-coefficients was used in deriving the second form in terms of populations per sub-level, n_i. The expressions for ϵ_ν and κ_ν can be combined to yield the line source function

$$S_\nu = \frac{\epsilon_\nu}{\kappa_\nu} = \frac{A_{21} N_2}{B_{12} N_1 - B_{21} N_2} = \frac{A_{21}}{B_{21}} \frac{1}{n_1/n_2 - 1} \tag{1.4}$$

Now, in thermodynamic equilibrium the level populations follow the Boltzmann distribution. In general, though, the level population distribution is not known and its determination in an arbitrary situation is the central problem in modeling radio-astronomical line emission in general and maser radiation in particular. However, although the population distribution is usually not known, it obviously can always be described by an expression like the Boltzmann relation if T is regarded as a free parameter. This is the motivation for defining the line excitation temperature T_x as

$$\frac{n_2}{n_1} = \exp(-\Delta E/kT_x). \tag{1.5}$$

The populations are then described by the Boltzmann distribution with the equivalent "temperature" T_x. With this definition and Einstein's relation for the A- and B-coefficients we get

$$S_\nu = \frac{2h\nu^3}{c^2} \frac{1}{\exp(h\nu/kT_x) - 1} = B_\nu(T_x). \tag{1.6}$$

The line source function is equal to the Planck function at the line excitation temperature.

With these results, the equation of radiative transfer

$$\frac{dI_\nu}{d\ell} = \kappa_\nu(S_\nu - I_\nu) \tag{1.7}$$

can be cast in the form

$$\frac{dB_\nu(T_b)}{d\tau_\nu} = B_\nu(T_x) - B_\nu(T_b) \tag{1.8}$$

by introducing the optical depth element $d\tau_\nu = \kappa_\nu d\ell$ and replacing the intensity and source term by the Planck functions of T_b and T_x, respectively. Now, the Planck distribution is a monotonically increasing function of temperature for any given frequency. Hence, for the radiation generated inside the source, the brightness temperature cannot exceed the line excitation temperature, namely

$$T_b \leqslant T_x, \tag{1.9}$$

because otherwise $dI_\nu/d\tau_\nu < 0$ and the medium is self-absorbing. Therefore, the brightness temperature, which is a measured quantity, provides a lower limit for the line excitation temperature in the source. Assuming that all the temperatures are in the Rayleigh-Jeans domain, eq. 1.8 becomes

$$\frac{dT_b}{d\tau_\nu} = T_x - T_b. \tag{1.10}$$

This form demonstrates explicitly that T_b cannot exceed T_x in the absence of external radiation. The solution of this equation is

$$T_b = T_x[1 - \exp(-\tau_\nu)] + T_e \exp(-\tau_\nu), \tag{1.11}$$

where T_e is the brightness temperature of an external source which may illuminate the cloud from behind. Although this is only a formal solution, since T_x is not known, it demonstrates explicitly that in the absence of background radiation T_b is smaller than T_x and approaches it only for optically thick lines.

Measured brightness temperatures provide lower bounds for the excitation temperatures in the emitting medium. Although in general line excitation temperatures need not coincide with the source kinetic temperature, they usually do have the same order of magnitude and in most cases are actually smaller. The discovery of OH brightness temperatures as high as 10^{10} K in some astronomical sources was therefore quite a spectacular event. Later on, the water vapor transition discovered at 22 GHz displayed brightness temperatures in excess of 10^{15} K. These enormous brightness temperatures obviously cannot bear any resemblance to the kinetic temperatures in the sources, since molecules dissociate at a few thousand degrees. In addition, the strong maser lines are always quite narrow. If their linewidths are interpreted as resulting from thermal broadening, the corresponding temperatures are usually no more than a few hundred degrees in general, in agreement with the values deduced by other means. Evidently, the reasoning that led from eq. 1.8 to eq. 1.9 does not always apply. Indeed, a hidden assumption was that all the temperatures involved are positive, as is usually the case. Suppose, however, that for one reason or another the populations of the two levels of a transition are inverted, so that $n_2 > n_1$ and T_x becomes negative. The Planck function of a negative argument is negative, and the absorption coefficient κ_ν (and hence also τ_ν) is also negative in this case. The equation of radiative transfer then does not restrict the brightness temperature any more. This is best illustrated by the version involving temperatures (eq. 1.10), which now becomes

$$\frac{dT_b}{d|\tau_\nu|} = |T_x| + T_b \qquad (1.12)$$

Obviously, T_b need not be smaller now than $|T_x|$. In fact, if negative excitation temperature and optical depth are inserted into the solution for the brightness temperature (eq. 1.11), the attenuation term $\exp(-\tau_\nu)$ becomes an amplification factor. An optical depth $|\tau_\nu|$ of more than 20 leads to amplification in excess of 10^8 and could explain the observed brightness temperatures.

One may wonder whether an experimental determination of T_x is possible at all, providing direct proof for population inversion. Unfortunately, eq. 1.11 includes two unknown quantities — the excitation temperature T_x and the optical depth τ. Both, obviously, cannot be determined from a single equation. Suppose, however, that an interstellar cloud happens to lie in front of a background continuum point source. Intensity can now be measured in two different directions, including and excluding the point source. This produces two brightness temperatures, with and without the term involving T_e. Since T_e can be measured at frequencies outside the line emission, this procedure provides two independent equations that can be solved for the two unknowns T_x and τ.

This clever technique was utilized by Rieu et al. (1976) to study an interstellar cloud in front of the extragalactic radio source 3C123. Their results for three of the OH ground-state transitions are displayed in figure 2 which shows the line spectra. The plotted quantity is the antenna temperature with the background subtracted so that the baseline corresponds to $T_a = 0$. In the so-called "off source" measurements, those in which 3C123 is outside the telescope beam, all three lines appear in emission. For the "on source" measurements, in the direction of 3C123, the plotted quantity is

$$T_b - T_e = (T_x - T_e)[1 - \exp(-\tau_\nu)] \qquad (1.13)$$

because of the baseline subtraction. Now the lines at 1665 and 1667 MHz appear in absorption, indicating that their excitation temperatures are lower than the brightness temperature of 3C123. In contrast, the 1720 MHz line emission is now stronger than before, so the cloud is actually *amplifying* the background radiation in this line instead of absorbing it. The maser amplification is rather weak, $|\tau|$ is only ~ 0.1. Nevertheless, this is one of the most spectacular maser effects observed in radio astronomy, since it directly demonstrates how a cloud can act as an amplifier for background radiation. The inferred excitation temperature is ~ -10 K. This experiment also demonstrates how maser action is sometimes recognized even though the intensity involved is not exceptional. In general, astronomical masers are identified through extreme brightness temperatures or unusual emission patterns.

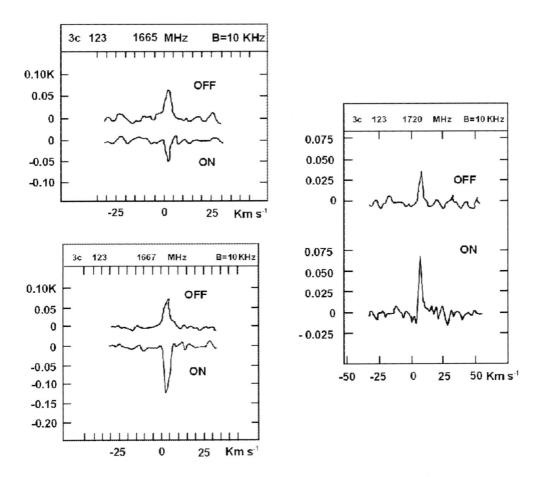

FIGURE 2. OH spectra toward 3C123 (from Rieu et al. 1976).

2. Sample Maser Sources

Amplified radiation is fundamentally different from thermal radiation. Some of the differences affect very basic properties of the data and must be recognized prior to any discussion of the maser observations.

2.1. *Peculiarities of Amplified Radiation*

Maser intensity greatly exceeds thermal emission, i.e., $I_\nu \gg S_\nu$. The source function can be neglected in the radiative transfer equation (1.7), which becomes†

$$\frac{dI_\nu}{d\ell} = \kappa_\nu I_\nu \tag{2.14}$$

From its solution, the amplified intensity at an arbitrary point ℓ along a given path obeys

$$I_\nu(\ell) = I_\nu(\ell_1) \exp \tau_\nu(\ell, \ell_1) \tag{2.15}$$

† Since the maser absorption coefficient and optical depth are negative, from here on κ and τ stand for the absolute values of these quantities.

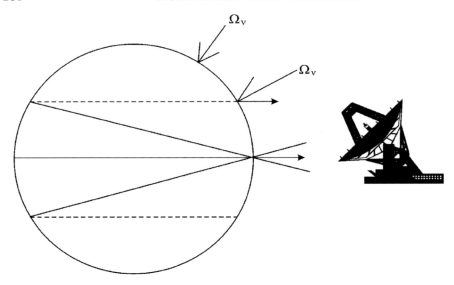

FIGURE 3. Beaming in a spherical maser.

where $\tau_\nu(\ell, \ell_1) = \int_{\ell_1}^{\ell} \kappa_\nu d\ell$ is the optical depth along the path from some fiducial point ℓ_1. The larger is τ the higher the intensity. Two immediate consequences are important for interpreting maser observations:

    • **Velocity Coherence.** Stimulated emission requires a perfect frequency matching between the parent photon and its interaction partner to within the natural line width (roughly the inverse life time of the transition). Since maser amplification is impossible for molecules whose transition frequency is Doppler shifted, it can occur only along paths that maintain good coherence in their line-of-sight velocity component. As a result, maser emission from sources such as rotating disks or expanding winds occurs preferentially along certain directions. Furthermore, since the interstellar medium is highly turbulent, maser sources tend to break-up into many small spots, each of which maintains by chance good line-of-sight velocity coherence.

    • **Beaming.** Maser radiation is highly beamed in all sources with an appreciable amplification no matter what their shape. The reason is that the intensity always increases with distance traveled in the source, and it is impossible to devise a geometry where all the rays are of equal length. At any given location in a source the ray path-length is different in different directions, whatever the geometry. Maser radiation is strongest along the direction that gives the longest chord through the source.

Consider for example the quintessential isotropic maser, a sphere. At every point except the center, the radial direction provides the longest path through the source and the radiation is strongest in this direction. The emerging radiation is tightly beamed into a solid angle Ω_ν centered on the radial direction (figure 3); as the amplification increases Ω_ν decreases. Because of the beaming, a given point on the surface is visible only to observers located inside its beaming cone. Conversely, the only points on the surface visible to a given observer are those whose beaming cones contain that observer's location. The points furthest removed from the axis connecting a distant observer and the sphere's center and still visible to that observer are those whose beaming cones graze the observer's

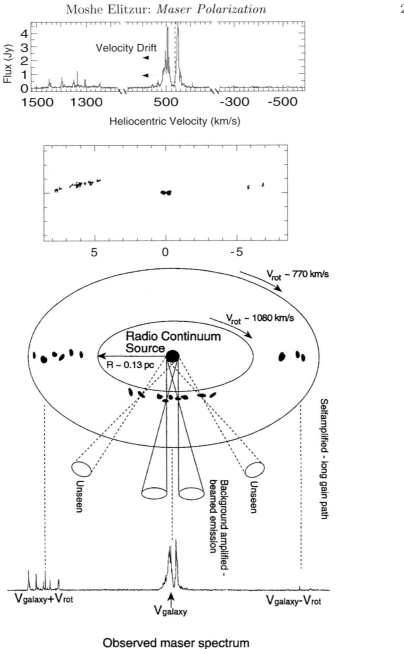

FIGURE 4. H$_2$O masers in NGC4258 (from Moran et al, 1999 and Bragg et al, 2000)

direction. Only a fraction of the source, indicated with dashed lines in the figure, is effectively visible, and the maser appears a lot smaller than it actually is. As the sphere increases, the amplification becomes stronger and the beaming angle smaller, increasing the disparity between the observed size and the actual physical radius.

2.2. *Black Hole in NGC4258*

The central role of beaming and velocity coherence is manifest in the H_2O maser emission from the Seyfert galaxy NGC4258, which provides the most compelling evidence that active galactic nuclei contain massive black holes (Moran, Greenhill & Herrnstein, 1999). The maser spectrum, shown at the top panel of figure 4, displays three groups of emission; one is centered on the systemic velocity (476 km s^{-1}), the other two shifted by velocities that range from ~ 770 km s^{-1} to ~ 1100 km s^{-1} on either side of the central component. There is no emission between those three groups. High resolution interferometry, shown in the mid panel, shows that the masers occupy an extremely narrow strip in the sky (the scale of this map is milli-arcsecond). Maser positions, too, are bunched in three distinct groups and each of them corresponds to a different spectral group. The blue- and red-shifted features are symmetrically located on the two sides of the systemic-velocity features—precisely the structure expected from a rotating disk.

The bottom panel shows an artist's sketch of the geometry. The masers are located in a rotating disk, viewed almost perfectly edge-on. The paths of longest line-of-sight velocity coherence are along the chords marked with dotted lines. The one in the radial direction corresponds to the systemic-velocity maser group, strongest because it amplifies the nuclear continuum radio radiation, the others to the components that are blue- and red-shifted by the rotation velocity. Although we observe only a few features, masers fill the entire disk and emit in all directions in its plane; those that are not visible to us are beamed in other directions.

A time sequence of interferometric mapping shows that, as expected, the systemic-velocity features do indeed move in the deduced rotation direction (see top panel). The precision afforded by the maser observations allows a complete construction of the source geometry and kinematics. The distance to NGC4258 is 7.2 ± 0.3 Mpc, the inner radius of the maser region is 0.14 pc, its outer radius is 0.28 pc and the disk thickness is less than 0.0003 pc. The maser velocities follow the Keplerian rotation law extremely tightly, determining the central mass as 3.9×10^7 $M_\odot$. The containment of such a large mass in such a small radius provides the best evidence yet for a black hole.

2.3. *Red Giant Winds*

The OH 1612 MHz masers in evolved stars provide another striking manifestation of the effects of beaming and velocity coherence. These stars are surrounded by circumstellar shells rich in molecules that emit maser radiation. Each maser molecule has a different set of energy levels, and so each radiates from the region in the stellar wind where conditions trigger its particular population inversion. SiO maser emission originates from just above the edge of the stellar atmosphere, H_2O from a larger shell with radius up to $\sim 10^{15}$ cm. OH main line 1665 and 1667 MHz masers are generated at $\sim 10^{15}$–10^{16} cm, 1612 MHz masers in a larger shell at $\sim 10^{16}$ cm. The intensity profiles of the latter are characterized by a distinct double-peak shape like the one shown in figure 5 for the star OH127.8. The peak separation is typically 20–50 km s^{-1}. This signature is so distinctive that it has been used to identify red giants at various locations in the Galaxy even when the optical emission from the star itself is obscured by intervening dust.

The double-peak profile arises naturally from the radial motion of the stellar wind. OH molecules in different sectors of the shell move in different directions and thus have large relative velocities. They cannot interact radiatively. Conversely, molecules on a given radial line move in the same direction at similar speeds; they are almost at rest with respect to one another. As a result, photons emitted by a molecule will affect only other molecules along the same radial direction, and maser amplification is possible only for radiation propagating inward and outward along a line through the center of the

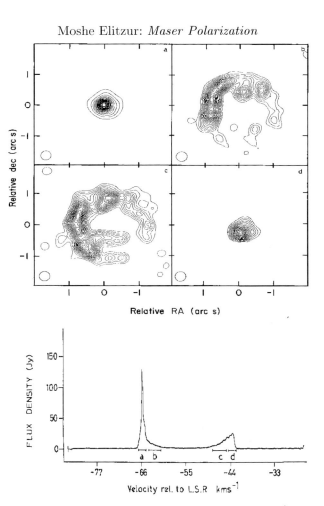

FIGURE 5. OH 1612-MHz maser profile and maps of the star OH127.8 (from Booth et al. 1981)

shell. The OH maser emission pattern resembles a hedgehog, with rays sticking out like spikes. At any given location, an observer can only detect the emission from the two regions on the shell along the line of sight to the central star. The signal from the front is blue-shifted, that from the back red-shifted. The velocity separation between the peaks is twice the shell expansion velocity and the mid-point corresponds to the stellar velocity. Each emission region is a small, circular section resembling a spherical cap.

Direct confirmation of the "front-back" explanation was provided in an elegant experiment by Booth et al. (1981) who performed precise interferometric observations of OH127.8 in various frequency intervals. In figure 5, the bracketed velocity intervals a—d in the 1612 MHz spectrum represent the ranges of velocity for which the spatial distribution of maser emission has been mapped with interferometry. The individual maps display the spatial distribution of OH emission in each of the corresponding velocity intervals. The contours outline points of equal intensity in the plane of the sky with positions measured as angular separations from a common origin. In each map the contour interval is 5% of the peak emission in that velocity interval and the lowest contour is ~ 10% of the peak. These maps show clearly that the radiation of the two peaks is emitted from compact, well defined caps that are also coincident in position along the

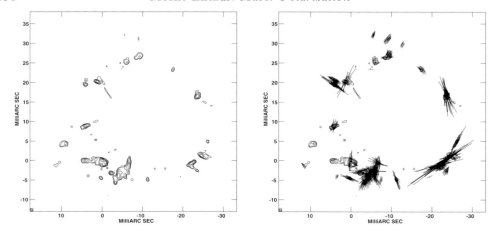

FIGURE 6. SiO maser emission in TX Cam. The intensity contour map (Stokes parameter I) is on the left, the linear polarization map on the right (from Kemball & Diamond 1997)

line of sight (maps a and d). By contrast, the radiation from the inner shoulders of the peaks covers a larger, circular region (maps b and c), as expected from an expanding shell. The incomplete and clumpy appearance of the shell reflects the maser region's deviations from pure spherical symmetry. Such irregularities are to be expected in a turbulent stellar wind.

2.4. *Red Giant Atmospheres*

In contrast with the distant location of OH, SiO maser emission originates from the immediate stellar vicinity, inside the dust formation zone and the onset of the wind. Thanks to modern interferometric capabilities and the high-brightness and small dimensions of masers, positions of individual maser spots can be determined to within a fraction of milli-arcsecond (mas), spot separations at the level of 0.01 mas. This enables a glimpse of stellar surfaces, a feat impossible for optical astronomy which cannot resolve the surface of any star other than the Sun.

A dramatic display of these capabilities is provided by the SiO maser imaging of the Mira variable TX Cam (figure 6). The observations were performed with the VLBA in the $v = 1$, $J = 1$–0 transition of SiO (Kemball & Diamond 1997). The left panel shows an intensity map of the maser emission. The ring-like structure indicates that the masers reside in a shell, presumably centered on the star, and emit preferentially along tangential directions because they provide the longest amplification path-lengths. The observed emission ring has a radius ~ 4.8 AU and width ~ 0.7 AU, implying that the masers probe the conditions in a spherical shell located within $\sim$ two stellar radii between the photosphere and the dust formation radius. Kemball & Diamond are producing similar maps at two-week intervals and have already accumulated a continuous coverage that exceeds the period of this star, which is 80 weeks. These maps enable study of surface activity similar to that afforded by Sun spots.

The right panel shows the polarization map, superimposing the linear polarization vectors over the intensity contours. The directions of plotted vectors indicate the plane of the electric vector and their lengths are proportional to the linearly polarized intensity. From this polarization map, Kemball & Diamond conclude that the magnetic field is poloidal. A close examination of the polarization map shows evidence for sharp 90° turns of the polarization direction, a behavior we discuss below in section 5.2.2.

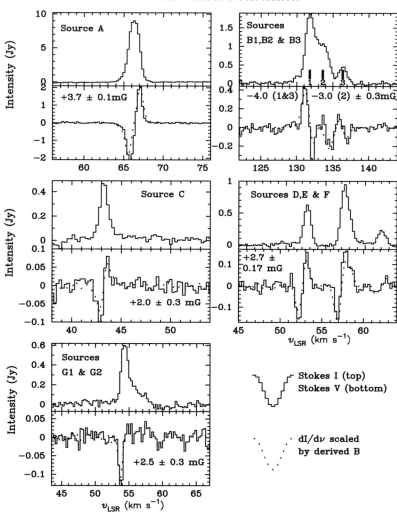

FIGURE 7. Profiles of the Stokes parameters I (upper panels) and V (lower panels) for various OH 1720 MHz masers at the Galactic center region Sgr A (from Yusef-Zadeh et al. 1996)

2.5. *Supernova Remnants*

A new class of masers emitting in the OH 1720 MHz transition was recently discovered in supernova remnants. These masers display strong circular polarization whose V-profiles show the anti-symmetric S-shape typical of Zeeman pattern when the components are separated by a small fraction of the thermal width. These profiles are shown in figure 7 for various OH 1720 MHz masers at the Galactic center region Sgr A (Yusef-Zadeh et al. 1996). Profiles of the Stokes parameter I are shown in the upper panels, V in the lower panels. The dotted lines superposed on the V spectra are the scaled derivatives of the corresponding I spectra. The scaling factors were used to determine the magnetic fields listed in the figure. The theory behind this procedure is discussed below in section 5.2.5.

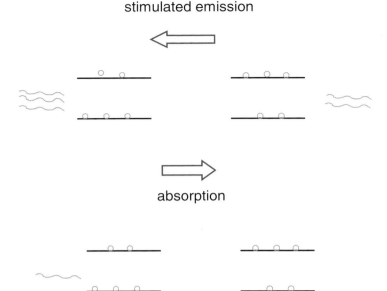

FIGURE 8. Schematic description of stimulated and spontaneous emission.

3. Fundamentals of Maser Emission

The radiation from maser and non-maser sources differs in many fundamental ways, and this section discusses those differences. Following common practice, all non-maser line emission is called "thermal". This covers every case in which the populations are not inverted even when they are not in thermal equilibrium or even excited by collisions.

3.1. *Spontaneous and Stimulated Emission*

At the core of the maser effect is the process of stimulated emission, first identified by Einstein (1917) in his classic paper on the theory of line radiation. His reasoning was based on the invariance under time reversal of all the fundamental laws of physics. If a movie of any microscopic process is run in either direction, each showing will depict an equally probable sequence of events. The top segment of figure 8 shows two frames from such a movie, involving particles (circles) populating the two levels of a transition and photons (wavy lines) whose frequencies match precisely the level energy separation. The difference between the two frames is the number of photons (one less on the right) and the distribution of particles between the energy levels. When the left frame is considered the movie's first and the right one its last, the sequence describes absorption. The reverse sequence, occurring in nature with equal probability, is called *induced* or *stimulated emission.*

Although stimulated emission was introduced invoking the photon concept, this is not necessary. Because line radiation involves discrete energy states, its proper description must utilize quantum mechanics for the atomic system. But since the radiation wavelength is many orders of magnitude larger than particle dimensions, there is no need to quantize also the radiation field. Most studies of the interaction of matter with maser radiation, notably since Lamb (1964) whose approach was adopted for astronomical masers by Litvak (1970) and Goldreich, Keeley & Kwan (1973; GKK hereafter), employ a hybrid, semi-classical approach: The energy levels are eigenstates of the system Hamiltonian, treated by quantum theory. Transitions between the levels are caused by interactions with the radiation field, which is described by standard classical electromagnetic waves and treated as a perturbation. The transition rates for both absorption and stimulated emission are obtained from the product of the (quantum) matrix element of the transition dipole moment with the (classical) intensity of the electromagnetic radiation field. Since this is a complete description, not merely a classical analog, an important consequence is that there are no properties of the radiation generated in stimulated emission that are peculiar to the quantum theory; we must be able to fully deduce all of them from purely classical electromagnetic concepts.

A left-to-right view of the bottom sequence in figure 8 depicts the absorption of a single photon by the system particles. From invariance under time reversal, the reverse process must occur with an equal probability. This right-to-left sequence describes the *spontaneous emission* of a single photon by a particle making the transition to the lower level in the absence of any external radiation. But such a transition cannot occur in standard treatments of quantum theory because the energy levels are stationary states of the system Hamiltonian, completely stable in the absence of external perturbations. This transition occurs only when quantization of the electromagnetic field is taken into considerations and can be interpreted as scattering off vacuum fluctuations.

In contrast with stimulated emission, spontaneous emission is a purely quantum process. It has no classical analog and must be analyzed in terms of the photon description of the radiation field.

3.1.1. *Induced Photons*

The process of stimulated emission is sometimes described as a photon absorption followed by the emission of two photons into the phase space cell of the absorbed photon. Because of energy and momentum conservation, the induced photon has the same frequency and direction as the parent photon. However, contrary to some widespread misconceptions, *the induced photon does not have the same*

- **Phase:** The argument of the oscillatory behavior of any wave is its phase $\phi = \phi_0 + \omega t - \mathbf{k} \cdot \mathbf{x}$, where ω is the angular frequency and $\mathbf{k}$ is the wave vector. While an electromagnetic wave has a phase, a photon does not. The uncertainty principle states that

$$\Delta E \Delta t \geqslant \hbar. \tag{3.16}$$

Consider photons that couple to a transition whose energy separation is $\hbar \omega$. The uncertainty in the energy of a state containing n such photons is $\hbar \omega \Delta n$, leading to the uncertainty relation between phase and photon number

$$\Delta n \Delta \phi \geqslant 1. \tag{3.17}$$

The phase of a state with a well-defined number of photons is completely undetermined. Conversely, wave functions for the radiation field with definite phases require superposition of states with different photon numbers ranging all the way to infinity (Glauber 1963). A "phase" for the induced photon is meaningless.

- **Polarization:** The induced photon polarization is not necessarily equal to that of the parent photon. It is determined instead by the change in magnetic quantum number m of the interacting particle. A $\Delta m = 0$ transition couples to photons linearly polarized along the quantization axis while $\Delta m = \pm 1$ transitions

$$\Delta m = 0 \qquad\qquad \Delta m = \pm\, 1$$

FIGURE 9. Polarization properties of different Δm transitions.

couple to photons that are right- and left-circularly polarized in the plane perpendicular to the quantization axis (figure 9). Consider first the interaction of a linearly polarized parent photon with particles in the upper level of spin $1 \to 0$ transition. When the interacting particle is in the $m = 0$ state it executes a $\Delta m = 0$ transition and the induced photon, too, is linearly polarized. Consider next, though, the case of a particle in one of the $|m| = 1$ states. Because a linearly polarized photon can be described also as a coherent mixture of two circularly polarized photons, it will now induce a $|\Delta m| = 1$ transition. The induced photon is now circularly polarized, although the interaction amplitude is reduced.

3.2. *Coherence and Incoherence in Astronomical Masers*

One of the hallmarks of laboratory laser radiation is the phase coherence across its wave front—different waves are oscillating in unison, their amplitudes reaching maximum together. For such a coherence to exist, the phase difference $\Delta\phi$ between different waves must be less than unity in the entire source. In laboratory lasers this coherence is maintained by specially designed resonant cavities and is not a prerequisite for the amplification process per se. Large dimensions and broad linewidths make such coherence impossible in astronomical masers. The phases of waves that start exactly the same at any given time and position will always diverge away from each other during subsequent travel in the source.

Consider first two waves that have exactly the same frequency ω, and thus the same wave number $k = 2\pi/\lambda$, traveling along slightly different directions separated by a small angle θ. The phase difference these waves will accumulate along a distance ℓ is

$$\Delta\phi = k\Delta\ell \simeq k\ell(1 - \cos\theta) \simeq \pi\frac{\ell}{\lambda}\theta^2. \qquad (3.18)$$

The dimensions of astronomical masers are at least a few pc and ℓ/λ always exceeds $\sim 10^{14}$, therefore $\Delta\phi$ is small only so long as $\theta \lesssim 10^{-7}$. However, although maser radiation is beamed, the beaming angles are never that small. Reasonable estimates give for the beaming angles $\theta \sim 10^{-2}$–10^{-1}, much too large to maintain a small $\Delta\phi$.

Consider next two waves propagating in exactly the same direction but with slightly different frequencies, separated by $\Delta\omega \ll \omega$. During a time interval Δt such waves accumulate a phase difference

$$\Delta\phi = \Delta\omega\Delta t. \qquad (3.19)$$

Line widths of astronomical masers always obey $\Delta\omega \gg 10^4$ s^{-1}, much larger than typical collisional and radiative rates. The phase differences generated across the radiation band-

	Thermal Source	Maser
Emission process:	spontaneous emission	stimulated emission
Process inherent nature:	purely quantum	classical analogue
Effect of propagation:	absorption	amplification
Coherence:	completely incoherent	line-of-sight velocity coherence
Visible region:	only edge of the source	radiation has sampled the entire length

TABLE 1. Fundamental properties of thermal vs maser sources

width between successive interactions are always much larger than unity, and coherence is impossible.

Since the radiation is phase incoherent across the wave front, many properties of laboratory laser radiation that are widely perceived to be fundamental characteristics of this phenomenon are missing in astronomical masers. At any point in the source, maser photons are generated by the interaction of particles whose velocities are distributed at random with radiation that arrives randomly at that point. Still, at a given frequency and along a given ray, the different particles involved in successive interactions are all in perfect tune with the photon wave vector (to within the natural line width), resulting in remarkable coherence, as evidenced by the extreme brightness temperatures. Recall that the maser intensity at an arbitrary point ℓ obeys $I_\nu(\ell) = I_\nu(\ell_1) \exp \tau_\nu(\ell, \ell_1)$ (eq. 2.15). Since every point in the maser can be considered the input source for every subsequent point, the entire maser is effectively coupled.

In contrast, no induced photons are detected from thermal sources. There the interaction with matter results in net absorption because of the excess population in the lower level. The photons leaving thermal sources were generated in spontaneous emissions within $\sim$ one optical depth from the surface. The fundamental differences between thermal and maser sources are summarized in the accompanying table 1.

4. Phenomenological Maser Theory

Another fundamental difference between maser and thermal line radiation is their effect on the populations of the levels that generated the radiation in the first place. Level populations are determined in optically thin sources by the collision rates and A-coefficients of relevant transitions. Increasing the optical depths, in thermal sources the only effect of interactions with the trapped radiation is to accelerate the approach of level populations to thermal equilibrium. Maser radiation, in contrast, has a rather drastic effect on the level populations.

A study of this effect requires considerations of levels other than the maser levels. Particle exchange between the two levels of any transition does not produce population inversion, the prerequisite for maser action. Inversion can only occur as a result of particle cycling through other levels. A description of the maser effect requires rate equations that include population exchange with other levels that do not directly interact with the maser radiation. Consider a much simplified model for the populations of the two maser levels in which the interaction with all other levels is summarized by phenomenological pump terms p_1 and p_2 and a loss rate Γ, assumed equal for both levels for simplicity (figure 10). The total pump rate into the maser system is $p = p_2 + p_1$ and an inversion occurs if $\Delta p = p_2 - p_1 > 0$, in which case the inversion efficiency is $\eta = \Delta p/p$. To

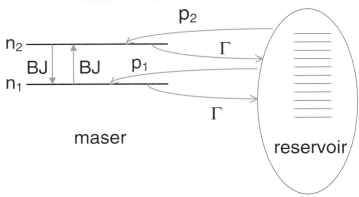

FIGURE 10. The interaction between a maser system and its reservoir.

further simplify matters, neglect spontaneous decays and collisional population exchange between the maser levels, and assume the same statistical weight for both; removing these assumptions only increases the complexity of resulting expressions without any fundamental change in the results. Then in steady-state

$$p_2 = \Gamma n_2 + BJ(n_2 - n_1)$$
$$p_1 = \Gamma n_1 - BJ(n_2 - n_1), \tag{4.20}$$

where B is the Einstein B-coefficient of the maser transition and $J = \int I \, d\Omega/4\pi$ is the angle-averaged maser intensity. Because maser radiation is beamed,

$$J = \frac{\Omega_b}{4\pi} I \tag{4.21}$$

where Ω_b is the beaming angle and I is the intensity along the beam axis. From the rate equations, the overall population of the maser system is

$$n = n_2 + n_1 = \frac{p}{\Gamma}, \tag{4.22}$$

fully determined by the interaction with all other levels and unaffected by the maser intensity. The population difference is

$$\Delta n = n_2 - n_1 = \frac{\Delta p}{\Gamma + 2BJ}. \tag{4.23}$$

4.1. *Unsaturated Maser; $BJ \ll \Gamma$*

Consider first masers with $BJ \ll \Gamma$, called *unsaturated* for reasons that will become clear shortly. The population inversion is then

$$\Delta n_0 = \frac{\Delta p}{\Gamma} = \eta n, \tag{4.24}$$

fully determined by the pumping scheme. The interaction with the maser radiation has no effect on the level populations, since it is negligible. The maser system enjoys a free, uninterrupted exchange with the reservoir. Particles are pumped into the system at the rate p and leave at the rate Γ before they had a chance to interact with the radiation.

The maser particles are practically oblivious to the radiation, the occasional radiative interaction that they have hardly affects the population inversion. But these sparse interactions have a dramatic effect on the radiation itself. The maser absorption coefficient, obtained by inserting Δn into the expression from equation 1.3, becomes $\kappa_0 \propto \Delta p/\Gamma$.

Since κ_0 is constant, the solution of the radiative transfer equation (1.7) is

$$I = S\left[\exp(\kappa_0 \ell) - 1\right]. \tag{4.25}$$

The amplified term dominates decisively once $\kappa_0 \ell \gtrsim 1$ and the intensity then increases exponentially with distance traveled in the source. The inverted population provides an amplifying medium because stimulated emissions outnumber absorptions. The exponential growth of the intensity reflects the shower of induced photons that can be generated by a single seed photon; the power series expansion of the exponential factor describes successive generations of induced photons. Since κ_0 is proportional to the pump rates, the intensity of an unsaturated maser responds exponentially to variations in pumping conditions. Such a maser can be expected to display erratic time variability.

In an unsaturated maser τ increases linearly with pathlength, therefore the intensity grows exponentially. The exponential growth reflects an amplification rate that remains uniform in the entire source because the radiation has no effect on the level populations. From its very description, the maser must be weak. Unsaturated masers display impressive growth but cannot be very strong since the radiative rate must remain negligible in comparison with other rates.

4.2. Saturated Maser; $BJ \gg \Gamma$

Exponential amplification cannot go on forever. Once the radiation grows to the level that BJ becomes comparable to Γ, the interaction with the maser radiation begins to affect the level populations. Each absorption adds to the inversion, each induced emission subtracts from it. And since induced emissions outnumber absorptions, the population inversion begins to decrease, an effect called *saturation*. Introduce

$$J_s = \frac{\Gamma}{2B}, \tag{4.26}$$

called the *saturation intensity*. From equation 4.23, the population inversion can be written as

$$\Delta n = \Delta n_0 \frac{J_s}{J + J_s}. \tag{4.27}$$

The interaction with the maser radiation reduces the inversion once $J > J_s$. The saturation intensity in astronomical masers is typically $\sim 10^5$–10^7 times larger than the source function, the intensity generated in spontaneous decays. Saturation becomes a factor when $\tau \gtrsim \ln 10^5 = 11$.

When $J \gg J_s$, the rate for interaction with the maser radiation greatly exceeds the loss rate and particles pumped into the maser system remain essentially trapped there; the particle just moves from one level to the other as a result of interactions with the line photons, never getting a chance to leave the maser system back to the reservoir. The populations of the maser levels become equalized because the particles are merely going up and down between them. Indeed, when $J \gg J_s$ the inversion becomes $\Delta n = \Delta n_0 J_s/J \to 0$; the longer a particle is trapped in the maser system, the less the memory of the original pumping event that landed it there. From eq. 1.3,

$$\kappa = \kappa_0 \frac{J_s}{J + J_s} \underset{J \gg J_s}{\simeq} \kappa_0 \frac{J_s}{J} \to 0. \tag{4.28}$$

Saturation reduces the amplification coefficient and the maser growth rate decreases. Consider, though, the generation rate of maser photons. The net rate per unit volume is the difference between the number of stimulated emission and absorption events, and from equation 4.27

$$BJ\Delta n = \frac{1}{2}\Delta p \frac{J}{J+J_s} \xrightarrow[J\gg J_s]{} \frac{1}{2}\eta p.$$ (4.29)

While the amplification rate is decreasing, the photon production rate is *increasing* toward its maximum possible value, $\frac{1}{2}\eta p$, hence the term saturation. In a strongly saturated maser, every pumping event results in the production of a maser photon with the maximal pump-determined efficiency. The extra factor of $\frac{1}{2}$ enters because a particle spends on the average half its time in the upper level when trapped in the maser system. In contrast, the photon production rate in an unsaturated maser is reduced from this maximum by a factor J/J_s (< 1).

Saturated masers produce as many photons as possible per unit volume, and the photons control the level populations. The actual magnitude of the inversion is irrelevant. It is merely adjusted by the radiative interactions to ensure just the right number of losses back to the reservoir, as required by steady state.

4.3. Maser—the Amplifier/Converter

The processes that control particles and photons undergo a complete role reversal during saturation. In unsaturated masers the pump rates control the population inversion, which in turn controls the radiation growth. In saturated masers the pump controls the photon production rate, the radiation in turn controls the inversion. Some insight into these fundamental differences can be gained by rearranging the maser radiative transfer equation 2.14 into the two equivalent forms

$$\frac{dI}{d\ell} = \kappa_0 I \times \frac{J_s}{J+J_s}$$

$$= C\,\eta p \times \frac{J}{J+J_s}$$ (4.30)

where $C = \frac{1}{2}h\nu\,(4\pi/\Omega_b)$. The first form invokes equation 4.28 for the maser amplification coefficient, the second the explicit expressions for the level populations (eq. 4.23) and the beaming angle Ω_b (eq. 4.21). While both forms are completely equivalent, their physical interpretation is quite different. The factor $\kappa_0 I$ in the first expression is a standard amplification term—propagating maser intensity increases in proportion to itself. The amplification is modified by the efficiency factor $J_s/(J+J_s)$ which is unity in an unsaturated maser, decreasing in saturated masers inversely with the degree of saturation J/J_s. The second form shows that the intensity growth per unit length is determined by the efficiency with which the maser converts pumping events to maser photons modified by the factor $J/(J+J_s)$. This efficiency factor reaches unity only when the maser is strongly saturated, and such a maser acts as linear converter of pumping events to maser photons. Unsaturated masers operate at a much lower efficiency, though, since their conversion efficiency is reduced by the degree of saturation J/J_s.

The reason for the different behavior in the different saturation regimes is simple. The production of maser photons in stimulated emission requires both a particle in the upper level and an interacting photon. A particle pumped into the upper level of an unsaturated maser leaves the system before it had a chance to participate in a stimulated emission—this is what the condition $BJ \ll \Gamma$ states. Most pumping events go wasted with regard to the radiation field because their only outcome is to get a particle in and out of the maser system without any maser interaction. The unsaturated maser produces exponential amplification precisely because it is an inefficient converter. On the other hand, once saturated, the maser converts pumping events to maser photons as efficiently as possible. Although the population inversion and the amplification are reduced by

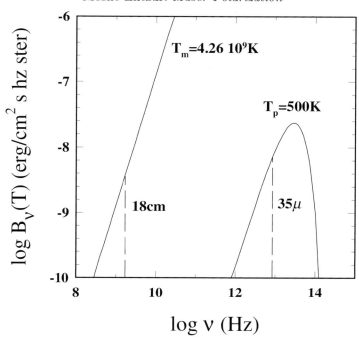

FIGURE 11. Brightness temperatures of pump and maser radiation for wavelengths applicable for the ground state of OH.

saturation, the net stimulated-emission rate increases toward its limit. Because the conversion efficiency of pump events to maser photons no longer depends on position in the source, the luminosity increases only linearly with dimensions instead of the exponential growth in unsaturated masers. This lower growth rate and the decrease in population inversion should not be confused with the fact that maser emission during saturation reflects the highest possible efficiency; for a given pumping scheme, a saturated maser always emits more than an unsaturated maser.

Determining the degree of saturation in any given source is difficult and usually must rely on indirect arguments. Nevertheless, observations seem to indicate that maser sources may be broadly divided into two classes: (1) large and relatively low density ($\lesssim 10^4$ cm^{-3}) clouds which lead to unsaturated, very weak ($\tau \lesssim 1$) maser action, and (2) compact and dense regions (densities in excess of at least 10^5 cm^{-3}) with very strong masers, usually saturated. It appears that sources in between, i.e., strong unsaturated masers with $1 < \tau \lesssim 10$, are the exception rather than the rule. This rough dichotomy can perhaps be explained with a plausibility argument. Consider the situation when τ varies linearly with some underlying property of the source, such as the density. Maser output would then vary linearly at small values of this physical parameter, where the unsaturated exponential amplification behaves linearly with τ, and also at large ones, where the maser saturates. The transition from one type of operation to the other occurs over a short interval of parameter space due to the fast exponential variation at the strong ($\tau > 1$) but unsaturated region. Under such circumstances, the apparent scarcity of such sources simply reflects statistics, since there are fewer of them.

4.4. *The Essence of the Maser Effect*

Some further insight into the maser effect can be gained from radiative pumps. It is easy to show that the pump rate of a radiatively pumped maser is equal to the number of pump photons absorbed per unit time (e.g. Elitzur 1992). Figure 11 displays the consequences for the OH ground-state 1612 MHz maser, which is pumped by radiative excitations of rotational states. The figure plots certain portions of the Planck function at two different temperatures. The dashed line under the $T_p = 500$ K Planckian marks its intensity at 35 μm, the wavelength of an OH rotational pump transition. The length of this line therefore corresponds to the number of pump photons. The other dashed line, at 18 cm, corresponds to the number of maser photons produced from the pump photons when the overall conversion efficiency is 50%. The brightness temperature required to pass a Planckian through the tip of this line is now $T_m = 4.26 \times 10^9$ K. A modest pump temperature is boosted many orders of magnitude to a spectacular maser brightness temperature even though the number density of maser photons is actually smaller than that of pump photons. The reason is simple. Both wavelengths are in the Rayleigh-Jeans domain, where $B_\nu(T) = 2kT\nu^2/c^2$. The factor ν^2 reflects the photon phase-space density and is widely different at the pump and maser frequencies ν_p and ν_m, respectively. This leads to a large temperature enhancement factor $(\nu_p/\nu_m)^2$, the primary reason for the spectacular brightness temperatures displayed by maser radiation. The inverted population enables the maser to efficiently shift photons from high to low frequencies, where their number density is much higher than that allowed in thermal equilibrium. This point is the essence of the maser effect.

While the number of photons is roughly conserved in the pumping cycle, their energy is degraded by ν_m/ν_p. The very same factor responsible for the spectacular brightness temperature also ensures that the energy flux carried in the maser line can never be as important as that in the pump-cycle lines in the overall energy balance of the maser source. Maser emission plays no role in energy considerations. However, detection instruments are based on photon counting and without the maser effect line emission from many sources would be undetectable.

4.5. *Linear Masers*

The transition to saturation and its effect on the maser structure can be studied only in the context of specific geometries. The simplest one involves the linear maser model where there are only two rays of radiation moving left and right. The only source of seed photons is spontaneous decays, no external radiation enters the maser.

Denote positions in the maser by the coordinate z, which varies in the interval $[-\ell, \ell]$, and the intensity of the rightward moving stream $I_+(z)$. From symmetry, the intensity of the opposite stream is $I_-(z) = I_+(-z)$. At the center $I_+(0) = I_-(0) = J(0)$, at all other points the angle-averaged intensity becomes $J(z) = \frac{1}{2}[I_+(z) + I_+(-z)]$. As a result, the absorption coefficient involves non-local effects introduced by the saturation factor J/J_s. Consider first a maser sufficiently small that $J \ll J_s$ everywhere (the maser in unsaturated). Since $\kappa = \kappa_0$ in that case there are no non-local effects and radiative transfer can be solved at once. If the maser is sufficiently strong that the source function S can be neglected in comparison with $J(0)$, the intensities of the two streams obey

$$I_\pm(z) = J(0)\exp(\pm\kappa_0 z). \tag{4.31}$$

Evidently, the outward moving stream completely dominates in each half of the maser; linear maser emission is essentially uni-directional. Consider now a succession of models with increasing ℓ while all the other parameters are held fixed. At a certain length the end intensities $J(\pm\ell)$ exceed the saturation parameter J_s. When the maser length is

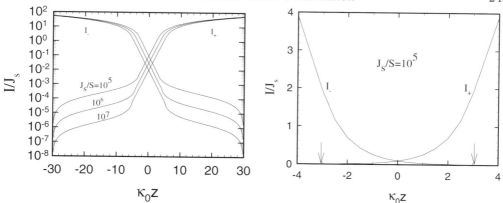

FIGURE 12. Intensities in linear masers.

further increased, the material added beyond the saturation length is subject to radiative intensity higher than J_s, resulting in saturated maser action. The maser develops a three-zone structure: an unsaturated core where $J < J_s$ and two saturated exterior zones where $J > J_s$.

The left panel of figure 12 shows the intensities in masers of equal length $\ell = 30$ and representative values of J_s/S. The maser central region, $|\kappa_0 z| \lesssim 5$, displays unsaturated exponential growth, outside this core the dominant stream grows only linearly. The logarithmic scale, necessary for the display of the full solution, is somewhat misleading in the prominent exposure it provides to negligible intensities, so the right panel plots the central region of the $J_s/S = 10^5$ maser on a linear scale. The saturation boundary points, $\kappa_0 z = \pm 3.05$ in this case, are marked with arrows. It is evident that the outward moving stream completely dominates as it exits the core. In spite of the large variation in source function among the three solutions, outside the unsaturated core the dominant intensities of the three models are essentially the same. To a good degree of approximation, each dominant stream can be considered as if it originated from the edge of the unsaturated core with intensity $2J_s$, independent of S. While S cannot be completely neglected, it is largely irrelevant.

Why does one stream dominate so decisively? After all, both the source function and the absorption coefficient are independent of direction. The reason is that the amplified quantity is not the source function but the maser's own radiation. A particle that enters the upper level requires an interaction partner to produce another photon. At every location other than the center the intensity of one stream is higher, increasing its probability of interaction. Photons are preferentially produced into the direction of the dominant stream and its dominance grows further. Maser amplification is the ultimate case of the strong robbing the weak to get stronger.

In three-dimensional geometries, the same properties are reflected in the high degree of beaming displayed by maser radiation. A particle in the upper level interacts preferentially with the strongest rays, increasing their dominance and reducing the beaming cone; the stronger the amplification, the tighter is the beam. The beaming relation (equation 4.21) can be written as

$$I = J \frac{4\pi}{\Omega_b} \,. \tag{4.32}$$

Since J increases linearly with the source size, I increases much faster because of the tightening of the beaming angle.

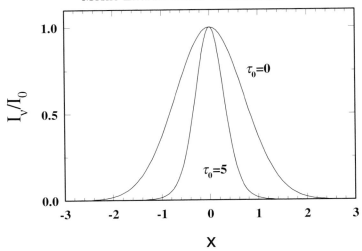

FIGURE 13. The effect of unsaturated amplification on the intensity profile.

4.6. *Line Narrowing*

Unsaturated maser intensity grows in proportion to $\exp[\kappa_0(\nu)\ell]$. Since $\kappa_0(\nu)$ is sharply peaked at the line-center frequency ν_0, the amplification there is stronger than at the line wings and the amplified line is narrower than the input line. As an example, consider amplification when κ_0 and the input signal are both Doppler-shaped with the some width. That is,

$$I_{in} = I_0 \exp(-x^2), \qquad \tau_\nu = \tau_0 \exp(-x^2), \tag{4.33}$$

where $x = (\nu - \nu_0)/\Delta\nu_D$ is the dimensionless frequency shift from line center. The intensity of the amplified line is then

$$I_\nu = I_0 \exp[-x^2 + \tau_0 \exp(-x^2)]. \tag{4.34}$$

Figure 13 displays this frequency distribution before ($\tau_0 = 0$) and after amplification with $\tau_0 = 5$. The line narrowing effect is evident. Note, though, that both profiles are normalized to unity at the line center. The central intensity of the amplified line is $e^5 \simeq 150$ times higher than that of the unprocessed signal.

Narrowing does not continue indefinitely, saturation rebroadens the line. This effect is best understood in terms of the converter part of eq. 4.30. The frequency dependence comes from the pump rate p, the beaming angle Ω_b and the conversion efficiency factor $J/(J + J_s)$. In a saturated maser the latter is unity and only the first two remain. The spectral distribution of the pump rate follows the same Doppler profile as κ_0. In linear masers, which do not have a meaningful beaming angle, this is the only relevant dependence on frequency and the line rebroadens during saturation to its thermal Doppler width. The same applies to filamentary masers, whose beaming angle is controlled by the frequency-independent aspect ratio of the filament. The situation is different in planar (e.g. disks) and three-dimensional (e.g. spheres) masers where Ω_b is controlled by the amplification and thus is tighter where the amplification is larger (i.e., line center). The result is that

$$\text{saturated maser:} \quad I_\nu \propto [\kappa_0(\nu)]^\alpha, \qquad \text{where } \alpha = \begin{cases} 1 & \text{linear (filament)} \\ 2 & \text{planar (disk)} \\ 3 & \text{3D (sphere)} \end{cases} \tag{4.35}$$

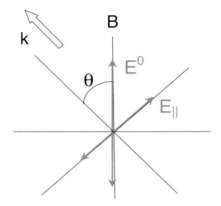

FIGURE 14. Polarization in $\Delta m = 0$ spontaneous emission.

A Doppler shaped profile $\kappa_0 \propto \exp(-x^2)$ results in a geometry-dependent spectral distribution $I_\nu \propto \exp(-\alpha x^2)$.

5. Polarization

From Maxwell's equations, the electric field of an electromagnetic wave is always perpendicular to the propagation direction. The tip of the field vector traces a well-defined figure in the transverse plane, usually an ellipse, a property referred to as *polarization*. This seemingly simple transverse condition provides a rather peculiar constraint when one attempts to reconcile it with the properties of the matter that emitted the radiation. The radiating system is characterized by directions that are intrinsic to the source and have nothing to do with the observer. Yet wherever that observer is located, the detected electric field is always orthogonal to the line-of-sight to the source.

The resolution of this puzzling dilemma is simple for emission from a classical oscillator. The oscillation frequency ν defines a characteristic wavelength $\lambda = c/\nu$ and the longitudinal electric field decays within a distance of a few λ; only the transverse component propagates. However, this reasoning cannot be extended to the quantum process of spontaneous emission. In this case the system emits a photon, which does not have a longitudinal part to begin with. Instead, the photon must be polarized in the plane perpendicular to its wave vector $\mathbf{k}$ (the direction of propagation) even though the orientation of the electric field associated with the transition is determined by the quantization axis, which has nothing to do with the direction of $\mathbf{k}$. Figure 14 shows the geometry for a $\Delta m = 0$ spontaneous decay. The quantization axis is denoted by B and the electric field generated in the transition is always parallel to this axis and has amplitude E^0. The photons propagate in the direction marked by the double arrow at an angle θ from B, the corresponding axis is denoted k. The electric field of the propagating radiation can be decomposed along two orthogonal axes in the plane of the sky (transverse to $\mathbf{k}$) as

$$E_\parallel = E^0 \sin\theta, \qquad E_\perp = 0. \tag{5.36}$$

Here $E_\parallel$ is the projection along the axis in the plane that contains both B and k (parallel to the projection of B on the plane of the sky) and $E_\perp$ is the component along the axis perpendicular to that plane. From the transverse condition, the component along the direction of propagation vanishes, i.e., $E_k = 0$. But how to reconcile this with the geometry, which gives $E_k = E^0 \cos\theta$? The answer is that we cannot as long as we apply

	σ^+	π	σ^-
	$\frac{1}{4} I^+$	$\frac{1}{2} I^0$	$\frac{1}{4} I^-$
$T_\parallel$	$2\cos^2\theta$	$2\sin^2\theta$	$2\cos^2\theta$
$T_\perp$	2	0	2
T_r	$(1+\cos\theta)^2$	$\sin^2\theta$	$(1-\cos\theta)^2$
T_l	$(1-\cos\theta)^2$	$\sin^2\theta$	$(1+\cos\theta)^2$
I	$2(1+\cos^2\theta)$	$2\sin^2\theta$	$2(1+\cos^2\theta)$
Q	$-2\sin^2\theta$	$2\sin^2\theta$	$-2\sin^2\theta$
V	$4\cos\theta$	0	$-4\cos\theta$

TABLE 2. Polarizations for fully resolved Zeeman pattern, $\nu_B \gg \Delta\nu_D$

classical reasoning. The resolution of this conflict is rooted in the quantum nature of spontaneous emission, which has no classical analog. Because of the uncertainty principle, only one component of any vector can be determined whenever the magnitude of that vector is known, the other two remain completely undetermined; recall the properties of angular momentum. The constraint $E_k = 0$ can be ignored in spontaneous emission, quantum mechanics can be counted on to take care of this transverse condition.

5.1. *Fully Resolved Zeeman Pattern; $\nu_B \gg \Delta\nu_D$*

In addition to providing a quantization axis, a magnetic field also shifts the energy levels of magnetic sub-states, resulting in a Zeeman pattern. When the field is sufficiently strong that the Zeeman shift ν_B exceeds the linewidth $\Delta\nu_D$, radiation is produced in pure Δm transitions centered on the appropriate Zeeman frequencies. For the $\Delta m = 0$ transition we have just derived the components of the radiation electric field. Denote by I^0 the intensity associated with the amplitude E^0 ($I^0 \propto |E^0|^2$). Then the intensities $T_\parallel$ and $T_\perp$ measured by an antenna with linear polarization response oriented parallel and perpendicular to the B-axis, respectively, are

$$T_\parallel = I^0 \sin^2\theta, \qquad T_\perp = 0, \qquad T_{r,l} = \frac{1}{2} I^0 \sin^2\theta. \qquad (5.37)$$

We have listed also the intensities $T_{r,l}$ obtained from the corresponding electric field amplitudes $E_{r,l} = 2^{-1/2}(E_\parallel \pm iE_\perp)$. These are the intensities that would be measured with right- and left-circular instrumental response. From these results, the Stokes parameters of $\Delta m = 0$ spontaneous emission are†

$$I = T_\parallel + T_\perp = T_r + T_l = I^0 \sin^2\theta$$
$$Q = T_\parallel - T_\perp = I^0 \sin^2\theta$$
$$U \propto 2\mathrm{Re}\left(E_\parallel E_\perp^*\right) = 0$$
$$V = T_r - T_l = 0 \qquad (5.38)$$

It is straightforward to repeat these calculations for $\Delta m = \pm 1$ spontaneous emission. The results are summarized in table 2 for the classical Zeeman pattern. In that case there are three spectral lines centered on $\nu_0 + \nu_B\Delta m$ ($\Delta m = 0, \pm 1$), where ν_0 is the line frequency in the absence of a magnetic field, with $I^0(\nu) = I^+(\nu + \nu_B) = I^-(\nu - \nu_B)$.

† We follow here the standard convention $Q = T_\parallel - T_\perp$. The maser literature contains many papers (notably GKK) that employ the opposite sign convention for Q.

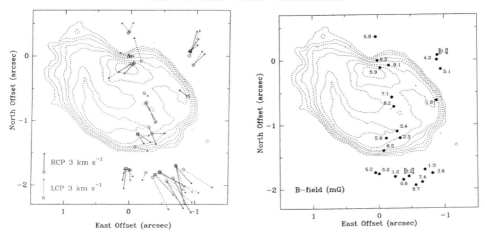

FIGURE 15. OH maser polarization in the sources W3OH (from Bloemhof et al 1992)

Quantities listed in the first column are obtained for each transition from the product of the intensity heading the transition column with the appropriate trigonometric factor. The parameter U vanishes for all transitions with this choice of axes.

The polarization properties are easy to understand. In the case of π-components the generated electric field oscillates along the magnetic field lines, therefore the polarization is always purely linear, parallel to the magnetic axis whatever the viewing angle. The intensity vanishes as the propagation direction approaches the magnetic axis on account of the transverse condition. In the case of σ-components, the generated electric field vector describes a circle in the plane perpendicular to the magnetic axis. For propagation along the field lines ($\theta = 0$) the circle is viewed face-on and the polarization is purely circular ($|V| = I$). When the propagation is at an oblique angle to the field, the circle is viewed in projection and the polarization is elliptical. Finally, the ellipse degenerates to a line for propagation perpendicular to the field ($\theta = \pi/2$), since the circle is viewed edge-on. The polarization is then linear, perpendicular to the magnetic field lines.

The spectral components of fully resolved Zeeman patterns are fully polarized ($I^2 = Q^2 + U^2 + V^2$). Each component can be considered an independent, isolated radiative transition that couples to a single sense of polarization. Therefore, the results apply also to maser radiation even though they were derived for spontaneous emission. Indeed, these are the maser polarization properties derived by GKK, although from an entirely different approach. The only difference between the thermal and maser cases is the disparity between the π and σ maser intensities, reflecting their different growth rates (Elitzur 1996). The two have equal intensities for propagation at $\sin^2 \theta = 2/3$, i.e., $\theta = 55°$. At $0 \leqslant \theta < 55°$ the σ-components have a higher intensity, at $55° < \theta \leqslant 90°$ the π-component is stronger.

5.1.1. *Maser Observations*

Thus far, the only molecular emission from astronomical sources to display the polarization properties of fully resolved Zeeman patterns involves the main-line OH masers in HII regions. The necessary large Zeeman splitting requires both a large Landé g-factor and a substantial magnetic field, a combination met only for these masers. Most interstellar molecules have closed electronic shells and their response to magnetic fields involves the nuclear angular momentum, resulting in g-factors of order 10^{-3}. OH is one of the few paramagnetic molecules with $g \sim 1$ and it requires magnetic fields of order milligauss for $\nu_B > \Delta\nu_D$. Magnetic fields in interstellar regions increase with the density and the only OH sources dense enough to sustain such fields are found around compact HII regions.

From the polarization solution, a maser spot should display two emission features at the same location but separated in frequency, each predominantly either left- or right-circularly polarized. The pairing of such Zeeman components can be difficult because it is impossible to decide a-priori whether a frequency shift is caused by the Zeeman or Doppler effects. The most detailed polarization measurements were done for W3(OH) by Garcia-Barreto et al. (1988) and Bloemhof, Reid & Moran (1992). Figure 15 shows the results of the latter study. The left panel displays the proper motions of various polarized features superposed on a map of the source. Each arrow marks the direction of proper motion for the corresponding feature, the arrow length is proportional to the velocity magnitude. Full-line arrows denote features predominantly right-circularly-polarized, dashed arrows are for left-circular polarization. The presence of oppositely-polarized pairs with identical position and proper motion, as predicted by the polarization solution, is evident. The right panel lists the magnetic fields derived at the various maser spots from the frequency splitting of their Zeeman pairs. The polarization of all features is adequately explained if Faraday rotation is responsible for removal of some linear polarization. Garcia-Barreto et al. find that there are no features in W3(OH) that might be identified as π-components, even accounting for possible Faraday rotation. Such preponderance of σ-components is expected for magnetic fields aligned within less than $55°$ from the line of sight because of the disparity between the growth rates of the π- and σ-components.

5.1.2. *Polarization Filters*

The pairing of Zeeman features is a difficult observational task because the two members of each pair must be identified at the same position, a challenging spatial-resolution problem compounded by spectral blending and accidental line-of-sight coincidences. Although the number of isolated, circularly polarized features without a matched companion has been reduced over the years, it has not been eliminated altogether. The occurrence of such "orphan" features indicates that the overall left-right symmetry of a homogeneous source must be broken. An ingenious filter mechanism that accomplishes this was proposed independently by Cook (1966) and Shklovskii (1969). The idea is as follows: The frequency of a $\Delta m = \pm 1$ transition is shifted in a magnetic field by the amount $\pm\nu_B$. When the magnetic field has a gradient along the ray path, the frequency of each Zeeman component varies continuously and must be compensated by some other shift to maintain the frequency coherence required for maser operation. Such a matching frequency shift can be provided, for example, by a large velocity gradient. The Doppler and Zeeman shifts can then be matched, but only for one of the polarization modes, if at all. This results in a polarization filter that amplifies only one sense of polarization, independent of the pump mechanism.

The proposal that velocity and magnetic gradients are correlated need not be as far-fetched as it might seem at first. The degree of ionization in interstellar regions is sufficient to ensure magnetic flux freezing, which leads to a coupling of the motions and the magnetic field.

5.2. *Overlapping Zeeman Components;* $\nu_B \ll \Delta\nu_D$

Although every electromagnetic wave is fully polarized, real radiation fields are usually unpolarized. The polarization ellipses of the different wave components that make up the radiation field are usually randomly oriented and the overall electric field vanishes.

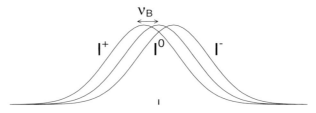

FIGURE 16. Intensities of thermal Zeeman components when $\nu_B \ll \Delta\nu_D$.

The average of the squared values of the field amplitudes does not vanish, though, and provides the radiative intensity.

The situation can be likened to thermal motions of gas particles. Each particle has a well defined velocity, but the overall velocity (though not the velocity square) averages out to zero. This analogy becomes even more appropriate when the radiation field is described in terms of photons. The radiative intensity, the average of the squared values of the field amplitudes, gives the photon density and never vanishes. But while single photons are fully polarized (the spin projection on the transverse plane has a well defined value), their polarizations are usually randomly oriented and average out to zero.

When the source is isotropic the polarization must vanish because there is no preferred direction to orient the individual polarization ellipses or photon spins. The introduction of an overall quantization axis defines a preferred direction and can result in net polarization. When that direction corresponds to the axis of a magnetic field that fully resolves the Zeeman pattern, each spectral component is fully polarized because each of its photons is polarized the same. When the magnetic field decreases so that $\nu_B < \Delta\nu_D$, the field still provides a quantization axis but different polarizations begin to blend across the line and the overall polarization is reduced.

5.2.1. *Thermal Emission*

Consider first thermal radiation fields. They are produced in spontaneous emission and must be considered in the photon picture. The various Δm transitions produce spectral components that have equal intensities but are centered on frequencies slightly shifted from each other as shown in figure 16. The intensity I^0 of the $\Delta m = 0$ component is an even function of the dimensionless frequency shift $x = (\nu - \nu_0)/\Delta\nu_D$, centered on $x = 0$. Introduce

$$x_B = \frac{\nu_B}{\Delta\nu_D} \ll 1. \tag{5.39}$$

Then $I^{\pm}(x \pm x_B) = I^0(x)$.

The three components are produced independent of each other and the overall radiation field is an incoherent superposition of them. At every frequency across the line the Stokes parameters of the three components simply add up because they merely count the different photon numbers. Within each component the transverse condition is obeyed independently and the photons are polarized as described in table 2. Therefore,

$$I = I^0 + I^+ + I^- = 2I^0$$

$$V = V^0 + V^+ + V^- = (I^+ - I^-)\cos\theta = \frac{dI}{dx}\, x_B \cos\theta$$

$$Q = Q^0 + Q^+ + Q^- = \left[I^0 - \tfrac{1}{2}(I^+ + I^-)\right]\sin^2\theta = -\frac{d^2I}{dx^2}(x_B \sin\theta)^2 \tag{5.40}$$

The final expression in each case is the leading order result from a series expansion in x_B of $I^{\pm}(x)$ around $I^0(x)$.

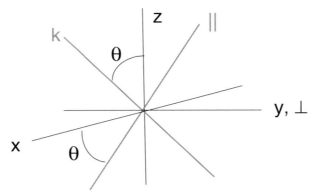

FIGURE 17. Geometry of the wave and quantization frames.

The overall radiation field is polarized because it is the superposition of three *different* polarized intensities. Even though the intensity of each Δm transition is the same, because of the Zeeman shifts the intensities are slightly different at every given frequency. These differences are controlled by the parameter x_B, and so is the polarization. The circular polarization is proportional to x_B because it arises from the odd terms in the expansion in powers of x_B. The linear polarization involves even expansion terms and is of order x_B^2 only. This small linear polarization is always perpendicular to the magnetic axis because it is produced by the σ-components.

When $x_B \to 0$, the polarization disappears. Although the radiation field is comprised of three fully polarized components, they are mixed in just the right amounts that the overall polarization vanishes. This is quite different from the unpolarized radiation in a thermal source without any quantization axis where the photon spins are completely randomly oriented, although the net result is the same.

5.2.2. *Overlapping Electromagnetic Waves*

Since maser radiation is generated in stimulated emission, its properties must be understood in terms of classical electromagnetic waves interacting with particles in quantized energy levels. The wave electric field must be handled in terms of its amplitude instead of its intensity (the square of the amplitude, which corresponds to photon number density). The geometrical setup involves two independent coordinate frames, one for the particle quantization the other for the wave propagation (figure 17). Particle quantization is defined with respect to the cartesian x-y-z coordinate frame, with z the quantization axis. Vector components of the electric field taken in this frame are uniquely associated with specific Δm transitions, the cartesian components are equivalent to the components $E^{\Delta m}$ ($\Delta m = 0, \pm 1$) defined as

$$E^0 = E_z, \qquad E^\pm = 2^{-1/2}(E_x \pm iE_y). \tag{5.41}$$

This notation reflects the fact that the z-component of the field couples only to $\Delta m = 0$ while the x- and y-components couple only to $\Delta m = \pm 1$. Each $E^{\Delta m}$ is the amplitude of the electric field that couples to the corresponding Δm transition.

The wave propagation is at an angle θ from the quantization axis in the x-z plane. The wave coordinate system is obtained from the quantization frame through rotation by an angle θ around the y-axis. The rotated z-axis is in the direction of wave propagation and is denoted k. The rotated x-axis corresponds to one possible direction of linear polarization, and is denoted $\parallel$ (the polarization is parallel to the quantization axis). The common y-axis corresponds to the other independent direction of linear polarization, and is denoted $\perp$. Since the electric field is a proper vector, it can be decomposed in this frame too. The transformation between the components in the two frames involves straight-forward, standard geometry. In terms of the quantization-frame vector components, the components of the electric field in the wave frame are

$$E_{\parallel} = \frac{1}{\sqrt{2}} \left(E^+ + E^- \right) \cos\theta + E^0 \sin\theta$$

$$E_{\perp} = \frac{i}{\sqrt{2}} \left(E^+ - E^- \right)$$

$$E_k = -\frac{1}{\sqrt{2}} \left(E^+ + E^- \right) \sin\theta + E^0 \cos\theta \qquad (5.42)$$

The transverse condition states that E_k must vanish irrespective of the direction of propagation. Now we cannot rely on quantum considerations, this condition must be obeyed as a geometrical constraint on the classical vector components of the electric field.

Since we wish to describe a maser radiation field, the magnitudes of its vector components in the quantization frame, $|E^{\Delta m}|$, reflect the interaction with the particles. From the discussion in section 4 we can expect these amplitudes to be controlled by the pump rates into magnetic sub-states and reflect their magnitudes—which have nothing to do with the propagation direction. Therefore, the magnitudes $|E^{\Delta m}|$ cannot be expected to guarantee the transverse condition. However, apart from their magnitude, the field vector components posses another important property. Describing electromagnetic waves, they contain an oscillating part, commonly expressed as a complex exponential whose argument is the wave phase $\phi_0 + \omega t - \mathbf{k} \cdot \mathbf{x}$. While their oscillatory part is the same, the phases of the three components can be shifted from each other by arbitrary amounts. Consider as an example the situation when the magnitudes of the three amplitudes $|E^{\Delta m}|$ are equal to each other. Denote this common magnitude E and select the arbitrary origin of time such that phases are initialized at E^0. Then the three vector components in the quantization frame are

$$E^0 = E, \qquad E^{\pm} = E\, e^{i\phi^{\pm}} \qquad (5.43)$$

with $\phi^{\pm}$ the undetermined phase shifts of the $\Delta m = \pm 1$ vector components. Inserting these expressions back into equation 5.42, the transverse condition $E_k = 0$ becomes

$$e^{i\phi^+} + e^{i\phi^-} = \sqrt{2}\cot\theta\,, \qquad (5.44)$$

a constraint on the phase differences between the three vector components of the radiation field. The real and imaginary parts of this complex equation provide two constraints on the two phase shifts. The solution is

$$\phi^+ = -\phi^-\,, \qquad \phi \equiv |\phi^{\pm}| = \arccos\left(2^{-1/2}\cot\theta\right). \qquad (5.45)$$

While one σ-component leads the π-component by the phase difference ϕ, the other one must trail by the exact same amount. Proper solutions require $2^{-1/2}\cot\theta \leqslant 1$, otherwise the phase shift ϕ is not physical. The interference dictated by the transverse condition can be maintained for three $E^{\Delta m}$ of equal magnitude only for propagation at

$$\sin^2\theta \geqslant \,^1\!/_3\,, \qquad (5.46)$$

i.e., $\theta \geqslant 35°$. Other values of the ratio $|E^0|/|E^{\pm}|$ make smaller propagation angles possible.

This completes the definitions of the three vector components $E^{\Delta m}$, both magnitudes and phases. From equation 5.42, the transverse components of the electric fields of propagating waves are

$$E_\parallel = E / \sin \theta$$
$$E_\perp = \pm E \left(2 - \cot^2 \theta\right)^{1/2} . \tag{5.47}$$

Note that $E_\parallel > E$, reflecting the constructive interference maintained by the phase coherence among the different components $E^{\Delta m}$. The lower bound in eq. 5.46 ensures that $E_\parallel$ does not blow up and that $E_\perp$ remains well defined. The transverse components immediately yield the polarization solution

$$q = \frac{Q}{I} = -1 + \frac{2}{3 \sin^2 \theta}$$
$$u = \frac{U}{I} = \pm \frac{2}{3 \sin^2 \theta} \left(3 \sin^2 \theta - 1\right)^{1/2}$$
$$v = \frac{V}{I} = 0 \tag{5.48}$$

The lower bound on propagation directions (eq. 5.46) protects against the unphysical $q > 1$ as well as imaginary u. We have found polarization for a superposition of three overlapping radiation fields with *equal intensities*, an impossible feat in the thermal case. The reason is that now we have considered the amplitudes themselves, instead of the intensities formed out of their squares, and the polarization reflects the specific phase relations among them. The polarization arises because the independent constraints imposed by the particle interactions and the transverse condition must be reconciled simultaneously. Particle interactions produce three independent fields, corresponding to the three different Δm transitions. The transverse condition dictates that only two independent fields propagate in any given direction, the longitudinal combination of the original fields must vanish. The phase relation, and the polarization, reflect the correlation that must exist to eliminate the longitudinal component of the field.

The circular polarization vanishes because of the complete left-right symmetry of the problem. The radiation is fully linearly polarized, $q^2 + u^2 = 1$, as could be expected—we have merely deduced the vector structure of propagating electromagnetic waves and such waves are always fully polarized. At any given direction there are two possible wave configurations corresponding to a role switch of the two σ-components as to which one trails and which one leads the π-component. These two arrangements for $\phi^\pm$ give two values of u at each θ, corresponding to two polarization position angles

$$\tan \chi = \frac{E_\perp}{E_\parallel} = \pm \left(3 \sin^2 \theta - 1\right)^{1/2} . \tag{5.49}$$

Both modes are equally probable, reflecting the complete symmetry in reflection about the x-z plane. The radiation field contains equal number of waves with either sign of the parameter u. Therefore, although $u \neq 0$ for any wave, u vanishes in the average over all the components of the radiation field. Only q survives this average, since it is the same for all waves propagating in a given direction, and the overall radiation field is only partially polarized. The polarization is either parallel (when $q > 0$) or perpendicular ($q < 0$) to the projection of the quantization axis on the plane of the sky.

This polarization solution was first derived by GKK from an entirely different approach. Their maser analysis assumed equal pump rates for the different m-states, an assumption reflected in the equal amplitudes $|E^0| = |E^+| = |E^-| = E$ taken here. The connection between these two ingredients is clarified in the next section. The solution is displayed in figure 18. The top panel shows the variation of q with propagation angle θ, the bottom panel shows the phase difference ϕ. Propagation at $35°$ requires all three wave components to oscillate in phase. The phase difference increases with θ, the σ-components

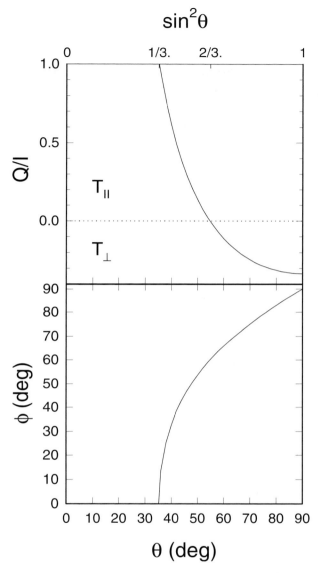

FIGURE 18. Maser polarization and the phase difference $\phi = |\phi^{\pm}|$ when $\nu_B \ll \Delta\nu_D$.

oscillate precisely out of phase to the π-component for propagation at 90°. The linear polarization changes sign at $\sin^2\theta = \frac{2}{3}$ ($\theta = 55°$), where it vanishes. At smaller angles q is positive, corresponding to polarization along the quantization axis, at larger angles it is negative, corresponding to polarization perpendicular to it. The transition angle $\theta = 55°$ between positive and negative q corresponds to a 90° flip in the polarization direction. Such flips are commonly observed in SiO masers. An example is shown in figure 19, an expanded view of one of the maser features from the maps displayed in figure 6. A natural explanation is a slight change in direction of the magnetic field, straddling the two sides of the transition angle $\theta = 55°$.

It is important to note that the linear polarization usually does not exceed 33% because $|q| > \frac{1}{3}$ only for the limited range of propagation directions $35° \leqslant \theta \leqslant 45°$. Propagation

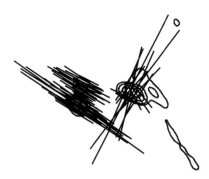

FIGURE 19. 90° flip of linear polarization in TX Cam (cf figure 6)

at $\theta > 45°$ gives only $|q| \leqslant \frac{1}{3}$, along the field when $45° \leqslant \theta < 55°$ and orthogonal to it when $\theta > 55°$. The direction of the magnetic axis is generally not known a-priori. It can be determined with certainty from the linear polarization only when it exceeds 33%, in which case the field projection on the plane of the sky is parallel to the polarization.

This polarization solution is entirely different from the one for thermal radiation. Whereas the thermal polarization arises from the superposition of different intensities, this one involves equal intensities but well defined phase relations among the amplitudes. The only assumption in its derivation was the existence of a quantization axis in the source. The physical process behind this axis was never specified, in principle it need not be a magnetic field. Indeed, no magnetic interactions entered into the analysis and the linear polarization depends only on propagation angle. It is entirely independent of x_B and would be the same for any physical mechanism that provided a good quantization axis.

5.2.3. *Maser Polarization Formalism*

The last section presents what can be called a kinematic derivation of the structure of polarization consistent with the fundamental physical processes that generate maser radiation. Equation 5.48 lists the only polarization structure consistent with the constraints that govern an interacting mixture of quantized particles and classical electromagnetic waves that have equal amplitudes in the quantization frame. We now proceed to the dynamics of the problem and study the mechanism that actually drives the system toward this solution. This requires a formalism to describe the components of the mixture, the electromagnetic waves and their interaction with the particles.

Electromagnetic Waves: It is convenient to describe the polarization state of electromagnetic waves by introducing the four dimensional space of Stokes parameters. Every electromagnetic wave is a vector in this space, defined by its own set of Stokes parameters

$$\mathcal{S} = (I, Q, U, V) = I(1, \mathbf{\Pi}), \qquad \text{where } \mathbf{\Pi} = (q, u, v). \tag{5.50}$$

The wave polarization structure is defined by the 3-vector $\mathbf{\Pi}$ of its normalized Stokes parameters (figure 20). Since individual electromagnetic waves are fully polarized, their polarization vectors obey $|\mathbf{\Pi}|^2 = q^2 + u^2 + v^2 = 1$.

During propagation in empty space, the wave Stokes parameters remain unchanged. Only particle interactions can modify the 4-vector $\mathcal{S}$. Since $\mathbf{\Pi}$ must remain a unit vector, the only possible effect of the interactions on the polarization vector of each electromagnetic wave is to rotate it.

Particle Interactions: The interaction with polarized maser radiation involves a straightforward extension of the phenomenological theory presented in section 4, adding magnetic sub-states to the maser levels. For simplicity we consider here a maser operating between an upper level with spin 1 and a lower level with spin 0. The upper level is populated from the reservoir by the three pump rates $p_{1,m}$ ($m = 0, \pm 1$), the lower level by the single pump rate p_0. The loss rate for both levels again is the same, Γ.

As before, the unsaturated level populations are $n_0 = p_0/\Gamma$ and $n_{1,m} = p_{1,m}/\Gamma$. As the maser radiation grows, the populations are modified by interactions with it. With equation 1.3, the level populations define three absorption coefficients $\kappa^{\Delta m} \propto n_{1,m} - n_0$ ($m = 0, \pm 1$). It is convenient to introduce the average $\kappa^1 = {}^1\!/_2 \left(\kappa^+ + \kappa^- \right)$ and form the three independent combinations†

$$\begin{aligned}
\kappa_m &= {}^1\!/_2 \left[\kappa^1 (1 + \cos^2 \theta) + \kappa^0 \sin^2 \theta \right] \\
\kappa_l &= {}^1\!/_2 \left(\kappa^0 - \kappa^1 \right) \sin^2 \theta \\
\kappa_c &= {}^1\!/_2 \left(\kappa^+ - \kappa^- \right) \cos \theta.
\end{aligned} \qquad (5.51)$$

The absorption coefficient κ_m is an angle-dependent mean of the three $\kappa^{\Delta m}$ while κ_l and κ_c involve differences that couple to linear and circular polarizations, respectively. The radiative transfer equation for the Stokes parameters is

$$\frac{d\mathcal{S}}{d\ell} = \kappa_m \mathcal{S} + \mathcal{R}\mathcal{S}, \qquad \text{where } \mathcal{R} = \begin{pmatrix} 0 & \kappa_l & 0 & \kappa_c \\ \kappa_l & 0 & 0 & 0 \\ 0 & 0 & 0 & 0 \\ \kappa_c & 0 & 0 & 0 \end{pmatrix} \qquad (5.52)$$

The matrix $\mathcal{R}$ resembles a standard rotation matrix. Indeed, it is easy to show from this equation that the polarization vector varies according to (Litvak 1975)

$$\frac{d\mathbf{\Pi}}{d\ell} = [\mathbf{\Pi} \times \boldsymbol{\kappa_p}] \times \mathbf{\Pi} \qquad (5.53)$$

where $\boldsymbol{\kappa_p} = (\kappa_l, 0, \kappa_c)$ is another 3-vector formed from the off-diagonal absorption coefficients (figure 20). This result is for single electromagnetic waves, which have $|\mathbf{\Pi}| = 1$; in the case of arbitrary polarization vectors, the right hand side contains an additional term proportional to $1 - \mathbf{\Pi}^2$. As expected, the effect of radiative transfer is to rotate the polarization vectors of individual electromagnetic waves.

5.2.4. *Stationary Maser Polarization*

The rotation velocity of an individual polarization vector is $\mathbf{\Pi} \times \boldsymbol{\kappa_p}$. Because of its dependence on the polarization vector itself, it is different for different electromagnetic waves and changes as $\mathbf{\Pi}$ is rotating. Since waves are launched with arbitrary initial polarizations that subsequently rotate at different rates, the radiation field can be expected to remain unpolarized unless there is a stationary configuration whose polarization vector does not rotate. When such a configuration exists, all the polarization vectors remain locked there once they have entered it and that becomes the polarization of the overall radiation field. Maser polarization can be determined as the solution of the stationary polarization equation

$$\frac{d\mathbf{\Pi}}{d\ell} = 0, \qquad \text{i.e.} \quad q' = u' = v' = 0 \qquad (5.54)$$

where the prime denotes derivative along the path. Since $q = Q/I$, $q' = 0$ is equivalent to $Q'/Q = I'/I$, and similar relations hold for U and V. Therefore stationary polarization

† This formalism was developed by Litvak (1975). The presentation here follows the notations of Elitzur (1996).

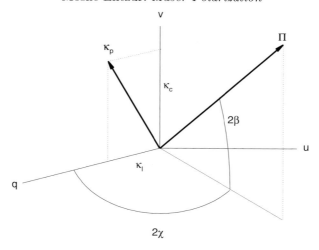

FIGURE 20. The polarization vector $\mathbf{\Pi}$ and the vector $\boldsymbol{\kappa_p}$ that controls its radiative transfer in the space defined by the normalized Stokes parameters q, u and v.

is defined by

$$\frac{I'}{I} = \frac{Q'}{Q} = \frac{U'}{U} = \frac{V'}{V} = \lambda, \tag{5.55}$$

where λ is some undetermined common factor, namely

$$\frac{d\mathcal{S}}{d\ell} = \lambda \mathcal{S}. \tag{5.56}$$

All four Stokes parameters of stationary polarization grow at the same rate, the Stokes 4-vector maintains the same direction, growing in proportion to itself. Comparison with equation 5.52 shows that the Stokes vector of stationary polarization obeys

$$(\lambda - \kappa_m)\,\mathcal{S} = \mathcal{R}\mathcal{S}, \tag{5.57}$$

it is an eigenvector solution of the radiative transfer matrix with λ the eigenvalue.

Two types of eigenvector solutions exist (Elitzur 1996), a result easily understood from either figure 20 or equation 5.53. *Type 1* stationary polarization occurs when $\mathbf{\Pi} \parallel \boldsymbol{\kappa_p}$, which gives

$$q = \frac{\kappa_l}{\kappa_p}, \quad u = 0, \quad v = \frac{\kappa_c}{\kappa_p}. \tag{5.58}$$

Since $\kappa_p{}^2 = \kappa_l{}^2 + \kappa_c{}^2$, type 1 solutions have $q^2 + v^2 = 1$ and describe fully polarized radiation. *Type 0* stationary polarization occurs whenever

$$\kappa_l = \kappa_c = 0; \tag{5.59}$$

obviously the polarization does not rotate in this case. Because the off-diagonal κ_l and κ_c determine all possible differences among the three basic absorption coefficients $\kappa^{\Delta m}$, their vanishing implies $\kappa^+ = \kappa^- = \kappa^0$. Type 0 polarization is obtained when the absorption coefficients of all three Δm transitions are equal to each other, namely, the population is the same in all three magnetic states of the upper level. Now, all absorption coefficients, including κ_l and κ_c, contain dependence on q and v, therefore the last two equations determine type 0 and type 1 polarizations only implicitly. Expressions for the complete absorption coefficients follow from solutions of the level population equations in steady state. When inserted in eqs. 5.58 and 5.59, the resulting equations for q and v can become rather involved.

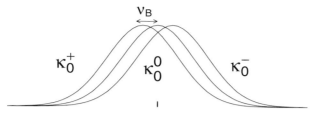

FIGURE 21. Unsaturated absorption coefficients in magnetic field with $\nu_B \ll \Delta\nu_D$.

In isotropic sources the pump rates are independent of magnetic quantum numbers. The populations of magnetic sub-states of each level are equal to each other, the off-diagonal absorption coefficients are identically zero and the polarization rotation trivially vanishes. No polarization can be generated in this case. Polarization requires anisotropy in the amplifying medium, which can be introduced in two general schemes. Pumping by a stream of either particles or radiation introduces a preferred direction, the stream axis. Few cases of such anisotropic pumping have been advocated in astronomical masers. More common is the action of a magnetic field. This introduces a good quantization axis whenever the gyro-rate for magnetic precession exceeds collisional- and radiative-interaction rates, which is the case in virtually all astronomical masers. The commonly studied model for masers in a magnetic field involves equal pump rates for the magnetic sub levels whose energies are shifted by the Zeeman effect. The solutions of the level population equations are somewhat involved; they are described in Elitzur (1996) and will not be reproduced here. The results enable study of the eigenvector problem. It turns out that type 1 provides the relevant polarization solution when $\nu_B > \Delta\nu_D$, producing the results described in section 5.1. When $\nu_B < \Delta\nu_D$, the relevant solution is type 0. Keeping terms to second order in x_B (< 1) gives

$$q = -1 + \frac{2}{3\sin^2\theta}\left(1 + \tfrac{8}{3}\,x_B^2 + \frac{2x_B^2}{J/J_s}\right)$$
$$v = \frac{16xx_B}{3\cos\theta}\left(1 + \frac{3}{4J/J_s}\right) \tag{5.60}$$

Although we have solved for stationary polarization, the result contains terms that vary with the intensity and thus is not stationary. The solution becomes stationary only when the dependence on maser intensity disappears. This requires intensity sufficiently large that the terms containing J/J_s can be neglected.

Fully stationary polarization is impossible during early stages of maser growth. The reason is obvious. Equal pump rates produce equal populations in each magnetic sub-state of the upper level, leading to unsaturated absorption coefficient $\kappa_0^{\Delta m}$ of equal strength. But because of the Zeeman shifts, the three are centered on slightly different frequencies as shown in figure 21, a situation similar to that for the intensities of thermal radiation (figure 16). At any given frequency the three absorption coefficients are different from each other and the type 0 polarization condition cannot be fulfilled. But these differences are only of order x_B and can be overcome by radiative interactions when $J/J_s \gtrsim x_B$, making it possible for stationary polarization to set in. Keeping only leading-order terms, the maser stationary polarization for isotropic pumping in a magnetic field with $x_B \ll 1$ is

$$q = -1 + \frac{2}{3\sin^2\theta}, \qquad v = \frac{16xx_B}{3\cos\theta}. \tag{5.61}$$

Selection of this solution requires a unique cooperation between the particles and radiation. The early stages of maser growth are marked by polarization rotation, caused by unequal magnetic sub-state populations of interacting particles at any given frequency. Maser growth affects the populations, equalizing them among the magnetic sub-states and eliminating the polarization rotation.

The type 0 linear polarization was found previously from the superposition of electromagnetic waves with equal amplitudes $|E^{\Delta m}|$ (equation 5.48). Because of the magnetic field, this linear polarization is accompanied by circular polarization of order $x_B = \nu_B/\Delta\nu_D$. When the transition frequency varies, the Doppler width $\Delta\nu_D$ varies proportionately while the Zeeman splitting ν_B remains fixed. Therefore x_B is inversely proportional to frequency and the circular polarization decreases with the transition frequency when all other properties remain fixed. SiO in evolved stars displays polarized maser emission in numerous rotational transitions within excited vibration states. As the rotation quantum number increases the transition frequency increases too and the circular polarization is expected to decrease. Observations show this indeed to be the case (McIntosh, Predmore & Patel 1994). In contrast, the linear polarization displays the opposite trend, increasing with rotation quantum number although all of its other properties remain the same—different lines display detailed similarities between fractional polarization profile, polarization position angle, and rotation of position angle across the profile (McIntosh & Predmore 1993). Since the solution linear polarization is independent of transition wavelength, this is the expected behavior in the presence of Faraday depolarization because this effect is proportional to λ^2. The low rotation states are more severely affected because of their longer wavelengths and the linear polarization can be expected to decrease toward lower angular momenta, as observed. Although detailed calculations of Faraday rotation have yet to be performed for $x_B \ll 1$, McIntosh & Predmore find this to be the most plausible explanation of the data.

Equation 5.61 gives the mathematical solution of an eigenvalue problem. The physical relevance of this solution is obviously confined to the domain

$$q^2 + v^2 \leqslant 1. \tag{5.62}$$

This condition imposes constraints on the directions of propagation. Physical values for the linear polarization preclude propagation too close to the magnetic axis while circular polarization eliminates perpendicular propagation, except at line center. These constraints are shown in figure 22, a display of the phase space for maser polarization solutions in the x_B—θ plane. They are plotted with the full lines labeled by frequency shift $x = (\nu - \nu_0)/\Delta\nu_D$ when $x_B < 1$. The domain for each x contains all the domains for larger values of x, and polarized maser propagation is allowed only inside the θ–x_B region enclosed by the corresponding boundary. At some frequencies these bounds are supplemented by secondary bounds, marked by dashed and dotted lines, that are of less importance. The lower branch of each full boundary is the lower bound on θ from the linear polarization, the upper branch is the upper bound from the circular polarization. At any fixed x_B, the interference generating the stationary type 0 solution is possible only for the finite range of propagation angles between the two branches. As x_B increases, the two bounds approach each other, causing the allowed region of propagation directions to shrink until eliminated. First the solution disappears at the line wings, as x_B increases the removed frequencies move closer to line center. Finally, type 0 solution is no longer possible when the line center is eliminated. The radiation is unpolarized across the entire line for all propagation directions when $x_B \gtrsim 0.3$. Further increase in x_B causes separation of the Zeeman components, ushering in the type 1 polarization solution of section 5.1. Zeeman separation starts at line center, spreading to the wings with further increase in x_B. At frequency shift x, complete Zeeman separation requires $x_B > 1 + x$

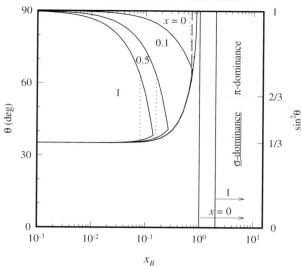

FIGURE 22. Phase space for maser polarization in a magnetic field (from Elitzur 1996)

and that is the applicability domain of the type 1 solution. Propagation is then allowed in all directions. The thin dotted line in this region marks the boundary between directions where π- and σ-components dominate.

Apart from the plotted constraints, the maser must meet a number of additional requirements to enable the polarization solution. In particular, the neglect of source terms, which was employed here, is justified only after certain initial growth. In combination with the other constraint on growth imposed by the requirement of stationary polarization, the system parameters must obey

$$(S/J_s)^{1/2} < x_B \lesssim J/J_s . \tag{5.63}$$

Every pumping scheme determines a threshold x_B and the maser remains unpolarized when the magnetic field is so weak that the Zeeman splitting is below this threshold. Furthermore, pumping schemes of "weak" masers have small J_s/S and such masers remain unpolarized irrespective of magnetic field strength. In this case, the lower bound on x_B becomes so large that the maser saturates before it had time to settle into the stationary polarization solution.

5.2.5. *Circular Polarization Profiles*

The V and I Stokes parameters of thermal radiation obey the relation

$$V(\nu) = b \, I'(\nu) \tag{5.64}$$

when the Zeeman components overlap (eq. 5.40). Here $b = \nu_B \cos\theta$ and the prime denotes derivative with respect to ν. The measured ratio V/I' is constant across the line profile, its value can be used to determine the magnetic field along the-line-of-sight.

Unsaturated masers do not obey such proportionality. Consider for example the unsaturated amplification of polarized thermal radiation. In this case all Stokes parameters are amplified by the same exponential of $\tau(\nu) = \kappa_0(\nu)\ell$, so that

$$I(\nu) = I_{\rm th}(\nu) \, e^{\tau(\nu)}$$
$$V(\nu) = V_{\rm th}(\nu) \, e^{\tau(\nu)} = b \, I'_{\rm th}(\nu) \, e^{\tau(\nu)} \neq b \, I'(\nu). \tag{5.65}$$

The derivative of the amplified term is not equal to the amplification of the derivative

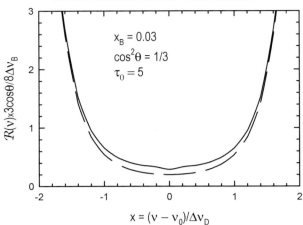

FIGURE 23. Variation of V/I' in an unsaturated maser with the parameters listed. When the maser saturates, this ratio becomes flat (from Elitzur 1998).

because the amplification itself depends on frequency (τ is peaked at line center). Instead of a constant, the ratio $\mathcal{R}(\nu) = V/I'$ rises toward the line wings. Figure 23 displays the $\mathcal{R}$ profile for self-amplified radiation in an unsaturated maser with the parameters listed. The two curves present analytical approximations for upper and lower bounds on the polarization rotation. The two bracket the possible range of the actual solution and the difference between them is practically insignificant.

When the maser saturates, $\mathcal{R}$ becomes flat. This result is best understood from the converter form of eq. 4.30. The only difference between the various Stokes parameters of the polarization solution in a saturated maser is the specific combination of pump rates into magnetic states that they involve. Since the unsaturated absorption coefficients are proportional to the pump rates and since $\kappa_0^{\pm}(\nu \pm \nu_B) = \kappa_0^0(\nu) \equiv \kappa(\nu)$ (cf figure 21), in a saturated maser

$$\frac{V}{I} \propto \frac{\kappa_0^+ - \kappa_0^-}{\kappa} \propto \nu_B \frac{\kappa'}{\kappa}. \tag{5.66}$$

But a saturated maser also obeys $I \propto \kappa^\alpha$ (eq. 4.35), therefore $I'/I = \alpha \kappa'/\kappa$ and equation 5.64 follows. The only difference from the thermal case is the proportionality constant b, which is $8\nu_B/(3\alpha \cos\theta)$ in a saturated maser (Elitzur 1998).

The flat shape of the ratio V/I' in a saturated maser is markedly different from its unsaturated profile, as is evident from figure 23. Profile analysis of this ratio provides a decisive, quantitative test of the degree of maser saturation. Based on such analysis, the circular polarization detected in OH 1720 MHz maser emission from supernova remnants is the first direct evidence for maser saturation.

5.2.6. *Limitations and Outstanding Issues*

The theory presented here was developed for an idealized maser. The results depend in a crucial manner on the assumption of a constant direction for the quantization axis. They provide the maximal polarization that can be produced in a source that maintains a uniform direction for the magnetic field. Any curvature in the field lines along the propagation direction results in θ variations that destroy the phase coherence between the π- and σ-components, reducing the degree of polarization. In particular, Alfven waves introduce ripples in the field lines that destroy the polarization whenever the Alfven wavelength is shorter than the amplification length. Furthermore, linear polarization may also be reduced by Faraday rotation. As a result, the information that can be extracted from polarization alone is limited because the same polarization can be produced in a number of different ways. For example, the same linear polarization can be attributed either to the maximal polarization at an appropriate angle θ or to a higher degree of polarization that was degraded either by curvature in the field lines or by Faraday depolarization.

Another fundamental assumption is that the only degeneracy of the maser levels involves their magnetic sub-states. When any of the levels includes additional degeneracy, so that magnetic sub-states of different levels overlap, the tight constraints responsible for the stationary solutions no longer apply and the polarization can be expected to disappear†. Indeed, the energy levels of both H_2O and methanol involve hyperfine degeneracy and both masers are generally only weakly polarized. Exceptions do exist, though, and H_2O masers sometime display high polarization, notably during outbursts such as one that occurred in Orion (Garay, Moran & Haschick 1989; Abraham & Vilas Boas 1994). This may involve the excitation of a single hyperfine component, in which case the general solutions derived here are applicable.

Finally, it is important to note that the theory presented here is still incomplete. The maser radiation field is an ensemble of waves, each of them launched with random polarization. Subsequent amplification through particle interactions is accompanied by rotation of each polarization vector and we have identified the stationary modes that do not rotate. However, we have not shown how the radiation field actually evolves into this solution, which is considerably more difficult. Indeed, for any statistical distribution it is always simpler to identify the stationary limit than to demonstrate how this limit is actually approached. A demonstration of the approach to Maxwellian of a particle velocity distribution or to Planckian of a photon distribution are considerably more difficult than the derivation of either functional form. The evolution of such ensembles requires numerical simulations of the type frequently performed in studies of plasma and laboratory lasers. This approach is not necessary for the analysis of polarization in thermal radiation, where the four Stokes parameters of different waves are uncorrelated. In contrast, for waves partaking in maser amplification the growth rates of the Stokes parameters depend on the Stokes parameters themselves and a full simulation of the ensemble evolution cannot be avoided. Such simulations have not yet been attempted for astronomical maser radiation. In addition to their inherent significance for demonstrating the approach to stationary polarization, these simulations are essential for a complete analysis of Faraday depolarization when $\nu_B \ll \Delta\nu_D$.

REFERENCES

ABRAHAM, Z. & VILAS BOAS, J.W.S. 1994, A&A 290, 956

J. ANDERSEN 1998, ed. *The AGN—Megamaser Connection*, in *Highlights of Astronomy, vol 11B*, IAU 23rd General Assembly (Kluwer)

BLOEMHOF, E.E., REID, M.J. & MORAN, J.M. 1992, ApJ 397, 500

BOOTH, R.S. ET AL. 1981, Nature 291, 382

BRAGG, A.E., GREENHILL, L.J., MORAN, J.M. & HENKEL, C.M. 2000, ApJ, 535, 73

CLEGG, A.W. & NEDOLUHA, G.E. 1993, eds. *Astrophysical Masers*, (Springer-Verlag)

COHEN, J. 1989, Rep. Prog. Phys. 52, 881

COOK, A.H. 1966, Nature 211, 503

---

† By example, consider the imaginary limit in which the hyperfine splitting of the OH molecule vanishes and the four ground-state lines are blended into one (figure 1).

EINSTEIN, A. 1917, Physikalische Zeitschrift, 18, 121. An English translation is available in the book *The Old Quantum Theory*, D. Ter Haar 1967 (Pergamon: Elmsford, N.Y), p. 167.

ELITZUR, M. 1992a, *Astronomical Masers*, (Kluwer)

ELITZUR, M. 1992b, ARAA 30, 75

ELITZUR, M. 1996, ApJ 457, 415

ELITZUR, M. 1998, ApJ 504, 390

GARAY, G., MORAN, J.M. & HASCHICK, A. 1989, ApJ 338, 244

GARCIA-BARRETO, J.A., ET AL. 1988, ApJ 326, 954

GLAUBER, R.J. 1963, Phys. Rev. 131, 2766

GOLDREICH, P., KEELEY, D.A. & KWAN, J.Y. 1973, ApJ, 179 111 (GKK)

KEMBALL, A.J. & DIAMOND, P.J. 1997, ApJ, 481, L111

LAMB, W.E. 1964, Phys. Rev., 134, A1429

LITVAK, M.M. 1970, Phys. Rev., A2, 2107

LITVAK, M.M. 1975, ApJ, 202, 58

MCINTOSH, G. & PREDMORE, C.R. 1993, ApJ 404, L71.

MCINTOSH, G., PREDMORE, C.R. & PATEL. N. A. 1994, ApJ 428, L29

MORAN, J.M., GREENHILL, L.J. & HERRNSTEIN, J.R. 1999, J. Astrophys. Astron., 20, 165

REID, M.J. & MORAN, J.M. 1991, in *Galactic & Extragalactic Radio Astronomy*, eds. Verschuur, G.L. & Kellermann, K.I., (Springer-Verlag)

RIEU, N.Q. ET AL. 1976, A&A 52, 467

SHKLOVSKII, I.S. 1969, Astr. Zh. 46, 3 [Sov. Ast. AJ, 13, 1 (1969)]

YUSEF-ZADEH, F. ET AL. 1996, ApJ 466, L25

Interstellar magnetic fields and infrared-submillimeter spectropolarimetry

By ROGER H. HILDEBRAND

University of Chicago, Department of Astronomy and Astrophysics, Department of Physics, and Enrico Fermi Institute. 5640 S. Ellis Ave., Chicago, IL 60637, USA

The large-scale features of the magnetic field in the arms of the Galaxy have been traced by observations of polarized starlight, synchrotron emission, Zeeman splitting, and Faraday rotation. More recently, it has become possible to map fields in dense clouds by observations of polarized thermal emission from magnetically aligned dust grains. Observations at far-infrared and submillimeter wavelengths provide measurements of the field as projected on the sky at hundreds of points in individual clouds. In the polarization maps, especially when compared at several wavelengths, one finds examples of fields shaped by gravitational contraction, differential rotation, and compression. One also finds evidence for unresolved thermal structure and turbulence. To interpret the results one must understand the physical principles that relate emission, absorption, and scattering; and that relate polarization to the shapes and materials of the emitting dust grains. When these principles are applied to emission one finds that the degree of polarization in homogeneous clouds should be nearly independent of wavelength in the far-infrared and submillimeter portions of the spectrum. The steep polarization spectra actually observed can tentatively be understood if one assumes a heterogeneous temperature and radiation structure in which there is a correlation between temperature and grain alignment. The potential sources of systematic errors in polarization measurements are such that anyone entering the field must carefully review the appropriate observing and analysis techniques. With attention to the required techniques and with new instruments to be commissioned in the next few years it should become feasible to pursue scientific goals that have thus far been largely inaccessible. Among these goals are *a*) tests of models to explain the unexpected structure of the polarization spectrum in dense clouds; *b*) determination of the field strengths and characteristic sizes of turbulent domains in molecular clouds; *c*) resolution of the smallest-scale magnetic domains in molecular clouds; and *d*) explanation of the orthogonal fields observed in the Galactic center (poloidal fields in synchrotron-emitting regions and azimuthal fields in regions of thermal emission from dust). Another goal likely to be reached in the next decade is detection of the polarized component of the cosmic microwave background.

1. Introduction

1.1. *Discovery of the Galactic magnetic field*

The magnetic field of the Galaxy was discovered by accident. About 52 years ago W. A. Hiltner, then an assistant professor at Yerkes Observatory, had set out to look for polarization signals emitted periodically from eclipsing binary stars. As demonstrated years later, there is such an effect, but Hiltner never saw it. In the attempt, however, he and his associate, John Hall, of Amherst College, later of the US Naval Research Laboratory, built instruments specifically for astronomical polarimetry. Their collaboration somehow ended when they realized that they were on the verge of a big discovery. Publishing separately in the same issue of *Science* (Hiltner 1949; Hall 1949) — and who cares now which one rushed in with his manuscript a day earlier? — they announced that the light from hundreds of stars, whether binary or not, was polarized to the extent of a few per cent. It was a startling discovery. No theory had pointed to such an effect. The degree of polarization tended to increase with redenning and the position angles of the polarization tended to be parallel to the Milky Way. It thus appeared that the effect had to do with absorption of the starlight by non-spherical dust particles that had somehow become aligned with respect to the Galactic plane.

At just that time Enrico Fermi announced his theory for magnetic acceleration of cosmic rays, and since that gave a reason to postulate magnetic fields in interstellar space, it was natural to think of attributing the anisotropic absorption to *magnetic* alignment of dust particles. It appeared that polarization of starlight might provide a means, then the only means, of tracing interstellar magnetic fields. In referring to these events Otto Struve (1949) said that "The year 1949 will undoubtedly be recorded in the annals of science as that in which one of the most important astrophysical discoveries was announced".

There were other theories that attempted to account for the anisotropic absorption of starlight in the arms of the Galaxy. But the only one that has survived, albeit with modifications, was that involving paramagnetic damping of the rotation of grains spinning about their short axes leaving only the component of the spin that was along the direction of the field. Grains colliding with gas molecules cannot avoid spinning on average at a rate, $\langle \omega \rangle \gtrsim 10^5$ radians/sec, given by Brownian motion [$\langle (\frac{1}{2}) I \omega^2 \rangle = (3/2) kT$]. As we will see later, they probably spin much faster. Stresses within a grain (principally Barnett relaxation; see Lazarian & Roberge 1997) rapidly align the spin axis with the axis of maximum rotational inertia, that is, the short axis. The subsequent alignment of the spin with the magnetic field proceeds more slowly but eventually the short axis is along the field and that is the direction of minimum extinction (or maximum transmission). The direction of maximum *emission* is thus parallel to the long axis and hence perpendicular to the field. It is important to note that the grains precess rapidly about the field so that the mean E-vector of the emitted radiation is along the field even if the grains are only partially aligned.

There are non-magnetic effects that can align grains in certain anisotropic environments. For example, gas molecules streaming by dust grains at more than the sound speed will tend to spin the grains about axes perpendicular to the direction of the relative gas-grain velocity (Gold 1952). The long axes of the grains will then, on average, be in the direction of the flow. Such an effect could occur where there is a rapid outflow from a compact object, but on a larger scale the grains would be carried along with the gas so that any relative velocity would disappear. Streaming can persist on a large scale in the case of ambipolar diffusion, that is, when neutral gas flows past charged grains tied to magnetic fields. In that case, however, the streaming will tend to put the long axis of the grains parallel to the component of the relative velocity that is perpendicular to the magnetic field which is the same orientation as that produced by ordinary magnetic alignment: the two effects are complementary and give the same indication of the field direction (Roberge, Hanany, & Messinger 1995). It appears safe to assume that on large scales the E-vector of the radiation emitted from dust grains is perpendicular to the direction of the magnetic field. One should keep in mind, however, that this generality may not hold on smaller scales where there are strong non-isotropic conditions other than magnetic fields.

1.2. *Characteristics of the Galactic field*

Hiltner published polarization maps covering large areas of the Galactic plane. Figure 1 shows his results for two adjacent regions. In those early maps one could already see evidence for a coherent field in the plane of the Galaxy. Moreover, those maps were the basis for the first estimate of the strength of the field. Using the angular dispersion of the vectors together with the known interstellar gas density and values of the velocity

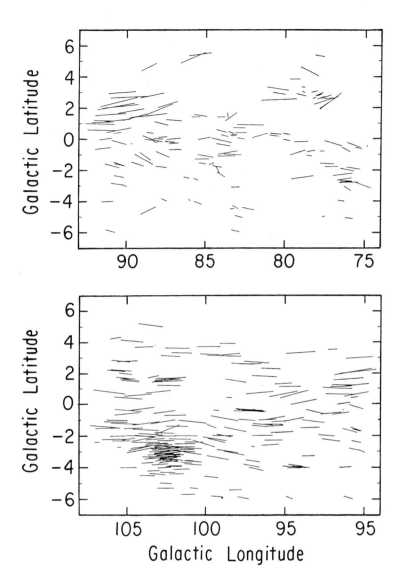

FIGURE 1. Starlight polarization in two adjacent areas of the Galactic plane (Hiltner 1951).

dispersion from the widths of atomic spectral lines, Chandrasekhar and Fermi (1953) estimated the strength of the magnetic field in the Galactic plane. Their argument was straightforward:

The velocity of a transverse hydromagnetic wave is given by

$$V = \frac{B}{\sqrt{4\pi\rho}} \tag{1.1}$$

where B is the strength of the field and ρ is the density of the diffuse matter. The transverse displacement, y, at a point, x, on a particular line of force can be described by

$$y = a \cos k(x - Vt). \tag{1.2}$$

Taking derivatives with respect to x and t, we obtain

$$\frac{\partial y}{\partial x} = -ak \cos k(x - Vt) \tag{1.3}$$

and

$$\frac{\partial y}{\partial t} = akV \sin k(x - Vt). \tag{1.4}$$

From these equations it follows that

$$V^2 \left\langle \left(\frac{\partial y}{\partial x} \right)^2 \right\rangle = \left\langle \left(\frac{\partial y}{\partial t} \right)^2 \right\rangle$$

The RMS velocity, v, in the y-direction is related to the velocity given by line broadening due to motion along the line of sight, by

$$\left\langle \left(\frac{\partial y}{\partial t} \right)^2 \right\rangle = \frac{1}{3} v^2. \tag{1.5}$$

The quantity $\frac{\partial y}{\partial x}$ is just the deviation of the line of force from a straight line as projected on the sky. Combining these expressions we have

$$B = \frac{v}{\left\langle \left(\frac{\partial y}{\partial x} \right)^2 \right\rangle^{\frac{1}{2}}} \sqrt{\frac{4}{3} \pi \rho}. \tag{1.6}$$

Using a value of $\left\langle \left(\frac{\partial y}{\partial x} \right)^2 \right\rangle$ from Hiltner's maps (predominantly for nearby stars) and values of v and ρ from other observations, Chandrasekhar and Fermi estimated a mean field of a few microgauss.

The characteristics of the field inferred from optical polarization measurements have been supplemented by observations of synchrotron emission (Dombrovsky 1954), Zeeman splitting (Vershuur 1969), and Faraday rotation (Manchester 1974). These techniques give complementary views of the Galactic field. Maps of polarized starlight and synchrotron emission both give the direction of the projected field as integrated along the line of sight and weighted, respectively, by the density of aligned dust grains or of relativistic electrons. Maps of Faraday rotation and Zeeman splitting give the strength of the field integrated along the line of sight and weighted, respectively, by the electron or atomic hydrogen density.

Observations by these techniques have led to the current picture of the magnetic field of the Milky Way: there is a coherent component parallel to the galactic plane, directed nearly along the spiral arms (Heiles 1996), and a fluctuating component of equal or greater magnitude which is structured on scales ranging from hundreds of parsecs to less than a parsec. Information about the disks of other galaxies is qualitatively similar but less detailed. Outside the plane of the disks there is considerable variation in the shapes of the fields from one galaxy to the next. As yet we know almost nothing about the fields in the inner few parsecs of external galaxies and in the center of the Milky Way we see a baffling co-existence of orthogonal fields.

Our knowledge of the Galactic field, still far from complete, has advanced considerably since 1949 when we were totally ignorant of its existence. We now understand the importance of the field to a wide range of Galactic phenomena. We know that magnetic fields affect the structure and energy balance of gas in galaxies; that they provide substantial vertical support of the gas layer; that turbulence is hydromagnetic in nature; that magnetic fields transport energy and modify the structure of interstellar shocks. It is also apparent that magnetic fields play a role in accelerating and confining cosmic rays. The cosmic rays produce ions in the interstellar gas, and that is what ties the gas to the field even where there is no other ionizing radiation. A consequence is that gravitational collapse of interstellar material tends to follow field lines producing clouds that are short along the field direction and longer in the transverse direction.

We turn now to investigations of magnetic fields in a different environment as observed in a different part of the spectrum.

1.3. *Magnetic fields in dense clouds.*

Magnetic phenomena on smaller scales, particularly on the scales of dense star-forming regions cannot be investigated by the techniques applied in the tenuous intercloud medium of the Galactic arms. Polarimetry of starlight can be extended to moderately dense regions by observations in the near infrared taking due care to distinguish between scattering and absorption. But in typical molecular clouds there are no visible background stars to show polarization by absorption; no relativistic electrons to emit synchrotron radiation; and no visible background pulsars to use for dispersion and Faraday rotation measures. Zeeman measurements still provide important information on the magnitude of the line-of-sight field but, in dense clouds, only as seen in absorption on the near sides of the objects under study.

It is just in those dense regions, however, that far-infrared polarimetry of the emitted radiation provides a unique view of magnetic phenomena. Because so little was known about the conditions required for grain alignment and about the magnetic fields and other conditions actually existing in dense clouds, it remained highly uncertain that polarized emission could actually be a useful tool until it was discovered that measurable polarization could be expected at almost any point where there is significant emission from dust. A balloon observation by Cudlip *et al.* (1982) showed polarization of $1 - 2\%$ in a beam centered on the Orion Nebula. And subsequent observations (e.g. Hildebrand, Dragovan, & Novak 1984; Hildebrand 1996; Dotson *et al.* 2000) showed measurable polarization throughout a large sample of molecular clouds and thermal streamers.

It is now feasible to map the fields as projected on the sky at hundreds of points in a single cloud. A far-infrared polarization map for an individual cloud is shown in the left panel of Figure 2. Since the dust grains spin about short axes parallel to the aligning field, and emit preferentially with the E-vector parallel to the long axis, the polarization of the emitted radiation is perpendicular to the field and the shape of that field as projected on the sky is most easily seen by rotating all the E-vectors through $90°$ as shown in the right panel of the figure. Here the vectors have all been drawn the same length to avoid any implication that a "B-vector" gives the strength of the field. Notice that the field lines are drawn towards the center of mass (Schleuning 1998). In other cases, the field lines bend around H II regions or are pulled into a pinwheel pattern by differential rotation.

In order to investigate the 3-dimensional structure of the field in a cloud, it is important, wherever possible, to make complementary Zeeman and linear polarization maps. In the cloud, M17, for example, Zeeman measurements by Brogan *et al.* (1999, 2000) show a strong line-of-sight field ($\sim 450~\mu$G) along a N-S line through the intensity peak where the polarization vectors of Dotson 1996 (and Dotson *et al.* 2000) are short; and a much weaker line-of-sight field ($\sim 50~\mu$G) in the eastern half of the cloud where the polarization vectors are long (Figure 3). We infer that the field is nearly along the line

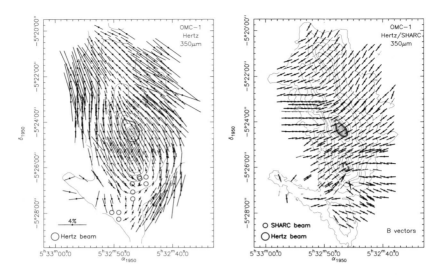

FIGURE 2. Polarization map of Orion at 350 μm (Dowell *et al.* 2001) made with the University of Chicago polarimeter, Hertz (Schleuning *et al.* 1997; Dowell *et al.* 1998) at the Caltech Submillimeter Observatory. The left panel shows E-vectors drawn to the scale indicated in the lower left corner. The right panel shows the same vectors drawn to uniform length and rotated 90° into the direction of the magnetic field. The contours in the right panel show 350 μm flux densities as measured with the photometer SHARC (Lis *et al.* 1998).

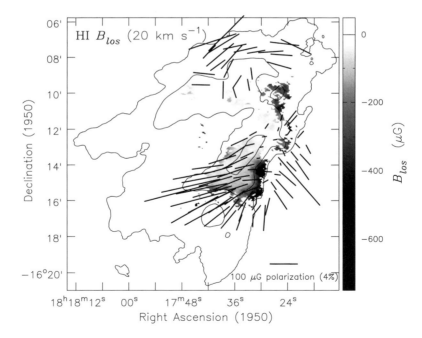

FIGURE 3. Gray-scale image of the M17 molecular cloud showing B_{los} from Zeeman observations of 20 km/s H I superposed with polarization vectors (Brogan *et al.* 1999, 2000). The 100 μm polarization vectors (Dotson 1996) have been rotated 90° to show $B_\perp$. The contours give the 21 cm continuum flux at levels of 0.3, 2.0, and 7.0 Jy/beam with 25 arcsec resolution.

of sight at the N-S line and turns toward the plane of the sky at points farther to the east.

The Zeeman measurements provide direct measurements of the mean field strengths in dense clouds. As the clouds collapse, the fields increase approximately as the square root of the gas density reaching values as large as ~ 10 mG. The rate of increase with density is more rapid in dense clumps (Vallée & Bastien 2000). Under certain conditions a comparison of the spectra of neutral and ionic molecular ion lines may provide another indicator of magnetic fields in dense clouds (Houde *et al.* 2000a, b).

1.4. *Emission and polarization spectra*

The polarization maps presented in this paper come from observations at 60 μm, 100 μm, 350 μm, and 850 μm. In almost all cases these wavelengths have been chosen for practical reasons with little or no thought to anything one could properly call the polarization spectrum; and before we venture into far-infrared spectropolarimetry it is worthwhile to review some of the practical constraints.

The most obvious practical concern is to observe at wavelengths where the dust is emitting. A typical emission spectrum for a molecular cloud peaks a little below 100 μm and that was a desirable wavelength to pick when the goal was simply to see whether there was measurable polarization throughout a cloud. To reach that part of the spectrum one must get to altitudes above ~ 14 km as was possible with the Kuiper Airborne Observatory and will again become possible with SOFIA, the next airborne observatory. At ground level, water vapor in the earth's atmosphere absorbs everything between ~ 30 μm and ~ 300 μm.

Before the Kuiper Observatory was retired in anticipation of SOFIA, a last opportunity for polarimetry in the far-infrared was used to observe a feature in the Galactic center known as the Sickle (see Section 5). The Sickle is a thin layer of warm gas with a peak emission at about 60 μm. Because it was faint in the far-infrared, the peak wavelength provided the best chance for an adequate signal. Moreover, because it was a small feature running across close strands of non-thermal emission visible at radio wavelengths it was desirable to have better spatial resolution than was possible at 100 μm. Accordingly we reconfigured the polarimeter to work at 60 μm (Dowell 1997). We were rewarded with a polarization map showing the strongest and most uniform polarization vectors we have found in any object. And as we will see in the last section the pattern of vectors showed a field in what seemed to be exactly the wrong direction.

After the last flight of the Kuiper Observatory we began observing on the ground because it was the only choice. It was not the place to try observing the Sickle but it was a very good place to work on the 350 μm polarization map of Orion (Figure 2). A disadvantage of working at that wavelength is that the atmospheric transmission, even at the altitude of the Caltech Submillimeter Observatory on the summit of Mauna Kea, only reaches as high as 50% on rare occasions. We chose 350 μm because it is one of several "atmospheric windows" of relatively good transmission, and because our limited resources forced us to use a detector array that was only marginal at 350 μm and no good at all at longer wavelengths. When the SCUBA photometer at the JCMT, an instrument with excellent detector arrays, was equipped to do polarimetry, the observations were most often made at 850 μm because the atmosphere is more transparent and more stable there than at shorter wavelengths.

These practical choices led to new types of scientific results. Maps of the cloud W3 (Figure 4) showed, in the center portion, a clockwise shift of $\sim 15°$ from 60 μm to 100 μm and another shift of $\sim 15°$ in the same direction from 100 μm to 350 μm. One could infer a corresponding shift in the direction of the magnetic field between the warm

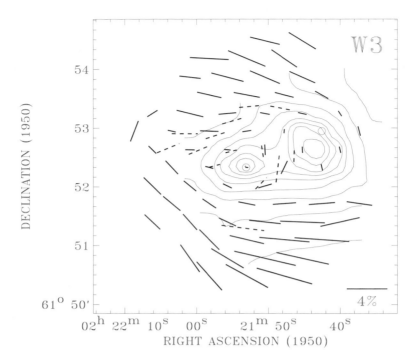

FIGURE 4. Polarization maps of the Galactic cloud, W3, at 60 μm (dashed vectors) and 100 μm (solid vectors). A further rotation of the vectors in the same direction appears at 350 μm (Schleuning *et al.* 2000). The contours show 350 μm flux densities.

core and cool outer regions of the cloud. Since the main temperature components of molecular clouds generally have peak intensities at ~ 50 μm to ~ 300 μm, it is in that range that multiwavelength observations are especially valuable in correlating measured features with source temperatures. It is a range that requires an observatory in the stratosphere or in space.

In some cases one sees no shift in position angle with wavelength but a significant change in the degree of polarization. It is that situation that confronts us for the first time with what one can properly call far-infrared spectropolarimetry. When we made histograms of the degrees of polarization in molecular clouds from 60 μm to 350 μm, we found, to our considerable surprise, a significant drop in the degree of polarization at the longer wavelength. Subsequent measurements at 850 μm from the SCUBA instrument at the James Clerk Maxwell telescope in Hawaii have shown that the spectrum rises again beyond 350 μm. This phenomenon and its interpretation will be the subject of Section 3.

To understand why a steep polarization spectrum was contrary to all expectations and to construct a theory to make sense of it we must first review the physics that relates grain cross sections to polarized emission. That will be the topic of the next section.

2. Physical principles of polarized emission

2.1. *Cross sections*

The effects of dust appear repeatedly in astrophysics. It obscures the center of the Galaxy; it reddens the light from stars; it interferes with measurements of the cosmic microwave background; it catalyzes chemical reactions; it stores the heavy elements; and it converts optical and UV photons into IR photons that escape from clouds taking energy with them. But what concerns us here has to do primarily with cross sections. To understand polarization one must, of course, be able to deal with cross sections of grains that are non-spherical. But there are important concepts that apply to a grain of any shape. In particular we will see that the discussion of cross sections will be greatly simplified by invoking three general principles: Babinet's principle, Kirchhoff's law, and the Kramers-Kronig relations.

We begin with the particular case of a black sphere in a beam of parallel light where the radius, r, is large compared to the wavelength, λ. In that case the amount of light absorbed will simply depend on the size of the object. That is, the absorption cross section, C_{abs}, will be the same as the geometrical cross section, $C_{geom} = \pi r^2$. But the sphere in question not only absorbs light it scatters it. And by Babinet's principle, the amount of scattered light must be equal to the amount absorbed. Any object in a parallel beam of light of wavelength less than the dimensions of the object will produce a diffraction pattern of scattered light. If we replace the object by a complementary screen, it too will produce a diffraction pattern. If the object is a sphere of radius r, its complement is a screen with a hole of radius r. The amount of light going through the hole in the second case is equal to the amount of light absorbed by the sphere in the first case. The amount of light absorbed by the screen is equal to the amount of light that was not absorbed by the sphere. When you remove both the sphere and the screen there is no diffraction pattern. The whole wavefront is simply the sum of the parts that would have been absorbed by the sphere or the screen. Hence the diffraction patterns of two parts must interfere so as to cancel one another, and that can be true only if, at every angle, the patterns have equal and opposite amplitudes and hence equal intensities. In other words, where $r > \lambda$ the cross section for scattering must be equal the cross section for absorption, $C_{scat} = C_{abs}$. The total cross section (what astronomers call the extinction cross section) must be twice the absorption cross section. That will prove to be a useful relationship in what follows.

Consider next the emission cross section, C_{em}. If you know the size and temperature of a black object, you can compute how much thermal radiation it will emit. If it is not black it will have an emissivity, Q_{em}, less than one and the the effective size of the object for radiating thermal energy will be what we call the emission cross section, $C_{em} = Q_{em}C_{geom}$. That is, C_{em} is the geometrical cross section that a blackbody would need to emit the same flux. The same (non-black) object will have an absorptivity (or "absorption efficiency") less than one and an absorption cross section $C_{abs} = Q_{abs}C_{geom}$. If a grain of any absorption efficiency and any temperature is placed in an oven it will come to equilibrium at the temperature, T, of the oven. The value of Q_{abs} will determine the rate at which it approaches equilibrium but not the equilibrium temperature. At equilibrium it will still absorb energy but it must emit energy at the same rate. The efficiency, Q_{em}, for emitting radiation must therefore be equal to Q_{abs}, the efficiency for absorbing radiation. That fundamental relationship, first enunciated by Gustav Kirchhoff in the nineteenth century, is known as Kirchhoff's law.

To simplify the discussions of Babinet's principle and Kirchhoff law we have spoken of the cross sections as if they were constants, and that is far from the truth. In the far-infrared the wavelengths are very long compared to the dimensions of a grain. Studies of scattering and absorption in the optical and UV give a wide range of grain radii with mean values $r \approx 0.08$ μm in the intercloud medium (Witt 1979). It is likely that there are some grains as large as a few microns, a size still small compared to far-infrared wavelengths. If a grain is not too far from spherical we can still speak of a "radius" given approximately by $r = \sqrt{C_{geom}/\pi}$). Treating a grain as a very small antenna it is easy to see why it should be an increasingly inefficient absorber as the wavelength increases far beyond the grain radius. Since the grains have a wide range of sizes they must have a wide range of cross sections and efficiencies even at a given wavelength. This problem, however, is simplified in the far-infrared by a relationship based on the Kramers-Kronig relations:

The factors that determine the absorptivity (or emissivity) are the wavelength, λ, and the complex dielectric function, $\epsilon(\lambda) = \epsilon_1(\lambda) + i\epsilon_2(\lambda)$. For $\lambda \gg r$, the relationship is

$$Q_{em} = Q_{abs} = \frac{8\pi r}{\lambda}\text{Im}\left\{-\frac{\epsilon(\lambda) - 1}{\epsilon(\lambda) + 2}\right\} \tag{2.7}$$

(e.g. Jackson 1961). I will not derive this expression, but the term in brackets will be familiar from elementary electrostatics. In the far-infrared, where $\lambda \gg r$, the grain is effectively in a uniform field. A dielectric sphere in such a field has a polarization proportional to the applied field multiplied by the term $\frac{\epsilon-1}{\epsilon+2}$, and the absorption is proportional to the imaginary part of that expression. The first thing to notice about this expression is that the quantity Q_{em}/r is independent of r hence it is not necessary to know a size distribution and a corresponding distribution in Q_{em}. It is sufficient to find mean values of r and Q_{em}.

If we multiply Q_{em} by πr^2 to obtain C_{em}, and expand the imaginary part, the expression becomes

$$C_{em} = \frac{2\pi v}{\lambda}3\text{Im}\left\{-\frac{\epsilon(\lambda) - 1}{\epsilon(\lambda) + 2}\right\}, \text{ or} \tag{2.8}$$

$$C_{em} = \frac{18\pi v}{\lambda}\frac{\epsilon_2(\lambda)}{[\epsilon_1(\lambda) + 2]^2 + [\epsilon_2(\lambda)]^2} \tag{2.9}$$

where $v = \frac{4}{3}\pi r^3$. The expression for non-spherical grains will take a similar form.

Now consider a cloud at a known distance D. Each grain will subtend a solid angle $\Omega = \pi r^2/D^2$ and will emit a flux $\Omega C_{em}B(\lambda, T)$ where $B(\lambda, T)$ is the Planck function at temperature T. If there are N grains per beam the total flux will be

$$F(\lambda) = N\frac{\pi r^2}{D^2}Q_{em}(\lambda)B(\lambda, T) = \frac{3V}{4D^2}\frac{Q_{em}(\lambda)}{r}B(\lambda, T) \tag{2.10}$$

Where V = total volume of dust $\approx N \times v$ (e.g. Hildebrand 1983). A measurement of $F(\lambda)$ and an estimate of T from the emission spectrum should thus give a value for the volume of dust, or, assuming a mean density, a value for the mass of dust, but only if one can evaluate Q_{em}.

The measurement of $Q_{em}(\lambda)$ at $\lambda \gg r$ consists in measuring the ratio $Q_{em}(\lambda)/Q_{ext}(UV)$ and multiplying by the value (2) of $Q_{ext}(UV)$ given by Babinet's principle. As we have seen, Babinet's principle explicitly relates C_{abs} and C_{scatt} and hence Q_{abs} and Q_{scatt} at short wavelengths. At long wavelengths Q_{scatt} disappears. The scattering cross section, C_{scatt}, decreases even more rapidly ($C_{scatt} \propto \lambda^{-4}$) than C_{abs}. Hence in the far-infrared one can, to good approximation, neglect C_{scatt} and assume $Q_{scatt} \longrightarrow 0$ and C_{ext}(far-IR) = C_{abs}(far-IR). The measurement of $Q_{em}(\lambda)/Q_{ext}(UV)$ has been discussed elsewhere (Whitcomb *et al.* 1981; Hildebrand 1983). One method is to find the equivalent ratio of optical depths, $\tau(\lambda)/\tau(UV)$ in a reflection nebula where some fraction of the UV radiation can escape. The long wavelength optical depth, $\tau(\lambda)$, is the ratio ($\ll 1$) of the observed brightness of a cloud to the brightness of a black body of the same temperature. The UV optical depth of the cloud is obtained by measuring the near-UV extinction of background stars. The result, then, is $Q_{em}(\lambda) = 2 \times \tau(\lambda)/\tau(UV)$. This result is subject to the inaccuracy of assigning a single temperature to the cloud. It is sufficient, however, to give approximate values of $Q_{em}(\lambda)$ [e.g. $Q_{em}(250\ \mu\text{m}) \approx 1/2666$; Hildebrand 1983].

A further inaccuracy in applying the expression $Q_{em}(\lambda) = 2 \times \tau(\lambda)/\tau(UV)$ is that the conditions under which Babinet's principle leads to the relationship $Q_{ext}/Q_{abs} = Q_{ext}/Q_{scatt} = 2$, are not fully satisfied where some of the wavelengths are longer than the dimensions of some of the dust grains. The actual value of the ratio is a matter of considerable interest. Its reciprocal, $Q_{scatt}/Q_{ext} \equiv \gamma$, is the "albedo". Measurements of the diffuse Galactic light give a value, $\gamma \approx 0.4$. Another measure of γ has been obtained from observations of a reflection nebula (Whitcomb *et al.* 1981). Any UV photon from an embedded star will either escape from the nebula or be absorbed and then be reemitted as far infrared radiation. The probability, p, of escape will then be given by the ratio of UV luminosity to the total luminosity,

$$p = \frac{L(UV)}{L(UV) + L(IR)}. \tag{2.11}$$

But p is also given by

$$p = \gamma^{\langle m \rangle}$$

where $\langle m \rangle$ is the mean number of times a photon scatters before it escapes. If the mean scattering angle is small, as indicated by the observations of Lillie & Witt (1976), then $\langle m \rangle \approx \tau$. Using measured values of p and τ one obtains again, a value, $\gamma \approx 0.4$.

2.2. *Dependence of polarization on cross sections*

The expression we have derived for $F(\lambda)$ assuming spherical grains is very nearly true for moderately elongated grains (where $r \equiv \sqrt{C_{geom}/\pi}$) but is clearly inadequate to describe polarization. One must take into account the differences in optical depths for components of radiation parallel to the different axes. If starlight becomes polarized it is because the component of light with the E-vector in one direction is absorbed more efficiently than that with the E-vector in an orthogonal direction.

Consider a beam of unpolarized starlight traveling through a cloud containing aligned grains, where the absorption, τ, is greatest for the component of the light with the E-vector in the x-directions and least for the component in the y-direction. The polarization, P_{abs}, by selective absorption in the cloud will be

$$P_{abs} = \frac{e^{-\tau_x} - e^{-\tau_y}}{e^{-\tau_x} + e^{-\tau_y}} = -\tanh[(\tau_x - \tau_y)/2] \approx -(\tau_x - \tau_y)/2. \tag{2.12}$$

(The minus sign indicates polarization by absorption.) Notice that the last expression on the right holds whenever $(\tau_x - \tau_y)/2$ is $\ll 1$: it is not necessary that the average value, $\tau \equiv (\tau_x + \tau_y)/2$, be $\ll 1$.

Polarization, P_{em}, by emission from the same cloud will be

$$P_{em} = \frac{(1 - e^{-\tau_x}) - (1 - e^{-\tau_y})}{(1 - e^{-\tau_x}) + (1 - e^{-\tau_y})} \approx \frac{(\tau_x - \tau_y)}{(\tau_x + \tau_y)}. \tag{2.13}$$

Notice, again, that the approximation holds when the *difference*, $(\tau_x - \tau_y)$, is $\ll 1$. Here the plus sign indicates that the direction of polarization is orthogonal to that for absorption. In these expressions we have not explicitly shown the wavelength-dependence of the optical depths or the corresponding wavelength dependence of the P's, but when we compare the expressions for P_{abs} and P_{em} we must be careful to do so. The relationship is

$$P_{em}(\lambda) = -P_{abs}(\lambda)/\tau(\lambda) \tag{2.14}$$

where $\tau(\lambda) \equiv (\tau_x + \tau_y)/2$. i.e. at a given wavelength the degree of polarization by emission is equal to the degree of polarization by absorption per optical depth. In fact, it is rarely possible to measure P_{abs} and P_{em} at the same wavelength. One measures P_{abs} in the optical or near-infrared in clouds that are tenuous enough so that one can see background stars. One measures P_{em} in the far-infrared or submillimeter in clouds that are dense enough to provide significant thermal emission. To make comparisons at different wavelengths one must make estimates of the λ-dependence of P_{em} using an assumed dielectric function for the grain material. To see how one makes such an estimate, or to see how one can use a measured spectrum, $P_{em}(\lambda)$, to infer the dielectric function, we must re-examine the expressions for C_{ext} and Q_{ext} taking into account the aspherical shape of the grains. We will first assume and then justify the assumption that the grains can be described approximately as oblate spheroids.

If a cloud were composed of identical oblate spheroids spinning about parallel axes perfectly aligned in the plane of the sky with the short (spin) axis, a, in the x-direction and the long (transverse) axis, b or c, in the y-direction, then one could simply replace τ_x with C_b and τ_y with C_a in the expressions for the P_{abs} and P_{em}. Because these conditions are not satisfied in a real cloud, we write the expression for $P_{em}(\lambda)$ in the form

$$P_{em}(\lambda) = \frac{C_b(\lambda) - C_a(\lambda)}{C_b(\lambda) + C_a(\lambda)} \Phi \tag{2.15}$$

where Φ is a factor to correct for the misalignment of the grains with the local field, the dispersion in direction of the field along the line of sight, and the inclination of the mean field to the plane of the sky. These components of Φ are all unknown but they do not depend on λ. Hence they do not influence relative values, $P_{em}(\lambda)/P_{em}(\lambda_0)$, where λ_0 is some reference wavelength at which the relative polarization is normalized. In summary, one can determine the shape of the polarization spectrum without knowing how well the grains are aligned. Values of P_{abs} are also subject to the unknown components of Φ, but again, one need not know Φ to compute relative values,

$$\frac{P_{abs}(\lambda)}{P_{abs}(\lambda_0)} = \frac{C_b(\lambda) - C_a(\lambda)}{C_b(\lambda_0) - C_a(\lambda_0)}. \tag{2.16}$$

Assuming spheroidal grains, the relationship of $C_b(\lambda)$ and $C_a(\lambda)$ to the shapes and dielectric functions is given by the expression

$$C_j(\lambda) = \frac{2\pi v}{\lambda} \frac{\epsilon_2(\lambda)}{\{L_j[\epsilon_1(\lambda) - 1]^2 + 1\}^2 + [L_j\epsilon_2(\lambda)]^2} \tag{2.17}$$

(e.g. Van de Hulst 1957) where C_j is cross section for radiation parallel to the jth principal axis of the grain (a or b or c), and L_j is a shape parameter for the jth axis given in terms of the eccentricity. Notice the similarity to the expression (eq'n 2.9) for the cross section, C_{ext}, of spherical grains. The L's for the three principal axes satisfy the relationship $L_a + L_b + L_c = 1$. For oblate spheroids (axes $a < b; b = c$) the L-values are defined by

$$L_a = \frac{(1 + f^2)}{f^2} \left[1 - \frac{1}{f} \arctan f \right] \tag{2.18}$$

where

$$f^2 = \left(\frac{b}{a}\right)^2 - 1 \qquad (2.19)$$

and

$$L_a + L_b + L_c = 1 \qquad L_b = L_c = \frac{1}{2}(1 - L_a). \qquad (2.20)$$

For prolate spheroids one can calculate the L-values using suitably modified expressions (e.g. Van de Hulst 1957). There is a simple relationship, however, between the degrees of polarization produced by oblate and prolate spheroidal grains spinning about their axes of maximum rotational inertia. The oblate grain will always present the same profile whereas the prolate grain will show an oscillating profile. If the wave amplitude, E, is resolved into components along the grain axes and the cross section is averaged over all angles about the spin axis ($C \propto \langle E_{proj}^2 \rangle$) one finds that oblate grains produce twice the polarization of prolate grains having the same ratio of axes (e.g. Appendix C of Hildebrand 1988). In the case of polarization by streaming where the spin axis tends to lie in a plane rather than along an axis, the degree of polarization for both oblate and prolate grains is reduced a factor of two below the values for magnetic alignment along an axis.

2.3. *Estimates of dielectric functions*

The shapes and dielectric functions appearing in these equations can best be investigated in portions of the spectrum near resonances where there are rapid changes in the cross sections (Aitken *et al.* 1988; Draine & Lee 1984). A striking resonance at ~ 9.8 μm corresponds to the stretching frequency of silicon-oxygen bonds. Another resonance, at about 20 μm corresponds to the bending of O-Si-O bonds. This is not surprising since silicon and oxygen are both abundant elements and silicates are common components of terrestrial rocks. Carbon, also abundant, must be present in some form and the observations are consistent with an admixture of graphite-like grains although the evidence is not as convincing as in the case of silicates. Draine & Lee (1984) use available photometry and spectropolarimetry to evaluate $\epsilon(\lambda)$ [hence $Q_{abs}(\lambda)$ and the shape, L_j]. They show that the observational results are consistent with a mixture of silicates with a smaller fraction of graphite where the materials have absorption efficiencies as shown in Figure 5. Other grain mixtures have been proposed (see next section) but so far none is clearly more reliable.

The absorption bands corresponding to the 9.8 and 20 μm resonances appear in many observations. Figure 6 by Aitken *et al.* (1988) is an example showing both the emission spectrum and the polarization spectrum.. The absorption at $\sim 20 \mu$m is less prominent because of the reduced optical depth with increased wavelength, but spectropolarimetry (middle panel) clearly shows enhanced polarization at both resonances. The bottom panel shows that Aitken *et al.* have verified that the position angle remains constant throughout the range used in the measurements; an essential qualification for interpreting any spectropolarimetry.

One can apply the expressions for P_{abs} and C_j to the results of mid-infrared spectropolarimetry to test the assumption that a typical grain is moderately oblate. In Figure 7 we show fits to the data of Aitken *et al.* (1988) for three different grain shapes assuming Draine's dielectric function for astronomical silicates. As is evident, the fits are very poor for extreme prolate grains (thin needles) and for extreme oblate grains (thin flakes) but relatively good for moderately oblate shapes. The best fit is obtained for a ratio of axes,

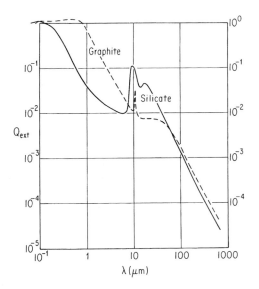

FIGURE 5. Absorption efficiencies for astronomical silicate and graphite as computed by
Draine (1985).

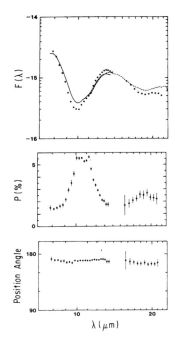

FIGURE 6. Spectra of AFGL 2591 as measured by Aitken *et al.* (1988). The solid and dashed
curves show the absorption spectra from IRAS and LRS respectively. The center pannel shows
the results of spectropolarimetry. The bottom pannel shows that the polarization spectrum is
not modified by a change in position angle with wavelength.

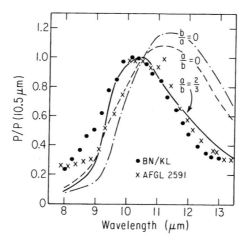

FIGURE 7. Results of spectro-polarimetry near the 9.8 μm silicate resonance for the objects AFGL 2591 (Aitken *et al.* 1988) and Orion (BN/KL; Aitken *et al.* 1985). The curves show calculated spectra for three grain shapes: needles ($b/a = 0$), flakes ($a/b = 0$), and moderately oblate spheroids ($a/b = 2/3$; Hildebrand & Dragovan 1995).

$a/b \approx 2/3$ (Hildebrand & Dragovan 1995). It is likely that there is a spread of axis ratios centered at about this value.

Spectropolarimetry at optical wavelengths shows a resonance corresponding to natural frequencies of whole grains rather than the frequencies of particular molecular bonds. This type of resonance, first seen by Serkowski (Serkowski 1973; Coyne, Gehrels, and Serkowski 1974; Serkowski, Mathewson, & Ford 1975), is best displayed when the polarization is normalized at the degree, P_{max}, and wavelength, λ_{max}, of maximum polarization. The "Serkowski Curve" is given by

$$\frac{P(\lambda)}{P_{\mathrm{max}}} = \exp\left[-K\ln^2\left(\frac{\lambda_{\mathrm{max}}}{\lambda}\right)\right] \tag{2.21}$$

where $K \approx 1.6\,\lambda_{\mathrm{max}}$. Values of λ_{max} are typically near 0.5 μm.

3. The far-infrared polarization spectrum

3.1. *Distributions in degrees of polarization*

We have seen that P_{abs} depends strongly on λ in the vicinity of the silicate resonances and that one can get an approximate fit to the observed polarization spectrum by applying the expressions for P_{abs} and C_j to silicate grains. One can use the same expressions and estimates of dielectric functions for several possible grain materials to predict polarization spectra in the far-infrared beyond any known resonance. The result is shown in Figure 8. The curves in this figure are computed for a particular grain shape, $a/b = 0.5$. The absolute values would be shifted upward for $a/b < 0.5$ and downward for $a/b > 0.5$ but there would be almost no change in the slope. Unless there is some abundant grain material with optical properties radically different from those we have considered, the curves remain almost flat in this range of wavelengths regardless of the shape and regardless of the material.

Because of the expectation that one would only find $P(\lambda) \approx$ constant, there seemed no incentive to investigate the far-infrared polarization spectrum. That is not to say

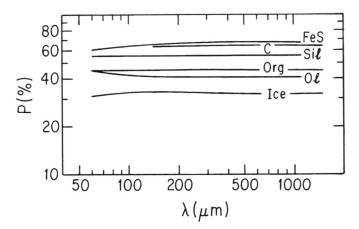

FIGURE 8. Computed polarization spectra for idealized clouds containing single types of oblate spheroidal grains (ratio of axes, $a/b = 0.5$) all at the same temperature and all perfectly aligned with spin axes in the plane of the sky (Hildebrand *et al.* 1999). The curves are derived using optical constants as given by Draine (1985) for "astronomical silicate" ("Sil") and graphite ("C") and by Pollack *et al.* (1994) for water ice ("Ice"), olivine ("Ol"), and troilite ("FeS"). The curve for orthopyroxene (not shown) is similar to that for olivine

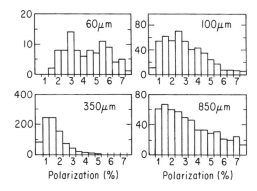

FIGURE 9. Distributions in degrees of polarization at four wavelengths for a sample of molecular clouds. The vertical scales give the numbers of events per $\frac{1}{2}\%$ bin in the range 0 to 7.5%. [60 and 100 μm data from the polarimeter Stokes on the KAO; 350 μm data from Hertz on the CSO (Hildebrand *et al.* 1999); 850 μm data from SCUBA on the JCMT (Greaves 2000)].

that there was no reason to make polarization maps at different wavelengths. It was well understood that a shift in the field direction between cloud components at different temperatures would be evident in comparisons of maps at different wavelengths. But there seemed little chance that the polarization spectrum would be interesting in itself. It was observations at different wavelengths chosen for purely practical reasons that forced a re-examination of the preconceived expectations. The first indication was simply a comparison of histograms in degrees of polarization at 60 μm, 100 μm, and 350 μm. More recently it has been possible to add a histogram of results from SCUBA (Greaves 2000) at 850 μm (Figure 9).

The results shown in this figure did not provide an adequate test of the flat-spectrum

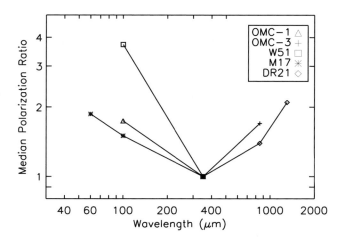

FIGURE 10. Polarization spectra for five molecular clouds normalized at 350 μm where data are available for all five (Vaillancourt 2001).

hypothesis because there was only partial overlap between the regions observed at the different wavelengths. To make valid comparisons one had to select those measurements where the observations at two or more wavelengths were in the same regions of the same clouds, where corresponding points were within one beam radius, and where there was no significant change in position angle with wavelength. Since the measurements were not made with such comparisons in mind only a small fraction of the accumulated data base could be used. Using this small sample the result (Figure 10) was qualitatively the same as that inferred from the histograms and altogether unlike the predicted spectra for idealized clouds (Figure 8).

An increasing polarization spectrum could be understood if the optical depth were great enough to reduce the degree of polarization significantly at the short wavelength end of the observed range and less so at the long wavelength end of the range (see section 3.4). But in most far-infrared observations that is not the case. Unless otherwise specified, we shall consider here only cases where $(\tau_x - \tau_y) \ll 1$.

The key to understanding the observed spectrum is to realize that interstellar clouds, and in particular the dense molecular clouds in which most of the data originated, are not homogeneous. They may, for example, include opaque clumps, transparent envelopes of embedded stars, and exposed surface material: and each such environment may be characterized by a different temperature and a different efficiency for grain alignment (e.g. exposure to radiation may be a critical factor for grain alignment). If the observed flux, F_{tot}, is the sum of n components, F_i, characterized by polarization efficiencies, P_i, then the polarization of the mixture will be

$$P_{mix}(\lambda) = \sum_{i=1}^{n} P_i X_i(\lambda) \tag{3.22}$$

where

$$X_i(\lambda) = \frac{F_i(\lambda)}{F_{tot}(\lambda)}, \tag{3.23}$$

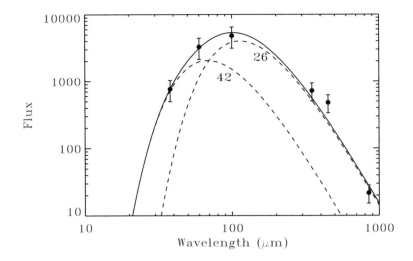

FIGURE 11. Flux spectrum at a point in Orion $\sim 1\frac{1}{2}$ arcmin north of BN/KL (30 arcsec beam). The dashed curves show a fit to the measured points assuming temperature components, 42 K and 26 K and spectra of the form $\nu^2 \mathcal{B}(\lambda, T)$. The mass ratio for the 42 K to 26 K components is 0.041 to 1 (Vaillancourt 2001).

and

$$F_i(\lambda) \propto V_i \frac{Q_i(\lambda)}{r} B(\lambda, T_i). \tag{3.24}$$

From the relationship $\partial P/\partial \lambda \approx 0$ we are led to the following rule: A significant rise or fall in the polarization spectrum (for $\Delta\tau \ll 1$, and $\lambda \gtrsim 50\mu$m) will occur if and only if there is emission from two or more populations of grains with contrasting polarization properties *and* contrasting emission spectra. The emission spectra will differ if the populations differ in the wavelength-dependence of (Q/r) or if there are differences in temperature (Hildebrand *et al.* 1999). Since none of the commonly assumed grain materials shows any significant wavelength-dependence of Q/r in the far-infrared, we are led to consider the effects of temperature variations.

3.2. *Temperature variations*

An emission spectrum like that shown in Figure 11 for a point in Orion is too broad to be consistent with emission from grains at a single temperature, but a satisfactory fit can be obtained assuming two discrete temperatures. In all probability a two-component model is oversimplified, and in fact we shall argue for a three-component or continuum model, but consider first what one should expect assuming that the 2-component model is a good approximation to the real situation. Let us further assume that for some reason only the warmer grains are aligned. As one can see from the figure, most of the flux at ~ 30 μm would be from grains in the warmer (aligned) component and most of the flux at 300 μm would be from grains in the cooler (unaligned) component. One would therefore expect a polarization spectrum that falls between 30 μm and 300 μm becoming level as one approaches the Rayleigh-Jeans portions of the spectra for both components. Notice that this model does not require different grain species but rather different grain environments.

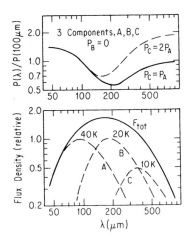

FIGURE 12. Polarization and flux spectra for a cloud with components at temperatures $T_A = 40$ K, $T_B = 20$ K, and $T_C = 10$ K where only components A and C are polarized (Hildebrand *et al.* 1999). The solid polarization curve (upper pannel) is drawn for relative polarization efficiencies $P_C = P_A$, $P_B = 0$. The dashed curve is drawn for $P_C = 2P_A$, $P_B = 0$. The flux density curves (lower panel) are drawn for peak flux densities in the ratio $1 : 1 : \frac{1}{2}$. The flux spectra for the individual components have the form $\nu B(\lambda, T)$.

To account for a spectrum that falls in the far-infrared and then rises again beyond ~ 350 μm, one can invoke a model where grains in three different environments are at different temperatures and are differently aligned. For example: (A) warm aligned grains near embedded stars; (B) cooler grains shielded from radiation; and (C) cold grains on a surface layer removed from internal sources of heat but exposed to radiation from the interstellar radiation field or nearby H II regions. The grains in (B) are near equilibrium with the local environment. Those in (A) and (C) are far from equilibrium and thus satisfy a condition for alignment established by Lazarian, Goodman, and Myers (1997). For such a cloud model the polarization and flux density spectra would have the forms shown in Figure 12 and thus qualitatively consistent with the observed spectrum. But before taking this or any other model seriously, one should determine whether both the flux density and polarization spectra agree with the model over a wide range of wavelengths and at many points. Such determinations have yet to be made with the required accuracy.

We have assumed thus far that within a cloud there may be domains characterized by different temperatures and different degrees of polarization. Now consider a cloud in which all grains are exposed to the same environment but in which different grain species are at different temperatures. Such differences could be expected in a tenuous cloud where all grains are exposed to the interstellar radiation field. Assume temperatures (graphite ~ 20 K; silicate ~ 15 K) and relative abundances (graphite $\frac{1}{4}$ by volume) as proposed by Draine & Lee (1984) and dielectric functions as computed by Draine (1985). The assumed dielectric functions result in nearly identical spectral indexes for the two species at $\lambda \gtrsim 50$ μm. The temperature differences are due to differences in UV and optical absorptivities. From the mid-infrared spectro-polarimetry, discussed in Section 2, we know that silicate grains can be aligned. If we assume that graphite grains are not aligned (perhaps because they are too small or because they have low paramagnetic susceptibility) then the fraction of the emission from aligned grains will increase with

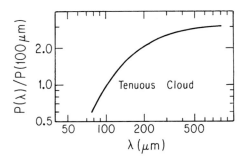

FIGURE 13. Polarization spectrum computed for a tenuous cloud ($\Delta\tau \ll 1$) exposed to the interstellar radiation field (Hildebrand *et al.* 1999). Individual grain species, A (aligned silicate) and B (unaligned graphite), are at temperatures, $T_A = 15$ K and $T_B = 20$ K (Draine & Lee 1984). The curve is based on the dielectric functions of Draine (1985) assuming a volume fraction $\frac{1}{4}$ for graphite.

wavelength and the polarization spectrum should have the form shown in Figure 13 (but only if the silicates and graphite are in separate grains).

3.3. *Grain alignment*

In developing models for both dense clouds and tenuous clouds we have assumed that stellar radiation — whether from embedded stars or from the interstellar radiation field — plays a role in aligning grains. The mechanism of grain alignment is a fascinating problem in itself (e.g. Lazarian, Goodman, & Myers 1997). All models of grain alignment must face the problem that damping torques produced by paramagnetic relaxation in ordinary grains in the weak fields of Galactic clouds cannot align grains as rapidly as impacts with molecules of the ambient gas can destroy the alignment. Purcell (1979) proposed that this difficulty could be overcome if the grains are not in equilibrium with the ambient gas but rather are spun up to suprathermal velocities by repeated impulses at special sites on the grain surfaces by processes such as the emission of photoelectrons or the formation and ejection of H_2 molecules. Such processes could plausibly account for alignment in the diffuse intercloud medium but in dense clouds, the accretion and evaporation of mantles would limit the survival of the special sites. Draine and Weingartner (1996) have proposed a variation of the Purcell model that does not depend on survival of the microstructure of the grain surface but rather on the overall shape of the grain. If a grain is asymmetric — if it is not its own mirror image — then when it is exposed to a radiation field it will be spun up like a pinwheel in the wind. Only a very slight asymmetry is sufficient. The spin-up does not hasten the paramagnetic damping: it simply provides more time for magnetic damping to act.

The temperatures of grains exposed to anisotropic radiation fields can be influenced by the orientation of the grains with respect to the source of radiation. In addition to spinning and heating the grains, as discussed in section 3.2, the radiation may selectively heat grains with spin axes along the direction of the radiation. That is, an oblate grain exposed to the radiation face-on will become warmer than one exposed edge-on. This effect will tend to steepen the polarization spectrum, especially at $\lambda \lesssim 100$ μm (Onaka 2000).

3.4. *Polarization vs optical depth*

In Section 2 we introduced the approximations, $P_{abs} \approx -(\tau_x - \tau_y)/2$, and $P_{em} = (\tau_x - \tau_y)/(\tau_x + \tau_y)$ for $\Delta\tau \ll 1$. One must remember that the condition $\Delta\tau \ll 1$ is not always satisfied. To describe the effect of non-negligible opacity, it is useful to rewrite the expression for P_{em} in terms of $\Delta\tau$. The result is

$$P_{em} = \frac{e^{-\tau}\sinh(\Delta\tau/2)}{1 - e^{-\tau}\cosh(\Delta\tau/2)} \qquad (3.25)$$

or

$$P_{em} = \frac{e^{-\tau}\sinh(P_0\tau)}{1 - e^{-\tau}\cosh(P_0\tau)} \qquad (3.26)$$

where $P_0 = \frac{\Delta\tau}{2\tau}$ is the polarization one would measure for $\tau \to 0$. If $P_0 \ll 1$ and if $\Delta\tau \ll 1$, then this reduces to the approximate expression

$$P_{em}(\lambda) \approx P_0[1 - \tau(\lambda)/2].$$

To first approximation $\tau \propto \lambda^{-\beta}$ where β is a constant of order 1 to 2. To this approximation the equation becomes

$$P_{em} = P_0\left(1 - \frac{K}{\lambda^\beta}\right)$$

where K is a constant. In dense cloud cores one often sees a drop in the degree of polarization which can be attributed in part to the increase in opacity. But a further drop may be caused by unresolved structure in the aligning field. The investigation of this unresolved structure is a matter of basic interest to be discussed in Section 5.

4. Observing techniques and analysis of results

4.1. *Source spectra and background flux*

The peaks of the emission spectra of Galactic clouds usually fall within a range, ~ 40 μm to ~ 300 μm, that is accessible only from the stratosphere or from space. The Stratospheric Observatory for Infrared Astronomy (SOFIA) will be especially valuable for observations in this range. The Rayleigh-Jeans side of the spectra can be observed in submillimeter "atmospheric windows" (e.g. 350 μm, 450 μm, 850 μm, 1300 μm) from ground-based telescopes at dry sites such as Mauna Kea and the South Pole (Figure 14). As we will see, reliable estimates of temperature components are essential to the interpretation of polarization spectra. It is difficult, however, to make accurate determinations of the flux spectra from which the temperatures can be derived. No one instrument covers the whole range of wavelengths shown in Figure 11. The data for this figure were obtained using several instruments at different sites and different epochs with different calibrators. We will return to this important limitation in Section 5.

The preponderance of the far-infrared or submillimeter flux entering a polarimeter at any ground or stratospheric site is not that due to the celestial source but rather to thermal emission of the sky and telescope. Moreover this background flux is noisy due to fluctuations in atmospheric transmission and emission. Procedures for removing the local background vary according to the nature of the observations, but all these procedures involve some kind of comparisons of signals at source points with signals at neighboring reference points chosen, where possible, to be off the source.

Let $F(x)$ be the flux from a point, x, on the source, and let $F(y1)$ and $F(y2)$ be the fluxes from two reference points at equal and opposite displacements from x. Assuming

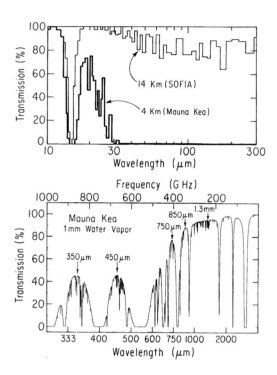

FIGURE 14. Atmospheric transmission at 14 km (SOFIA) and 4.2 km (Mauna Kea at 1 mm precipitable water vapor). The arrows in the lower pannel mark the centers of "atmospheric windows" commonly used in ground-based submillimeter observations.

that the background at x is given approximately by the average of the backgrounds at $y1$ and $y2$, the net flux will be

$$I(x) = G\{F(x) - [F(y1) + F(y2)]/2\} \qquad (4.27)$$

where G is a calibration factor which one need not evaluate unless the goal is an absolute flux measurement. (For arrays of detectors one must evaluate relative values of the G's to correct for the different efficiencies of the individual sensors.)

Because of fluctuations in atmospheric transmission and emission, the comparisons of flux densities at source and reference points must be made more rapidly than would be possible by moving the whole telescope. Instead one moves a "chopping mirror" somewhere in the optical path, usually the secondary mirror, to switch rapidly between two of the points, say x and $y1$. One integrates the difference of the signals between the two points and then moves ("nods") the telescope so as to switch between $y2$ and x. If the source was first in the "left" beam it will now be in the "right" beam and, if there is a net signal from the source, the sign of that signal will be reversed. A strip chart showing the difference signal between the two beams will then appear (ideally) as a square wave with the corners rounded because of the time required to nod between the two positions. But unless the source is very bright and the atmosphere unusually stable, the trace will look more like a succession of jagged peaks and valleys displaced from a wandering baseline as seen in Figure 15. We will return to this figure when we discuss

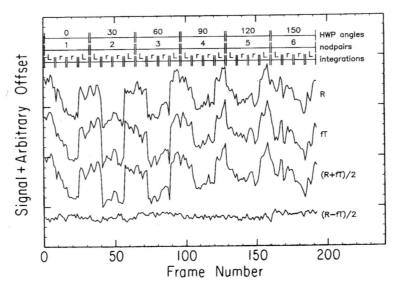

FIGURE 15. Strip chart record (Dowell *et al.* 1998) of frames accumulated during a single file (six steps of the half-wave plate) during an observation of the source IRC+10216 ($F = 30$ Jy/beam). Each point on a chart trace is one frame, the fundamental unit of stored data for two cycles of the chopping secondary at the CSO (0.6 s). The T-frames (one component of polarization) have been multiplied by a normalization factor, f, to bring them to the same scale as the R-frames (the other component). Correlated sky noise is evident in R and T as well as their sum (third trace: the total photometric signal). The correlated noise is removed by taking the difference $(R - fT)$ (bottom trace). In this example, both the source and the instrument have negligible polarization.

noise in polarization signals. The trace labeled "$(R + fT)/2$", however, is the sum of the signals for two orthogonal components of polarization and is thus exactly equivalent to an ordinary photometric signal $[\times \frac{1}{2}]$. This is an example of "sky noise" caused by changes in the background emission between the left and right beams. For an instrument with a single detector there is little one can do except to repeat the measurement for many cycles. But with an array of detectors one can take advantage of the fact that the atmospheric noise is correlated over the whole focal plane and can be removed by taking the difference between signals in the center pixels on the source and signals in the edge pixels if they are known with certainty to be off the source.

An alternative method permits removal of the background and sky noise while scanning "on the fly" without nodding. This procedure is especially valuable for sources that are larger than the array. If one scans, say from right to left, across a source while chopping in the scanning direction along a row of detectors, the signal from the source will appear in each detector first in the left beam and then in the right. A plot of the signals (left minus right) *vs* time will then have the form shown in Figure 16. The sky noise appears simultaneously in the whole row of detectors while the source moves through the detectors sequentially. The total signal, $S_d(t)$, in detector d at time t is

$$S_d(t) = G_d\{F[x_d(t)] + N(t)\} + C_d \tag{4.28}$$

where G = detector gain; F = source flux as a function of sky position, $x_d(t) = x_1(t - t_d)$, where t_d = time to scan from pixel 1 to pixel d; $N(t)$ = correlated sky noise; and C_d = constant offset (residual background or electronic offset).

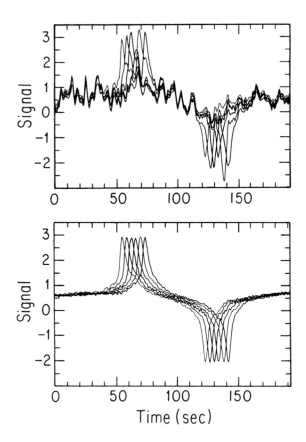

FIGURE 16. The signal, l-r, (arbitrary units) for a scan on-the-fly along a row of six detectors in the University of Chicago polarimeter Hertz. (Source: OMC-1. Scan rate: 5 arcsec/sec. Chop throw: 5.5 arcmin. Chop rate: 2.7 Hz). The raw signal from the source appears first in the left beam and then in the right beam of each detector with a time lag equal to the beam separation divided by the scan rate. The coherent noise appears in each detector with no time lag. The lower panel shows the signals after processing to remove the correlated noise.

The noise is removed by adjusting $N(t)$ to minimize the difference between the net signals registered by pixels along the scan. The resulting solution for $S_d(t) - N(t)$ is shown in the lower panel of the figure. To avoid inverting a large matrix, one can solve sequentially for $S_d(t)$, $N(t)$, C_d, and G_d.

4.2. Polarization signals

If one measures two components of polarization simultaneously, the sky noise will affect both components equally. The noise will therefore be removed from the polarization signal by taking the difference of the signals for the two components. We consider first the detection of a single component with a single detector; then two components with one detector for each component; and finally two components in two arrays of detectors.

If a fine grid of parallel wires (wire separation $\lesssim \lambda/5$ center to center) is placed across the path of the radiation ahead of a detector, the component parallel to the wires will be reflected: the component perpendicular to the wires will be transmitted and can be coupled to the detector. If the grid is rotated, the signal in the detector will vary with the angle of rotation, θ, according to the degree and angle of polarization; i.e. the grid will serve as a "polarization analyzer". The degree of polarization is

$$P = \frac{I_{\max} - I_{\min}}{I_{\max} + I_{\min}}. \tag{4.29}$$

The direction of polarization is given by the angle, $\phi(\max)$, of the E-vector at which the intensity, $I(\phi)$, is a maximum. Notice that the intensity is also a maximum at $\phi(\max) + n\pi$. The difference of $I(\phi)$ from its mean value, $\langle I(\phi) \rangle$, should vary as $\sin(2\theta)$.

Instead of rotating a grid, one can rotate a half-wave plate ahead of a fixed grid. A rotation of the half-wave plate through an angle θ rotates the plane of polarization by 2θ: i.e. rotating the half-wave plate through an angle 2θ is equivalent to rotating a grid by 4θ. Thus the variation in intensity is given by

$$I(\theta) - \langle I(\theta) \rangle = A \sin[4(\theta - \delta)] \tag{4.30}$$

and

$$\frac{I(\theta) - \langle I(\theta) \rangle}{\langle I(\theta) \rangle} = P \sin[4(\theta - \delta)] \tag{4.31}$$

where δ is a phase angle depending on ϕ, the angle of the E-vector on the sky; and on the rotation angle of the sky with respect to the instrument; the angle of the instrument rotator, and an angle depending on the orientation of the grid and half-wave plate within the instrument.

Because of the atmospheric noise the signal will not have the form given by this equation. For a moderately faint source typical signals will have forms like those shown in the top two panels of Figure 18, each recorded for a single component.

An optical design for simultaneous detection of two orthogonal components of polarization is shown schematically in Figure 17. In this design, a fixed wire grid following the half-wave plate is inclined at 45° to the optic axis. The "R-component" parallel to the wires is Reflected to one detector array and the other ("T-component") is Transmitted to the other array. The polarization signal is then

$$S(\theta) = \frac{R(\theta) - T(\theta)}{R(\theta) + T(\theta)} \tag{4.32}$$

Fluctuations in atmospheric transmission affect R and T by equal factors and hence leave $S(\theta)$ unchanged. Fluctuations in atmospheric emission (unpolarized) produce correlated excursions in R and T that are removed in the numerator, $R(\theta) - T(\theta)$, when taking the difference. These fluctuations are small compared to those in the denominator, $R(\theta) + T(\theta)$. The effectiveness of this scheme is shown in the bottom panel of Figure 18 and also in the bottom trace of Figure 15.

Note: In principle one could measure P by rotating an analyzer without chopping to obtain the polarized flux, $P \times F$, and then dividing by a value of the flux, F, obtained from a photometric map. But as we have said, the preponderance of the flux reaching the detectors is not from the source but from the local background, and that flux will become polarized by the instrumental polarization to produce a total polarized flux much greater than that due to the source.

4.3. *Instrumental polarization*

I will assume familiarity with the Stokes parameters, I, Q, U, and V and will discuss the problem of solving for these parameters in observations with a telescope and instrument both of which introduce components of polarization which must be removed in order to

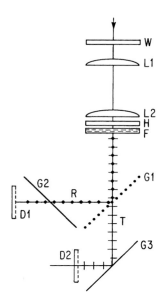

FIGURE 17. Schematic design of a polarimeter for simultaneous detection of two components of polarization. Radiation enters through a vacuum window, W, passes through lens, $L1$, at an image of the sky, and through a second lens, $L2$, at an image of the primary. $L2$ reimages the sky at the detector arrays, $D1$ and $D2$. A rotating half-wave plate, H, close to $L2$ rotates the plane of polarization. A spectral filter, F, defines the passband. A wire grid, $G1$, reflects component R parallel to the wires and transmits component T perpendicular to the wires. Component R is then transmitted by a second grid, $G2$, and enters detector array, $D1$. Component T is reflected by grid $G3$ and detected by array $D2$. One measures the quantity $(R-T)/(R+T)$ as a function of the angle of rotation, θ, of the half-wave plate.

find the polarization due to the source. It will be convenient to do the analysis in terms of the reduced Stokes parameters defined by

$$q = Q/I = P \cos 2\xi \qquad (4.33)$$

$$u = U/I = P \sin 2\xi \qquad (4.34)$$

where

$$\xi = \frac{1}{2} \arctan(u/q) \qquad (4.35)$$

is the angle of polarization with respect to a reference frame in the instrument. Notice that the quantity

$$q^2 + u^2 = P^2[\cos^2 2\xi + \sin^2 2\xi] = P^2 \qquad (4.36)$$

is invariant with respect to the orientation of the polarization. If ξ changes (e.g. by rotation of the sky with respect to the telescope) while the reference frame remains fixed in the instrument then points in the $q - u$ plane will move on a circle of radius P.

The observed polarization is due to the combined effects of the source, s, the telescope, t, and the polarimeter, p. The values of q and u for those components are additive when their sum is $\ll 1$ as in all cases considered here. Hence the measured Stokes parameters can be written as the sums

$$q = q_s + q_t + q_p \qquad (4.37)$$

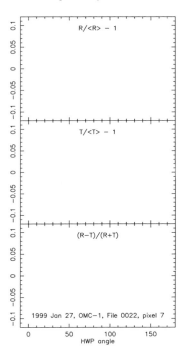

FIGURE 18. Elimination of sky noise by simultaneous detection of orthogonal components of polarization. The top two frames show the polarization signals derived from the two components individually (as would be necessary with a one-array instrument). The bottom frame shows the signal derived from the difference of the two signals divided by the sum.

and

$$u = u_s + u_t + u_p. \tag{4.38}$$

If the polarimeter is fixed with respect to the (alt-az) telescope and the sky rotates, then the measured Stokes parameters, q_i, u_i, at a point i, at sky rotation α_i, will be given by

$$q_i = P_s \cos 2(\xi_i - \alpha_i) + (q_t + q_p) \tag{4.39}$$

$$u_i = P_s \sin 2(\xi_i - \alpha_i) + (u_t + u_p), \tag{4.40}$$

and these are just the parametric equations of a circle of radius P_s, centered at a point $(q_t + q_p)$, $(u_t + u_p)$ representing the combined effects of the sky and telescope. We thus have the magnitude, $P_s = [(q_i - q_t - q_p)^2 + (u_i - u_t - u_p)^2]^{\frac{1}{2}}$ and the angle, $\xi_i = \frac{1}{2}\arctan[(q_i - q_t - q_p)/(u_i - u_t - u_p)]$ for the source polarization for a pixel on the axis of rotation. See Figure 19.

4.4. *Polarized Flux in Reference Beams*

The procedure we have outlined for removing background due to emission from local sources is valid only if the reference beams, $y1$ and $y2$, are, in fact, off the source. But that condition is often not satisfied for extended sources. Polarized flux in the reference beams is a potential source of large systematic errors and is the main reason to hesitate before crediting a polarization map. In the case of photometry one can at least determine that the flux at point x is a measured level above (or below) the average of that at $y1$ and $y2$. But in the case of polarimetry, it is not only the flux but also the degree and direction of polarization at $y1$ and $y2$ that is unknown. There is no justification for

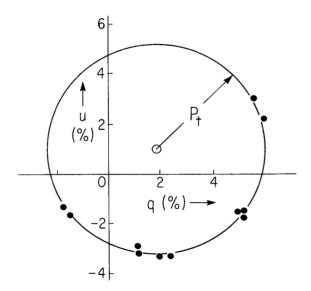

FIGURE 19. Points in the q-u plane as measured in the boresight pixel. The polarimeter has been rotated to follow the sky. The points trace a circle of radius P_t centered at $(q_i + q_p)$, $(u_i + u_p)$. If the instrument had remained fixed with respect to the telescope the points would have traced a circle of radius P_s centered at $(q_i + q_p)$, $(u_i + u_p)$.

assuming that the polarization is low where the flux is low. The opposite is more often true.

To solve the problem properly one must make long scans across the object while chopping and nodding and advancing the half-wave plate at each point. Alternatively one can make a series of scans on the fly, advancing the half-wave plate between each scan, recording the data in bins no larger than a resolution element, and analyzing the result as if done at discrete points. The procedure is most easily described for the case of chopping in the direction of the scan.

Consider a scan in the x-direction.

Let D = chopper throw,

$f(x)$ = true distribution of the desired quantity (I or Q or U)

$g(x)$ = measured distribution as given by

$$g(x) = f(x) - \frac{1}{2}[f(x + D) + f(x - D)]. \tag{4.41}$$

The relationship between $f(x)$ and $g(x)$ is shown graphically in Figure 20.

Now choose a point x_0 believed to be at least a distance D from the source. Measure $g(x_0 - D)$, $g(x_0)$, $g(x_0 + D)$, $g(x_0 + 2D)$, $\cdots$, $g(x_0 + nD)$. If point x_0 is, in fact, off the source by $\geqslant D$, then one should find $g(x_0 - D) = g(x_0) = 0$. If point $x_0 + nD$ is at least a distance D beyond the other side of the source then one should find $g(x_0 + mD) = g[x_0 + (m + 1)D] = 0$ when $m \geqslant n$.

Rearranging the above equation we have

$$f(x + D) = 2[f(x) - g(x)] - f(x - D), \tag{4.42}$$

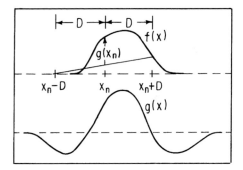

FIGURE 20. Scan of an extended object. The curve, $f(x)$, represents the true distribution of the desired quantity (e.g. I, Q, or U). The curve $g(x)$, represents the measured distribution for a chopper throw, D.

and applying this expression repeatedly starting from $f(x_0) = 0$, we have

$$f(x_0 + nD) = -2 \sum_{i=1}^{n-1} (n - i)g(x_0 + iD). \tag{4.43}$$

To limit the random walk of the errors, now adjust the measured values of $f(x_0 + nD)$ within the errors to satisfy the condition $f(x_0 + mD) = 0$ for $m \geqslant n$. A limitation to this procedure is that its sensitivity for features in $f(x)$ approaches zero whenever those features approach a periodicity of length D, $D/2$, etc. That problem can be overcome by measurements with different values of D.

The usual procedure in acquiring a source is to maximize the signal $I(x)$ as given by $g(x)$. If both reference beams are off the source then $f(x_{\max}) = g(x_{\max})$. If one then moves to a point, x', where $g(x')$ is, say, $\frac{1}{10}g(x_{\max})$, one can likewise assume $f(x') = g(x') = \frac{1}{10}g(x_{\max})$ but *only* if both beams are still off the source. If one or both beams are on the source, $f(x')$ may be $\gg g(x')$, and if one assumes that $g(x')$ is the value of $I(x')$ to use in the denominator of the expressions for q, u, and P, then errors in measuring the Stokes parameters - for example, errors due to the unknown degrees and directions of polarization in the reference beams — will be multiplied by the ratio $f(x')/g(x')$.

4.5. *Noise*

We conclude the discussion of polarization techniques with a summary of noise sources. The relative importance of these sources depends on weather, wavelength, source intensity, detector characteristics, chopping amplitude, and optics.

(1) *Detector/Amplifier Noise*: Typically white noise but increasing as $1/f$ at low frequencies. (Usually not a limiting factor with modern detectors.)

(2) *Photon Noise*: White component of noise from the sky due to statistical fluctuations in photon arrival times. (Generally below sky noise for observations within the earth's atmosphere for chopping amplitudes $\gtrsim 1$ arcmin, but the fundamental noise once sky noise is removed.)

(3) *Atmospheric Noise*: Here we must distinguish between atmospheric emission and transmission

A. *Emission* ("sky noise"): Background noise due to changes in atmospheric emission at source and reference points (adds equal increments to the two orthogonal components). This noise can be reduced by reducing the chopping amplitude. On extended objects, however, low chopping amplitudes exacerbate the problem of systematic errors due to unknown polarization in the reference beams. Sky noise is removed from the difference, $(R - T)$, in the numerator of the expression for the polarization signal by simultaneous detection of two orthogonal components of polarization.

B. *Transmission*: Changes in the received signal due to fluctuations in atmospheric opacity. These changes affect the signals R and T by equal factors and are effectively removed from the polarization signal, $(R - T)/(R + T)$, by simultaneous detection of two orthogonal components

(4) *Pointing Noise*: Noise generated by drifting of the source with respect to the center of a detector, especially in the case of bright compact sources. If the arrays for the two components are not accurately aligned, or if only one array is used, then the pointing error will introduce noise in the polarization signal wherever there are gradients in the flux density or polarization.

5. Far-infrared polarimetry in the next ten years

Attempts to look into the future have value at least in providing amusement for future generations when the future has become the past and the predictions become embarrassments. In this field, however, there is plenty to see within the time horizon of current graduate students and post doc's. Any embarrassments can be shared among contemporaries. Whenever a new phenomenon appears — something unexpected, dimly perceived, and explained only by speculation — there is surely work to do in much less than ten years.

5.1. *The Polarization Spectrum*

The polarization spectrum is such a phenomenon. Instead of the expected flat spectrum there is a spectrum that falls and then rises again. The spectrum is only roughly described in a single type of object, molecular clouds, and the explanation is still an untested hypothesis. Among the thousands of polarization measurements that have been recorded in the range $60\mu m$ to 1300 μm, fewer than 100 provide valid comparisons. In only a few objects can one compare existing measurements at different wavelengths at points within one beam radius where there is no change in position angle and where the measurements are demonstrably free of spurious effects due to unknown flux and polarization in reference beams. To be believable the comparisons should be at groups of adjacent points satisfying all these conditions.

An obvious step toward investigating the spectrum is simply to increase the store of relevant data by at least an order of magnitude by carrying out observations designed specifically for that purpose. To do it properly there should be new instruments for both airborne observations in the far-infrared and ground based observations at submillimeter wavelengths. A second step is to test the hypothesis that differences in polarizing power are correlated with differences in temperature (Vaillancourt 2001). That will require comparisons point by point of polarization spectra with emission spectra. The emission spectra must be measured with sufficient accuracy and at a sufficient number of wavelengths to permit analysis of the temperature components of the sources. If one focuses on molecular clouds the accuracy will be limited not by statistical errors but rather by systematic errors, especially in calibration. No single instrument has yet been developed to cover the whole range of wavelengths required for such an analysis. The existing data come from different instruments at different observatories and, most seriously, from observations based on different methods of calibration. The range of wavelengths must be sufficient to reach both the short- and long-wavelength sides of the emission spectra. No

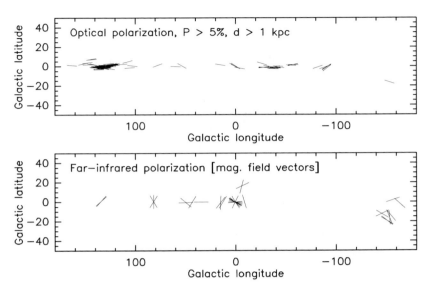

FIGURE 21. Two views of magnetic fields in the arms of the Galaxy (Dowell, Hildebrand & Jones 2001). Upper panel: polarization by absorption at optical wavelengths in the diffuse intercloud medium. The vectors are selected from the summary of Heiles (2000) for long lines of sight (> 1 kpc) and strong polarization (P > 5%). Lower panel: mean field directions for 27 Galactic clouds as seen in far-infrared emission.

temperature components can be identified from measurements entirely on the Rayleigh Jeans tail of all the components. Notice, however, that even when the majority of the flux is from warm components, the bulk of the emitting dust may be from cool components that are seen preferentially at long wavelengths. The number of spectrum measurements at each point in the source must be greater than twice the number of temperature components one would like to fit. Even if one assumes discrete temperature components with identical spectral indices, each component must be described by a temperature and an abundance.

5.2. Large-scale fields

We began these lectures by viewing Hiltner's maps of polarized starlight. Because those were among the first such maps, they were primarily for bright, nearby stars. The most detailed compilation of data on polarized starlight is that assembled by Matthewson and Ford in 1970 showing polarization vectors toward some 7000 stars. From this large sample one can select distant stars using distances as tabulated by Heiles (2000). Figure 21 (upper panel) is a polarization map for distant stars (> 1 kpc) showing strong polarization (> 5%). Here the effects of dispersion at smaller scales are suppressed by confusion along the line of sight and what one sees is the mean field in the tenuous intercloud medium.

Another view of fields in the plane of the Galaxy is provided by far-infrared/sub-millimeter measurements of the polarized emission from individual dense clouds. In those clouds one generally finds orderly fields with well-defined mean field directions. In the lower panel of the figure we show a large-scale map giving the mean directions for 27 Galactic clouds. Comparing this to the optical results one sees that the direction of the mean field in the intercloud medium is lost in the process of condensing into dense clouds.

Despite the apparently random orientation of the mean fields in individual molecular clouds, some order may appear when the observations are extended to large-scale cloud complexes. On a much smaller scale, it is already becoming feasible to investigate the turbulent component of the fields in dense clouds.

5.3. *Mean and turbulent fields*

From Hiltner's maps and from the far-infrared maps shown in this paper, it is evident that there are non-uniform components in both the tenuous intercloud medium and in dense clouds. In both cases turbulent components tend to be lost by confusion along the line of sight. Although the turbulent components are largely hidden, they may contribute or even dominate the total energy stored in the field; they may control the collapse of interstellar material into dense clouds and clumps and protostars; and may provide significant heating by dissipation of turbulence.

Two key numbers in assessing the role of magnetic fields in interstellar clouds are the ratio of magnetic energy, M, to kinetic energy, K, and the ratio of M to the gravitational energy, W. These ratios determine whether a cloud is in virial equilibrium and whether it can be magnetically supported. The ratio M/K is a measure of the extent to which magnetic fields control motions within the cloud.

Jones, Klebe, & Dickey (1992) have measured the degree of polarization, P_{abs}, *vs* optical depth, τ, as seen in absorption through the tenuous intercloud medium. Where $\Delta\tau$ is less than a few tenths, one would expect to find $P_{abs} \propto \tau$ for a purely uniform field and $P_{abs} \propto \sqrt{\tau}$ for a purely random field. The τ-dependence they find falls between these extremes and indicates an energy density in the turbulent component at least as great as that in the uniform component. The characteristic depth of the turbulent domains is $A_V \approx 1$ or $N \approx 3 \times 10^{21} \mathrm{cm}^{-2}$. It should be feasible to carry out a complementary investigation into the turbulent component in dense clouds by measuring polarized emission, P_{em}, *vs* far-infrared optical depth, $\tau(\mathrm{FIR})$. When $\tau(\mathrm{FIR}) \ll 1$, one would expect $P_{em}(\tau) \approx$ constant for a purely uniform field and $P_{em} \approx 1/\sqrt{\tau}$ for a purely random field. If the condition for $\tau(\mathrm{FIR}) \ll 1$ is not satisfied, the results must be corrected for opacity as given by equation (3.26). By observing the actual distributions in $P_{em}(\tau)$ it should be possible to determine the turbulent fraction and correlation lengths in clouds that are 10^4 times denser and contain fields that are 100 times stronger than those in the intercloud medium.

In Section 1 we reviewed the analysis of the Galactic field by Chandrasekhar and Fermi using only the angular dispersion of the polarization vectors. To the extent that the apparent dispersion is reduced by confusion along the line of sight such an analysis tends to give an overestimate of the field strength (Zweibel 1996). The analysis described here uses only the τ-dependence of P. It should be feasible, however to incorporate both the angular dispersion and the degrees of polarization in the analysis by comparing computer-generated models with polarization maps of greatly improved precision. The next generation of polarimeters should permit measurements with $> 15\sigma$ significance ($\Delta\phi < 2°$), a five-fold improvement, and $> 3\times$ better angular resolution than has heretofore been possible.

5.3.1. *Small-scale fields*

To sustain a hydromagnetic wave in which neutral particles are coupled to the field, the collision time between ions and neutrals must be shorter than the period of the wave. The minimum length, λ_A, satisfying this requirement is determined by the strength of the magnetic field, B, the number density of the interstellar material, n, and the ion

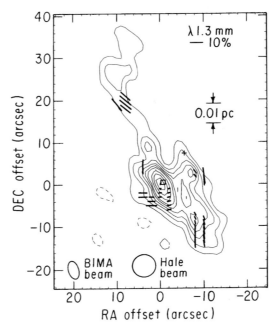

FIGURE 22. Polarization map of Orion-KL at 1.3 mm from observations at the BIMA interferometer (Rao *et al.* 1998). The peak flux at IRc2 (hollow square) is 2.4 Jy/beam. The ellipse and circle at the bottom show, respectively, the beams for BIMA (1.3 mm) and SOFIA (60 μm).

fraction, f_i. In rough approximation the relationship is

$$\lambda_A > 0.3\text{pc} \left(\frac{B_{\mu G}}{100}\right) \left(\frac{10^4}{n}\right)^{3/2} \left(\frac{10^{-7}}{f_i}\right) \tag{5.44}$$

(e.g. Hildebrand 1996). Where not hidden by confusion along the line of sight, features of this magnitude should be easily resolvable with current observing techniques. Situations in which confusion is not serious may be found where the emitting material is concentrated in regions only a few times larger than the features one wishes to resolve. An example (not necessarily an example of hydromagnetic effects) has appeared in polarimetry of the central $\frac{1}{2}$ arcmin of Orion (Rao *et al.* 1998) where one sees strong polarization with large shifts in angle between domains separated by ~ 10 arcsec (Figure 22). It is not yet clear whether this represents abrupt shifts in the field direction or an effect of outflows. In either case the Orion results call for further investigation. Continuing interferometry will be able to probe other clouds with and without outflows but because this technique is applicable only for bright, compact sources it will be difficult to cover larger areas. A polarimeter, "Hale", to be proposed for SOFIA will have sufficient resolution at 60 μm to resolve the structure in the Orion core and the sensitivity not only to cover a larger area but also to examine fainter clouds closer to the Earth. Moreover Hale's ability to observe over a range of far-infrared wavelengths down to ~ 53 μm will make it possible to isolate domains of different temperatures.

Interferometry will clearly be required to resolve circumstellar features. There again one can expect the emission to be dominated by the small regions under study.

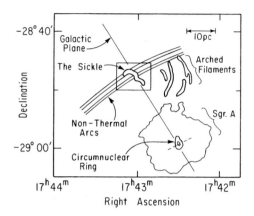

FIGURE 23. Sketch of features in the center of the Galaxy.

5.4. *Galactic Center*

Figure 23 is a schematic drawing of features near the Galactic center. Observations of synchrotron emission from the long non-thermal arcs indicate a strong field normal to the Galactic plane (Yusef-Zadeh, Morris, & Chance 1984; Tsuboi, Ukita, & Handa 1997). It was generally assumed that that was the prevailing direction of the field in the whole region. The polarization map of the circumnuclear disk could tentatively be explained as a magnetic accretion disk in which rotation has wound up an ambient poloidal field into a tourous (Werner *et al.* 1988, Hildebrand *et al.* 1993). But the fields subsequently measured in the Sickle (Figure 24), the arched filaments (Morris, Davidson, & Werner 1995; Davidson 1996), and the clouds surrounding the circumnuclear disk (Novak *et al.* 2000); and the large-scale field surrounding the circumnuclear disk and extending 150 pc along the ridge of emission (Novak *et al.* 2000, 2001; Figure 25) indicate a prevailing field along the Galactic plane (i.e orthogonal to the field in the arcs).

A possible interpretation is that the azimuthal field is the result of shearing by motions in the plane. But that interpretation does not address the contrasting picture of the field in regions of synchrotron emission. If the Sickle actually intersects the non-thermal arcs, as the evidence suggests (Yusef-Zadeh, & Morris 1988), then evidence for field reconnection (Serabyn & Morris 1994) might appear in observations with sufficient angular resolution to resolve the intersections at individual threads of the non-thermal arcs. The magnetic model for the circumnuclear disk could be tested by comparing high-resolution multi-wavelength polarimetry with temperature maps derived from multiwavelength photometry. The polarization maps of W3 (Figure 4) show, in principle, how that might be done (see discussion in section 1.4).

5.5. *The Cosmic Microwave Background*

We have discussed polarization due to absorption, scattering, and emission from interstellar dust, and polarization due to synchrotron emission from relativistic electrons. It is likely that within a few years there will be detections of polarization from another astrophysical source, Thompson scattering of photons in the early universe.

An introduction to the physical principals of this process can be found in "A CMB Polarization Primer" by Wayne Hu and Martin White (1997). In contrast to thermal fluc-

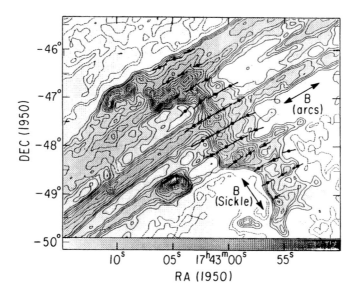

FIGURE 24. Polarization map (E-vectors) for the thermal feature known as the Sickle indicating a field direction orthogonal to that in the long non-thermal arcs (Dotson *et al.* 2000). The contours are from the radiographs of Yusef-Zadeh (1986).

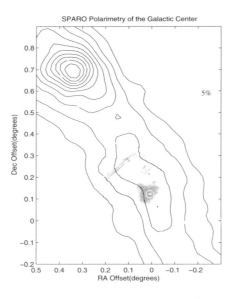

FIGURE 25. A large-scale map of the Galactic center at 450 μm (Novak *et al.* 2001; First results from the Northwestern University Polarimeter, SPARO). The ridge of 450 μm emission shown by the flux density contours (also from SPARO) follows the Galactic plane. The E-vectors of the polarization indicate a field parallel to the plane. The gray scale shows the radio emission surrounding SgrA* and along the non-thermal arcs. The field in the arcs is perpendicular to the plane.

tuations which may evolve, this phenomenon probes directly the epoch of last scattering. A search for anisotropies in the polarization will provide a direct test of the assumption that the large scale anisotropies we see today have evolved from small fluctuations in the early universe. From the observed patterns of polarization and from cross correlations between fluctuations in temperature and polarization it will be possible to learn not only the distribution of quadrupole anisotropies in the photon flux at $z \approx 1000$, and what fraction of the photons were re-scattered when the intergalactic medium was re-ionized, but also to distinguish between distortions associated with compression, vortical motion, and gravitational waves.

The expected intensity of the fluctuations on the polarized flux is only a few μK on scales of less than 1 degree. At this level the problem of avoiding systematic effects due to foreground sources is even more demanding than for measurements of temperature fluctuations. Nevertheless the problem may be manageable by taking into account the different spectral indexes of the various effects. (See *Microwave Foregrounds* 1999 for discussions of systematics.) Synchrotron emission should be the dominant background below ~100 GHz and dust emission should be dominant at higher frequencies. To isolate emission from the cosmic microwave background it will be necessary to use multifrequency measurements straddling this part of the spectrum.

Most of the material in these lectures has been obtained in collaboration with my students and former students. In particular, those who have worked on the polarization spectrum are Jacqueline Davidson, Jessie Dotson, Darren Dowell, Giles Novak, David Schleuning, and John Vaillancourt. I thank John Vaillancourt for help in preparing this manuscript, Jane Greaves for the 850 μm data shown in Figure 9, Crystal Brogan for permission to use Figure 3; David Aitken for permission to reproduce Figure 6; and Giles Novak for permission to show the preliminary results in Figure 25. Sections 2, 3, and 4 respectively are based in part on work published in *QJRAS* (Hildebrand 1983 and 1988), *ApJ* (Hildebrand *et al.* 1999), and *PASP* (Hildebrand *et al.* 2000). These lectures have been prepared with support from NSF Grants AST-9732326 and AST-9987441.

REFERENCES

AITKEN, D. K., BAILY, J. A., ROCHE, P. F., & HOUGH, J. M. 1985 *MNRAS* **215**, 815.

AITKEN, D. K., ROCHE, P. F., SMITH, C. H., JAMES, S. D., & HOUGH, J. M. 1988 *MNRAS* **230**, 629.

BROGAN, C. L., TROLAND, T. H., ROBERTS, D. A., & CRUTCHER, R. M. 1999 *ApJ* **515**, 304.

BROGAN, C. L., TROLAND, T. H., ROBERTS, D. A., & CRUTCHER, R. M. 2000 Submitted to *ApJ*.

CHANDRASEKHAR, S., & FERMI, E. 1953 *ApJ* **118**, 113.

COYNE, G. V., GEHRELS, T., & SERKOWSKI, K. 1974 *AJ* **79**, 581

CUDLIP, W., FURNISS, I., KING, K. J., & JENNINGS, R. E. 1982 *MNRAS* **200**, 1169.

DAVIDSON, J. A. 1996 In *Polarimetry of the Interstellar Medium,* eds. W. G. Roberge & D. C. B. Whittet, ASP Conference Series, vol. **97**. San Francisco. pp. 504 – 521.

DOMBROVSKY, V. A. 1954 *Dokl. Akad. Nauk SSSR* **94**, 1021.

DOTSON, J. L. 1996 *ApJ* **470**, 566.

DOTSON, J. L., DAVIDSON, J. A., DOWELL, C. D., SCHLEUNING, D. A., & HILDEBRAND, R. H. 2000 *ApJS* **128**, 335.

DOWELL, C. D., HILDEBRAND, R. H., & JONES, T. J. 2001 In preparation: To be submitted to *ApJ*.

DOWELL, C. D., HILDEBRAND, R. H., SCHLEUNING, D. A., VAILLANCOURT, J. E., DAVIDSON, J. A., DOTSON, J. L., & HOUDE, M. 2001 In preparation.

DOWELL, C. D., HILDEBRAND, R. H., SCHLEUNING, D. A., VAILLANCOURT, J. E., DOTSON, J. L., NOVAK, G., RENBARGER, T., & HOUDE, M. 1998 *ApJ* **504**, 588.

DRAINE, B. T. 1985 *ApJS* **57**, 587.

DRAINE, B. T., & LEE, H. M. 1984 *ApJ* **285**, 89.

DRAINE, B. T., & WEINGARTNER, J. C. 1996 *ApJ* **470**, 551.

GREAVES, J. S. 2000, Private communication.

GOLD, T. 1952 *MNRAS* **112**, 215.

HALL, J. S. 1949 *Science* **109**, 166.

HEILES, C. 1996 *ApJ* **462**, 316.

HEILES, C. 2000 *AJ* **119**, 923.

HILDEBRAND, R. H. 1983 *QJRAS* **24**, 267.

HILDEBRAND, R. H. 1988 *QJRAS* **29**, 327.

HILDEBRAND, R. H.. 1996 In *Polarimetry of the Interstellar Medium*, eds. W. G. Roberge & D. C. B. Whittet, ASP Conference Series, vol. **97**. San Francisco. pp. 254 – 268.

HILDEBRAND, R. H., DAVIDSON, J. A., DOTSON, J. L., FIGER, D. F., NOVAK, G., PLATT, S. R., AND TAO, L. 1993 *ApJ* **417**, 565.

HILDEBRAND, R. H., DAVIDSON, J. A., DOTSON, J. L., DOWELL, C. D., NOVAK, G., & VAILLANCOURT., J. E. 2000 *PASP* **112**, 1215.

HILDEBRAND, R. H., DOTSON, J. L., DOWELL, C. D., SCHLEUNING, D. A., & VAILLANCOURT, J. E. 1999 *ApJ* **516**, 834.

HILDEBRAND, R. H., & DRAGOVAN, M. 1995 *ApJ* **450**, 663.

HILDEBRAND, R. H., DRAGOVAN, M., & NOVAK, G. 1984 *ApJ* **284**, L51.

HILTNER, W. A. 1949 *Science* **109**, 165.

HILTNER, W. A. 1951 *ApJ* **114**, 241.

HOUDE, M., BASTIEN, P., PENG, R., PHILLIPS, T. G., & YOSHIDA, H. 2000a *ApJ* **536**, 857.

HOUDE, M., PENG, R., PHILLIPS, T. G., BASTIEN, P., & YOSHIDA, H. 2000b *ApJ* **537**, 245.

HU, W., & WHITE, M. 1997 *New Astron.* **2**, 323.

JACKSON, J. D. 1961 Introduction to dispersion Relation Techniques, in *Dispersion Relations*, chapter 1, ed. Screaton, G. R., Interscience, New York.

JONES, T. J., KLEBE, D., & DICKEY, J. M. 1992 *ApJ* **389**, 602.

LAZARIAN, A., GOODMAN, A. A., & MYERS, P. C. 1997 *ApJ* **490**, 273.

LAZARIAN, A., & ROBERGE, W. G. 1997 *ApJ* **484**, 230.

LILLIE, C. F., & WITT, A. N. 1976 *ApJ* **208**, 64.

LIS, D. C., SERABYN, E., KEENE, J., DOWELL, C. D., BENFORD, D. J., PHILLIPS, T. G., HUNTER, T. R., & WANG, N. 1998 *ApJ* **509**, 299.

MANCHESTER, R. N. 1974 *ApJ* **188,** 637.

MATHEWSON, D. S., & FORD, V. L. 1970 *MNRAS* **74**, 139.

Microwave Foregrounds, 1999 *Astronomical Society of the Pacific Conference Series*, Volume **181**, Ed A. de Olivera-Costa & M. Tegmark.

MORRIS, M., DAVIDSON, J. A. & WERNER, M. W. 1995 In *Proc. of the Airborne Astronomy Symp. on the Galactic Ecosystem: From Gas to Stars to Dust*, ed. M. R. Haas, J. A. Davidson, & E. F. Erickson (San Francisco: ASP), p. 477.

NOVAK, G., *et al.* 2001 In preparation: To be submitted to *ApJ*.

NOVAK, G., DOTSON, J. L., DOWELL, C. D., HILDEBRAND, R. H., RENBARGER, T., & SCHLEUNING, D. A. 2000 *ApJ* **529**, 241.

ONAKA, T. 2000 *ApJ* **533**, 298.

POLLACK, J. B., HOLLENBACH, D., BECKWITH, S., SIMONELLI, D. P., ROUSH, T., & FONG,

W. 1994 *ApJ* **421**, 615.

PURCELL, E. M. 1979 *ApJ* **231**, 404.

RAO, R., CRUTCHER, R. M., PLAMBECK, R. L., & WRIGHT, M. C. H. 1998 *ApJ* **502**, L75.

ROBERGE, W. G., HANANY, S., & MESSINGER, D. W. 1995 *ApJ* **453**, 238.

SCHLEUNING, D. A. 1998 *ApJ* **493,** 811.

SCHLEUNING, D. A., DOWELL, C. D. HILDEBRAND, R. H., & PLATT, S. R. 1997 *PASP* **109**, 307.

SCHLEUNING, D. A., VAILLANCOURT, J. E., HILDEBRAND, R. H., NOVAK G., DOTSON. J. L., DAVIDSON. J. A., & DOWELL, C. D. 2000 *ApJ* **535**, 913.

SERABYN, E., & MORRIS, M. 1994 *ApJ* **424**, L91.

SERKOWSKI, K. 1973 in IAU Symposium 52, *Interstellar Dust and Related Topics*, ed. J. M. Greenberg and H. C. van de Hulst (Dortrecht: Reidel), p. 145.

SERKOWSKI, K., MATHEWSON, D. S. & FORD, V. L. 1975 *ApJ* **196**, 261.

STRUVE, O. September 1949 *Sky and Telescope* pp. 274.

TSUBOI, M., UKITA, N., & HANDA, T. 1997 *ApJ* **481**, 263.

VAILLANCOURT, J. E. 2001 PhD thesis, University of Chicago, In preparation.

VALLÉE, J. P., & BASTIEN, P. 2000 *ApJ* **530**, 806.

VERSHUUR, G. L. 1969 *ApJ* **156**, 861.

WERNER, M. W., DAVIDSON, J. A., MORRIS, M., NOVAK, G., PLATT, S. R., & HILDEBRAND, R. H. 1988 *ApJ* **333**, 729.

WHITCOMB, S. E., GATLEY, I., HILDEBRAND, R. H., KEENE, J., SELLGREN, K., & WERNER, M. W. 1981 *ApJ* **246**, 416.

WITT, A. N. 1979 *Astrophys. Space Sci.* **65**, 21.

VAN DE HULST, H. C. 1957 *Light Scattering by Small Particles* (New York: Dover).

YUSEF-ZADEH, F. 1986 PhD thesis, Columbia University.

YUSEF-ZADEH, F., & MORRIS, M. 1988 *ApJ* **329**, 729.

YUSEF-ZADEH, F., & MORRIS, M., & CHANCE, D. 1984 *Nature* **310**, 557.

ZWEIBEL, E. G. 1996 In*Polarimetry of the Interstellar Medium, ASP Conference Series*, ed's. W. G. Roberge and D. C. B. Whittet Vol. **97**, pp 486–503.

Instrumentation for Astrophysical Spectropolarimetry

By CHRISTOPH U. KELLER

National Solar Observatory, 950 N. Cherry Ave., Tucson, AZ 85719, USA

Astronomical spectropolarimetry is performed from the X-ray to the radio regimes of the electromagnetic spectrum. The following chapter deals with instruments and their components that are used in the wavelength range from 300 nm to 20 μm. After introducing the terminology and formalisms that are used in the context of astronomical spectropolarimeters, I discuss the most widely used optical components. These include crystal and sheet polarizers, fixed monochromatic and achromatic retarders, and variable retarders such as liquid crystals and photoelastic modulators. Since polarimetric measurements are often limited by systematic errors rather than statistical errors due to photon noise, I deal with these instrumentally induced errors in detail. Among these errors, I discuss instrumental polarization of various kinds and chromatic and angle of incidence errors of optical components. I close with a few examples of successful, modern night-time and solar spectropolarimeters.

---

1. Introduction

1.1. *Scope of chapter*

Astronomical polarimetry is performed over a large fraction of the electromagnetic spectrum, from X-rays to radio waves. The following chapter is restricted to the optical range that can be observed from the ground, i.e. 300 nm to 20 μm. Far-infrared polarimetry is described by Hildebrand in this volume. Information on polarimetry in the X-ray and radio regimes can be found in Tinbergen (1996).

Furthermore, the following text focuses on instruments for spectropolarimetry, i.e. the instrumental aspects of polarimetry with a spectral resolution that resolves spectral lines, either with a spectrograph or a filter, i.e. a spectral resolution of $\lambda/\delta\lambda = R > 10,000$. However, it needs to be remembered that polarimetry using the same optical techniques is often performed at low spectral resolution. Finally, the text mostly deals with high-precision polarimetry, i.e. sensitivities of the order 10^{-3} or better, and most examples are from solar observations.

This introduction to instrumentation for astrophysical spectropolarimetry is based on a series of 5 lectures. The viewgraphs shown in those lectures can be found at www.noao.edu/noao/staff/keller.

1.2. *Brief history of optics and instruments for spectropolarimetry*

Over the centuries, many people have contributed to the development of instruments to measure the polarization of astronomical objects. Table 1 lists some of them with an emphasis on their contribution to instrumentation. In addition, it also lists some of the failed attempts to detect polarization in order to show how early scientists tried to use spectropolarimetry in astrophysics.

Historically important contributions to polarized light have been collected by Swindell (1975) and Billings (1990). Livingston (1993) included some historically relevant papers on astronomical polarimeters.

1669	Erasmus Bartholinus discovers double refraction in calcite
1690	Christian Huygens discovers extinction with two crossed calcites
1808	Etienne-Louis Malus discovers polarization from reflection
1812	David Brewster discovers angle of reflection where light is totally polarized and finds relation with index of refraction
1818	Augustin Jean Fresnel and Dominique Francois Arago find the transverse nature of polarization
1828	William Nicol invents the first calcite polarizing prism
1852	George Gabriel Stokes introduces the Stokes parameters
1852	William Bird Herapath and his student Phelps find polarizing crystals in the urine of a dog that was fed quinine when iodine was dripped into it
1858	E. Liais discovers linear polarization in the solar corona during an eclipse
1875	John Kerr discovers the birefringence of isotropic materials when an external electrical field is applied
1906	George E. Hale visually searches for linear polarization in sunspots without success
1908	George E. Hale finds circular and linear polarization in sunspots using photographic plates
1910	W.H. Wright finds no polarization due to the Zeeman effect in stellar hydrogen line spectra
1913	P.W. Merrill finds no polarization due to the Zeeman effect in stellar hydrogen line spectra
1928	Edwin H. Land develops the first sheet polarizer using herapathite crystals
1933	Hale and collaborators fail to obtain photo-electric measurements of the circular polarization due to the Zeeman effect in the sun
1941	R. Clark Jones introduces 2 by 2 complex matrices to describe the effect of a crystalline plate on the state of polarization
1943	Hans Mueller introduces four by four real matrices to describe the influence of depolarizing optics
1944	Bernard Lyot develops the birefringent filter
1946	S.M. MacNeille invents the thin-film polarizing cube beam splitter
1946	S. Chandrasekhar introduces Stokes parameters to astrophysics and includes polarization in radiative transfer
1947	Horace W. Babcock discovers circular polarization in 78 Vir
1953	K.O. Kiepenheuer and Babcock and Babcock use photo-electric spectropolarimeters to measure magnetic fields all over the sun
1958	Audouin Dollfus develops a modulating polarizer that reaches a sensitivity of 10^{-5} using a rotating waveplate
1966	M. Billardon and J. Badoz invent the photoelastic modulator
1969	James C. Kemp predicts that photoelastic modulators could measure the circular polarization of astronomical objects with an accuracy of 10^{-6}

TABLE 1. Timetable of important events in the development of astronomical spectropolarimeters.

1.3. *Books on polarimetry*

Various books have been written on polarimetry in general. The classical book by Shurcliff (1962) is compact and contains many historical references, but it is not in print anymore. Collett (1993) is extensive and easy to read, but the book is rather expensive. Clarke and Grainger (1971), Azzam and Bashara (1987), and Kliger et al. (1990) are some of the other books that deal with polarimetry. Only a few books deal specifically with astronomical polarimetry. The conference proceedings by Gehrels (1974) contain interesting papers on instrumentation, although many are nowadays of mostly historical interest. Two recent monographs dealing specifically with astronomical polarimetry are those by Tinbergen (1996) and Leroy (2000). Leroy's book is an introduction to polarization in astrophysics that avoids equations to remain readable for a wide audience. Tinbergen's text is aimed at a higher level but leaves many interesting equations to be derived as exercises.

2. Principles of optical polarization measurements

2.1. *Terminology*

While the Jones and Mueller formalisms are explained in detail by Landi Degl'Innocenti in this volume, the following paragraphs will briefly introduce the terminology and definitions used here and highlight issues associated with instrumentation. A formalism that is not reviewed here is the Poincaré sphere. While that approach has its advantages, it has become considerably less important with fast computers that can easily deal with large sets of Mueller matrices. An extended review of the Poincaré sphere formalism has been given by Ramachandran and Ramaseshan (1962).

2.1.1. *Stokes parameters, vectors, and Mueller matrices*

Mueller matrices describe the (linear) transformation between Stokes vectors (formed by grouping the four Stokes parameters into a single vector) associated with optical elements and surfaces, i.e.

$$\boldsymbol{I'} = \mathsf{M}\boldsymbol{I} , \tag{2.1}$$

where the Stokes vector consists of the following Stokes parameters

$$\boldsymbol{I} = \begin{pmatrix} I \\ Q \\ U \\ V \end{pmatrix} = \begin{pmatrix} I_1 \\ I_2 \\ I_3 \\ I_4 \end{pmatrix} . \tag{2.2}$$

The latter form of the Stokes vector components is useful when applying methods of linear algebra to Stokes vectors. Extensive examples of Stokes vectors are given by Shurcliff (1962).

Mueller matrices have the following form:

$$\mathsf{M} = \begin{pmatrix} M_{11} & M_{12} & M_{13} & M_{14} \\ M_{21} & M_{22} & M_{23} & M_{24} \\ M_{31} & M_{32} & M_{33} & M_{34} \\ M_{41} & M_{42} & M_{43} & M_{44} \end{pmatrix} . \tag{2.3}$$

A *normalized* Mueller matrix is obtained by scaling the matrix such that the upper left element is equal to one. Some useful examples of Mueller matrices are given in the following sections. Extensive lists of Mueller matrices are given by Shurcliff (1962) and Kliger et al. (1990).

When a beam of light passes through N optical elements, each described by a Mueller matrix M_i, the combined Mueller matrix M' of the whole assembly is given by

$$\mathsf{M}' = \mathsf{M}_N \mathsf{M}_{N-1} \cdots \mathsf{M}_2 \mathsf{M}_1 . \tag{2.4}$$

Note the reversed order of the Mueller matrices as compared to the order in which the light passes through the optical elements. This order is important since Mueller matrices do not commute in general.

Rotation of elements described by Mueller matrices are given by

$$\mathsf{M}' = \mathsf{R}(-\alpha)\mathsf{M}\mathsf{R}(\alpha) , \tag{2.5}$$

where α is the rotation angle, and the rotation matrix R is given by

$$R\left(\alpha\right) = \begin{pmatrix} 1 & 0 & 0 & 0 \\ 0 & \cos 2\alpha & \sin 2\alpha & 0 \\ 0 & -\sin 2\alpha & \cos 2\alpha & 0 \\ 0 & 0 & 0 & 1 \end{pmatrix}. \tag{2.6}$$

The formalism for rotating Mueller matrices given above cannot be applied blindly. In particular, one has to remember that this formalism assumes that we keep the same coordinate system for the incoming and outgoing beams. However, when reflections are considered, the convention is to change the coordinate system for the reflected beam as compared to the incoming beam (see Landi Degl'Innocenti in this volume). The Mueller matrix for an ideal reflection at normal incidence is given by

$$M = \begin{pmatrix} 1 & 0 & 0 & 0 \\ 0 & 1 & 0 & 0 \\ 0 & 0 & -1 & 0 \\ 0 & 0 & 0 & -1 \end{pmatrix}, \tag{2.7}$$

which indicates nothing but the change in the coordinate system for the Stokes vector. Therefore, the above formalism for calculating the Mueller matrix of rotated optical components cannot be directly applied when reflections are involved. In that case, it is often easier to consider the rotation of Stokes vectors (to which the rotation matrix R can be applied).

Let us consider an example. It is well known that successive reflections off two mirrors will not modify the polarization if the angles of incidence are the same for both reflections and if the surface normal of the second mirror is perpendicular to the plane of incidence of the first mirror. This principle is sometimes used to compensate the instrumental polarization introduced by a single reflection. If M_r is the Mueller matrix describing the reflection off a single mirror, a naive use of the above formalism would lead us to write the combined Mueller matrix of both mirrors as

$$M' = R(-90°)M_r R(+90°)M_r . \tag{2.8}$$

Using the diagonal matrix given above for normal incidence as the mirror reflection Mueller matrix M_r shows immediately that the above equation is incorrect and should be replaced by

$$M' = R(+90°)M_r R(+90°)M_r . \tag{2.9}$$

The reason for this is that the Mueller matrix for reflection implies a change in coordinate system that requires a change in the sign of the angle for the rotation matrix.

Finally, Stokes vectors and Mueller matrices operate on intensities and their differences, i.e. incoherent superpositions of light, they are not adequate to describe interference nor diffraction effects. However, they are ideally suited to describe partially polarized and unpolarized light.

2.1.2. *Jones vectors and matrices*

The Jones calculus is the adequate way to describe the coherent superposition of polarized light because it operates on amplitudes rather than on intensities. However, Jones vectors and matrices can only describe 100% polarized light because a monochromatic wave is always 100% polarized.

The electrical field vector of a monochromatic electromagnetic wave traveling along the z axis of a right-handed coordinate system can be decomposed into its x and y

components

$$\mathrm{Re}E_x \, , \mathrm{Re}E_y \, , \tag{2.10}$$

where E_x and E_y are complex quantities with an amplitude and a phase.

The Jones vectors contain the complex amplitude electrical field components in the form

$$\boldsymbol{E} = \begin{pmatrix} E_x \\ E_y \end{pmatrix} . \tag{2.11}$$

Note that amplitudes are not observed directly by detectors in the wavelength range considered here. Therefore, observables always depend on products of Jones vector components such as $|\boldsymbol{E}|^2$.

The transfer of 100% polarized light through an optical medium is described by 2 by 2 complex matrices. Combined Jones matrices describing a series of optical elements are equal to the matrix product of the individual Jones matrices in the same way as with Mueller matrices. Examples of various Jones matrices are given in the following sections. Extensive lists have been prepared by Shurcliff (1962) and Kliger et al. (1990).

The rotation of Jones matrices is given by

$$\mathsf{J}' = \mathsf{R}(-\alpha)\mathsf{J}\mathsf{R}(\alpha) \, , \tag{2.12}$$

where the rotation matrix R is given by

$$\mathsf{R} = \begin{pmatrix} \cos\theta & \sin\theta \\ -\sin\theta & \cos\theta \end{pmatrix} . \tag{2.13}$$

Any Jones matrix can be transformed into the corresponding Mueller matrix using the following relation (Azzam & Bashara 1987):

$$\mathsf{M} = \mathsf{B} \left(\mathsf{J} \otimes \mathsf{J}^* \right) \mathsf{B}^{-1} \, , \tag{2.14}$$

where * indicates the complex conjugate,

$$\mathsf{B} = \begin{pmatrix} 1 & 0 & 0 & 1 \\ 1 & 0 & 0 & -1 \\ 0 & 1 & 1 & 0 \\ 0 & i & -i & 0 \end{pmatrix} , \tag{2.15}$$

and $\otimes$ is the tensor product. For matrices B and C with N by N elements, the tensor product is a N^2 by N^2 matrix given by

$$\mathsf{A} = \mathsf{B} \otimes \mathsf{C} \, , \tag{2.16}$$

where

$$a_{i+(j-1)N,k+(l-1)N} = b_{i,k}c_{j,l} \, . \tag{2.17}$$

While the Jones matrix has 8 independent parameters, the absolute phase information is lost in the Mueller matrix, leading to only seven independent matrix elements for a Mueller matrix derived from a Jones matrix. A general Mueller matrix has 8 degrees of freedom, the additional degree being related to the transfer of unpolarized light.

2.1.3. *TE and TM waves, s and p polarization*

In the context of optics manufacturing and testing, a terminology different from Stokes vectors and Mueller matrices is used (see Fig. 1). We consider the plane of incidence formed by the normal to the surface and the direction of the incoming light. If the electrical field of the incoming wave is in the plane of incidence, then the magnetic field must be transverse to it, and this is therefore called a TM (transverse magnetic) or *p*-

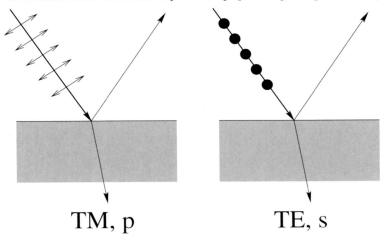

FIGURE 1. Definition of transverse magnetic (TM), transverse electric (TE), p, and s polarized waves. The definitions are only valid in the context of a light beam being reflected off or transmitted through an interface between two materials with different indices of refraction at non-normal incidence.

polarized wave (p for parallel). The other one is the TE (transverse electric) or *s-polarized* (s for 'senkrecht'= perpendicular in German) wave.

To fully characterize the influence of an interface between two materials (e.g. air to metal), not only the reflectivities for s and p polarization need to be considered, but also the phase delay between s and p polarized light introduced by the interface. In general, there will also be absorption for the transmitted light, i.e. the energies of the reflected and the transmitted beam are not equal to the energy of the incoming beam. Purely dielectric media do not absorb energy.

2.1.4. *Polarization sensitivity and accuracy*

In the following, we will distinguish between polarization *sensitivity* and *accuracy*. Polarization sensitivity describes the magnitude of a small polarization signal on top of a big background that can just be detected. An adequate measure of the polarization sensitivity is the standard deviation of a polarized spectrum that is known to have no spectral variation of the polarization signal. Polarization accuracy is the magnitude of the absolute error in the polarization measurement. Both are typically expressed as a fraction of the intensity. These definitions are in analogy to definitions used in photometry.

2.2. *Polarized ray-tracing*

To simulate the design of a polarimeter, we need to calculate the influence of the telescope, instrument, and polarimeter on the polarization of the incoming light. While initial simulations can assume a single ray passing through the optical components along the optical axis, more accurate simulations require that many rays are traced through the optics, similar to what is done for general optical design. The polarization aspect of ray-tracing can often be described by a series of Mueller matrices that can be combined into a single Mueller matrix for each ray corresponding to one point in the field of view and one point in the pupil. The Mueller matrices corresponding to all the points in the pupil can then be averaged into a single Mueller matrix (incoherent superposition) that is equivalent to the average over the point-spread function (PSF) in the final focus for a single point in the field of view.

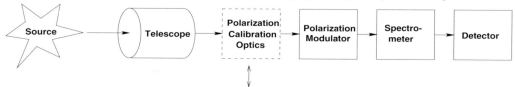

FIGURE 2. General layout of an astronomical spectropolarimeter: Light from a source passes through the telescope. A polarization calibration package can be inserted into the beam to calibrate the polarimeter and the rest of the instrument. During observations of the source, the calibration optics is not in the beam. The polarization modulator translates the polarization information into spatial and/or temporal variations. The beam(s) then pass through the spectrometer and are recorded by the detector.

A more accurate calculation requires that the various beams are combined coherently in the focal plane, something the Mueller calculus cannot describe. If one uses the Jones formalism in the equivalent way as the Mueller formalism above, one obtains the Jones matrix that corresponds to the center of the PSF. This method can be expanded to calculate the Mueller matrix at any position within the PSF for a given point in the field of view by calculating the Fourier transform of the Jones matrices in the exit pupil (Sánchez Almeida & Martínez Pillet 1992).

However, the latter approach is only valid for slow beams where one can assume that all rays that are averaged are almost parallel in the final focus. For fast beams, a more extensive way of polarized raytracing has to be used. Chipman (1995) developed such a method by extending the Jones vector approach to three complex components of the electric field. An optical element is described by a 3 by 3 complex matrix, and the rays are described by a complex vector with three components. While this approach is very general and can calculate complex and fast optics, it is normally not needed for astronomical instruments.

2.3. *Basic polarimeters*

In general, polarimeters consist of optical elements such as retarders and polarizers that change the polarization state of the incoming light in a controlled way (see Fig. 2). The detectors only measure intensities, at least for the part of the spectrum considered here. The various intensity measurements are then combined to retrieve the polarization state of the incoming light. Polarimeters differ mostly by the way that the polarization modulator works. In addition, a good polarimeter should also include optics for the polarization calibration, i.e. optics with very well known polarization characteristics that can be temporarily inserted in front of the polarization modulator.

2.3.1. *Rotating waveplate polarimeter*

A simple polarimeter can be built using a rotating waveplate and a linear polarizer (see Fig 3). The Mueller matrix calculus introduced above and the Mueller matrices for a retarder and a linear polarizer (see below) allow us to determine the intensity seen by the detector as a function of retardance δ and position angle θ of the rotating retarder. The intensity seen by the detector is given by

$$I' = \frac{1}{2} \left(I + \frac{Q}{2} \left((1 + \cos\delta) + (1 - \cos\delta)\cos 4\theta \right) + \frac{U}{2} \left(1 - \cos\delta \right)\sin 4\theta - V\sin\delta\sin 2\theta \right) .$$

$$(2.18)$$

From this equation, we can deduce the following:
- only the terms that depend on θ will lead to a modulated signal;

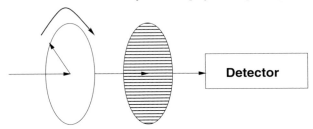

FIGURE 3. A simple polarimeter consisting of a rotating retarder and a fixed linear polarizer, which together make up the modulator package. The arrow on the retarder indicates the orientation of the fast axis (see below).

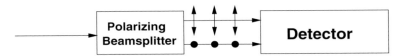

FIGURE 4. A simple (linear) polarimeter consisting of a polarizing beam-splitter that produces two displaced beams corresponding to orthogonal linear polarization states.

- to obtain equal modulation amplitudes in Q, U, and V, the retardation of the retarder should be close to $\delta = 127°$;
- Q and U are modulated at twice the frequency of V;
- the phase shift in modulation between Q and U is $90°$, and the corresponding frequency is twice that of Stokes V, which requires measurements at 8 angles to determine all 4 Stokes parameters.

In the following, the modulation scheme employed by a polarimeter that measures Stokes parameters sequentially will be referred to as *temporal modulation*.

2.3.2. *Polarizing beam-splitter polarimeter*

Another simple polarimeter for linear polarization contains only a fixed polarizing beam-splitter that produces two beams corresponding to orthogonal polarization states (see Fig. 4). The full linear polarization information can be deduced by rotating the whole polarimeter assembly. In the following, we will refer to such an arrangement of simultaneous measurements of two (or more) Stokes parameters as *spatial modulation*.

2.3.3. *Comparison of temporal and spatial modulation schemes*

Temporal and spatial modulation schemes have different advantages and disadvantages. The most important ones are summarized in Table 2.

Since none of these schemes has clear advantages over the other, but the two schemes are rather complementary, modern, sensitive polarimeters often combine the two modulation schemes to combine the advantages and minimize the disadvantages. Such a combined approach is described in detail in Sect. 5.

2.4. *Statistical errors of polarization measurements*

When designing a polarimeter, it is important to model and understand the performance that can be expected. Once clear performance goals have been established, the design can be optimized to maximize the performance. The literature shows various ways to define and maximize the efficiency with which the polarization is measured, and it often depends on the exact polarimeter that is used, therefore making it difficult to directly compare the efficiency of various polarimeters.

Modulation Scheme	Advantages	Disadvantages
temporal	negligible effects of flat field and optical aberrations	influence of seeing if modulation is slow
	potentially high polarimetric sensitivity	read-out rate of regular array detectors limits modulation frequency
spatial	off-the-shelf array detectors can be used	requires up to four times larger array detector
	high photon collection efficiency	influence of flat field
	allows post-facto image reconstruction	influence of differential aberrations

TABLE 2. Comparison of spatial and temporal modulation schemes.

Errors in polarization measurements are typically separated into two components: statistical errors (mostly due to the photon statistics and detector read-out noise) and systematic errors (also called instrumental errors). When designing a polarimeter, a trade-off between the two error components needs to be made. For instance, it does not make sense to build a polarimeter that has very small systematic errors but so low an efficiency that the statistical errors completely dominate.

Making a few assumptions, we can derive the expected statistical noise in the Stokes parameters, which we will try to minimize. For this analysis, we can assume that:

• there is a linear relation between the Stokes parameters of the incoming light and the signals that are measured;
• the noise in the various measurements is independent;
• the noise has a Gaussian distribution.

For a large number of photons, which are needed for accurate polarimetry, a Gaussian distribution is a good approximation to the Poisson statistic that describes photon noise.

We combine the measured intensities into a signal vector S, which is related to the incoming Stokes vector, I by the signal matrix X, where

$$S = X I \ . \tag{2.19}$$

X is a 4 by m matrix where m is the number of intensity measurements that contribute to the polarization measurement. For example, $m = 4$ for most systems that use liquid crystals, while $m = 8$ for a rotating retarder approach as outlined above. X is a function of the free parameters in the polarimeter design. Since the polarimeter optics can be described by Mueller matrices, each row of X corresponds to the first row of the Mueller matrix describing the particular intensity measurement.

To determine the Stokes vector I from the measurements S, X needs to be inverted. If

$$Y = X^{-1} \ , \tag{2.20}$$

or more generally for a non-square matrix using the Moore-Penrose generalized inverse (e.g. Albert 1972)

$$Y = \left(X^T X\right)^{-1} X^T \ , \tag{2.21}$$

then the standard deviations of the Stokes parameters, σ_{I_i}, are given by

$$\sigma_{I_i} = \sqrt{\sum_{j=1}^{4} \mathsf{Y}_{ij}^2 \sigma_{S_j}^2} \, , \qquad (2.22)$$

where σ_{S_j} is the standard deviation of the intensity in measurement j. In most cases, σ_{S_j} does not depend on the measurement number j.

A more extensive treatment of statistical errors in polarimetry has been given by del Toro Iniesta and Collados (2000). When designing a polarimeter, we generally wish to minimize the standard deviations of the deduced Stokes parameters.

2.5. *Data reduction*

In order to achieve high polarimetric sensitivity, there will always be a substantial amount of data reduction involved. Often the raw data (before calibrations have been applied) have errors on the order of a few percent in the fractional polarization, while the expected signal is one or two orders of magnitude smaller. Typical reduction steps include:

- subtraction of dark current and bias;
- division by flat field;
- calculation of fractional polarization Q/I, U/I, and V/I;
- subtraction of polarization bias;
- removal of polarized fringes;
- calibration with polarization efficiency (polarization flat field);
- if required, multiplication with calibrated intensity I to obtain V, Q, and U.

To deduce the best data reduction strategy, one needs to understand the relevant instrumental effects so that they can be removed during processing. All data reduction steps should be based on a physical model of the data collection process, i.e. on a theory of the observing process and a model of the instrument. Once a theory is available, one can solve it for the parameters that should be determined as a function of the measured quantities. This solution will then also help in identifying the necessary calibration observations.

3. Optical components for spectropolarimetry

3.1. *Polarizers*

A polarizer is defined as an optical element that produces (at least partially) polarized light when the input light beam is unpolarized. Therefore, a polarizer can be linear, circular, or in general, elliptical, depending on the type of polarization that emerges.

There is a large variety of polarizers that all have their respective advantages and disadvantages. Here we will discuss the types of linear polarizers that are most often used in astronomical polarimetry.

3.1.1. *Jones and Mueller matrices for linear polarizers*

A linear polarizer can be described by its transmittance of the electrical field in two orthogonal directions. The Jones matrix for a linear polarizer is then given by

$$\mathsf{J}_p = \begin{pmatrix} p_x & 0 \\ 0 & p_y \end{pmatrix} , \qquad (3.23)$$

where the real values $0 \leq p_x \leq 1$ and $0 \leq p_y \leq 1$ are the transmission factors for the x and y-components of the electrical field, i.e. $E_x' = p_x E_x$ and $E_y' = p_y E_y$. $p_x = 1, p_y = 0$

describes a linear polarizer in the $+Q$ direction, $p_x = 0, p_y = 1$ describes a linear polarizer in the $-Q$ direction, and $p_x = p_y$ describes a neutral density filter.

The corresponding Mueller matrix is given by

$$M_p = \frac{1}{2} \begin{pmatrix} p_x^2 + p_y^2 & p_x^2 - p_y^2 & 0 & 0 \\ p_x^2 - p_y^2 & p_x^2 + p_y^2 & 0 & 0 \\ 0 & 0 & 2p_x p_y & 0 \\ 0 & 0 & 0 & 2p_x p_y \end{pmatrix}. \tag{3.24}$$

From this equation, it is evident that an unpolarized incoming beam will always be linearly polarized. However, the emerging Stokes vector does not correspond to a completely polarized beam unless $p_x = 0$ or $p_y = 0$. If the polarizer only produces a partially polarized beam from unpolarized light, it is called a *partial* linear polarizer. Any real polarizer is always only a partial polarizer.

If the incoming beam is polarized, the emerging beam is, in general, elliptically polarized, even for a purely linear polarizer because of the non-zero diagonal terms $2p_x p_y$ in the Mueller matrix. A totally polarized beam will remain totally polarized even when passing an ideal partial linear polarizer, i.e. an ideal polarizer does not depolarize. However, real polarizers can produce minute amounts of unpolarized light from a fully polarized beam because of scattering within the polarizer. Nevertheless, this effect is small and can almost always be neglected.

When characterizing the quality of actual linear polarizers, two different parameters are used to describe the performance. k_1 describes the (intensity) transmittance of the polarizer for a fully linearly polarized beam whose angle is chosen such as to maximize the transmitted intensity. k_2 is the minimum transmittance as a function of the angle of the incoming linearly polarized beam. It is evident that $k_1 = p_x^2$ and $k_2 = p_y^2$ if we assume that $p_x > p_y$. The ratio of k_1 to k_2 is called the *extinction ratio*. k_1 and k_2 are often tabulated for various polarizers as a function of wavelength and can be determined from the transmittances for unpolarized light of parallel and crossed identical polarizers, which are given by

$$\begin{aligned} T_{\text{parallel}} &= \tfrac{1}{2} \left(k_1^2 + k_2^2 \right) \\ T_{\text{crossed}} &= k_1 k_2 \end{aligned}. \tag{3.25}$$

Finally, the Mueller matrix for a total linear polarizer at position angle θ is given by

$$M_p(\theta) = \frac{1}{2} \begin{pmatrix} 1 & \cos 2\theta & \sin 2\theta & 0 \\ \cos 2\theta & \cos^2 2\theta & \sin 2\theta \cos 2\theta & 0 \\ \sin 2\theta & \sin 2\theta \cos 2\theta & \sin^2 2\theta & 0 \\ 0 & 0 & 0 & 0 \end{pmatrix}. \tag{3.26}$$

3.1.2. Wire grid polarizers

Grids of parallel conducting wires with a spacing d of the order of the wavelength λ of the light act as a polarizer. Intuitively, one might expect that the electric field parallel to the wires is transmitted because it 'slips' through the wires. On the contrary, it is the plane of polarization perpendicular to the wires that is transmitted because the electric field component (of the electromagnetic wave) parallel to the wires induces electrical currents in the wires, which strongly attenuates the transmitted electric field parallel to the wires. The induced electrical current is such that the polarization parallel to the wires is reflected. It is thus possible to produce a polarizing beam-splitter with a wire grid polarizer, which reflects and transmits orthogonal linear polarization states. As a rule of thumb, if $d < \lambda/2$, then the polarization is strong. If $d \gg \lambda$, then the transmission of both polarization states is high, and thus the polarization of the transmitted beam is weak.

Wire grid polarizers were first used by Heinrich Hertz at radio wavelengths. They are used mostly in the infrared because the wire spacing becomes very small at visible wavelengths. Modern commercial wire grid polarizers are made by depositing a thin-film metallic grid pattern on a suitable infrared substrate. They are available for wavelengths larger than 1 μm. For longer wavelengths where appropriate substrates are not available, free-standing wire grid polarizers are employed.

3.1.3. *Polarcor*

Available since 1984, Polarcor made by Corning is a glass polarizer with high performance between 633 and 1550 nm. It is made from a borosilicate glass containing silver particles aligned along a common axis. The elongated, conducting silver particles act as small wires. Polarization occurs within 25 to 50 μm of each surface. In the UltraThin version, polarization occurs throughout the entire body of the glass. Unfortunately, the maximum diameter of Polarcor is currently limited to 30 mm.

3.1.4. *Dichroic crystals*

Dichroic materials preferentially absorb one polarization state. The behavior depends on the wavelength, i.e. the materials appear to have different colors depending on the angles of illumination and viewing. Dichroism arises from the anisotropy of the complex index of refraction (see Landi Degl'Innocenti in this volume). Examples of naturally occurring dichroic crystals are tourmaline and herapathite. In 1852, W.B. Herapath discovered a salt of quinine that had polarizing properties. He succeeded in making artificial crystals large enough to study under a microscope. Di-iodosulphate of quinine is now known as herapathite. However, it is generally difficult to produce uniform, large dichroic crystals, which is why this type of polarizer is only of historical interest.

3.1.5. *Polaroid-type polarizers*

In 1928, Edwin Land made a suspension of tiny herapathite crystals, which he spread as a thin layer between supporting sheets. The crystals also have a magnetic dipole moment, so that if a suspension is placed in a very strong magnetic field, they become oriented to form a uniform dichroic layer. If the aligned crystals are suspended in a polymer, they set. Land called this a J-type polarizer. However, this first type of sheet polarizer had problems with the finite lifetime because the crystals become disoriented over time.

The next-generation sheet polarizer, also invented by Edwin Land in 1938, is based on molecular dichroism. H-type sheet polarizers are made by heating and stretching a sheet of polyvynil alcohol (PVA) that is laminated to a supporting sheet of cellulose acetate butyrate. The PVA is then treated with an iodine solution. The difference between various H-type polarizers is in the amount of iodine absorbed by the PVA. The PVA-iodine complex is analogous to a short, conducting wire. This is the operating principle of H-type polarizers, which are still in use today. Of course, the axis of maximum transmission for linearly polarized light is perpendicular to the stretch direction. The commercial names for Polaroid sheet polarizers such as HN-38 identify the overall type (H), the color (N=neutral) and the approximate transmittance of a single polarizer for unpolarized light (38%).

Other types of Polaroid sheet polarizers include the K-type, which is close in performance to H-type polarizers but more stable under extreme environmental influences such as temperature and humidity. Again, PVA is the starting material, but this time

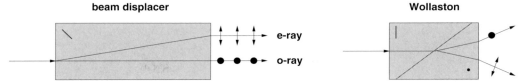

FIGURE 5. Beam displacers and Wollaston prisms are often used in astronomical polarimeters. A simple block of calcite with the optic axis at about 45° (indicated by the short line) splits a single ray into two parallel rays of opposite linear polarization. In the Wollaston, two calcite or quartz prisms with their optic axes at 0° and 90° deflect the two linear polarizations symmetrically.

hydrogen and oxygen atoms are removed by a chemical process to produce polyvinylene from some of the PVA molecules, and stretching aligns the long chain molecules.

Finally HR-type polarizers, developed at Polaroid between 1943 and 1951 by Blake, are based on a PVA-polyvinylene-iodine complex that works well from 0.7 to 2.3 μm.

The very thin polarizer material is laminated between sheets of plastic or glass. The plastic sheeting can be removed with organic solvents to obtain the polarizing foil alone.

3.1.6. *Crystal-based polarizers*

Crystals can have different indices of refraction for each axis:

$$n_x \neq n_y \neq n_z . \qquad (3.27)$$

Uniaxial crystals axis have only one axis that has a different index of refraction as compared to the other two axes:

$$n_x \neq n_y = n_z . \qquad (3.28)$$

The *optic axis* of a uniaxial crystal is the axis that has a different index of refraction. The *fast axis* is the axis with the smallest index of refraction since the speed of light along that axis is the fastest. When sending a ray of light through a uniaxial crystal, the single ray is generally split into two rays. The *ordinary ray* (or short *o-ray*) passes the crystal without any deviation, while the *extraordinary ray* (or *e-ray*) is deviated at the air-crystal interface. The two emerging rays have orthogonal polarization states. It is common to use the indices of refraction for the ordinary ray (n_o) and the extraordinary ray (n_e) instead of the indices of refraction in the crystal coordinate system.

It is outside of the scope of this chapter to deal with the optical calculations of crystals, which are rather complicated, even for uniaxial crystals. However, we will present some of the most important results here. An introduction to uniaxial crystal optics is given in Collett (1993). A more advanced treatment of crystal optics that is easy to read has been given by Wahlstrohm (1960).

The polarizing beam displacer or beam-splitter is the most simple crystal polarizer. It consists of a single block of calcite (or other uniaxial crystal with large birefringence) with the optic axis at about 45°. On the interface between air and calcite, the extraordinary ray is deflected while the ordinary ray just passes through (see Fig. 5). For calcite, the splitting is about 0.095 times the thickness. The two emerging rays are parallel and have orthogonal linear polarization states.

The Nicol prism, invented in 1828 by William Nicol was the first actual polarizer based on crystals. However, it has been superseded by considerably better designs (discussed below). It consists of two calcite prisms that are held together by Canada balsam. The angles are chosen such that the ordinary ray undergoes total internal reflection on the interface between the calcite and the Canada balsam.

Glan-Thompson polarizers also consist of two calcite prisms that are cemented together. The ordinary beam undergoes total internal reflection and is absorbed by black paint on the side of one of the prisms. There are two versions that differ in their prism angles and the index of refraction of the cement between the two crystal prisms. The long version has an acceptance angle of about 25°, while the short form has an acceptance angle of about 15°.

Glan or Glan-Foucault polarizers consist of two prisms made of calcite that are separated by an air gap. As all the other crystal-based polarizers, it is usable from about 300 nm to 2700 nm, but absorption of the deflected beam occurs above 2 μm. The field of view is between 13° and 7.5°, but this acceptance angle is rotationally symmetrical only at a single wavelength. Again, total internal reflection of the ordinary ray is responsible for the polarization. The transmission is not as high as for the Glan-Thompson polarizer because of reflection losses at the internal calcite-air interfaces.

Wollaston prisms are similar to beam displacers since they also produce two beams with orthogonal linear polarizations. However, Wollaston prisms do not displace the two beams but deviate them in opposite directions (see Fig. 5). They are made of calcite or quartz prisms with perpendicular optical axes that are cemented together. The usable spectral range is typically 300 nm to 2200 nm, the upper limit being given by the absorption of the ordinary beam. The polarization directions are parallel and vertical to the refracting edge. The e-ray in the first prism becomes the o-ray in the second prism and is bent toward the normal. The o-ray in the first prism becomes the e-ray in the second prism and is bent away from the normal. The angle of divergence is determined by the wedge angle, which is typically between 15° and 45°.

3.1.7. *Brewster-angle polarizers*

When calculating the polarization of reflected and transmitted beams for an air-dielectric interface as a function of angle of incidence (see the chapter by Landi Degl'Innocenti), it becomes evident that there is an angle α_B, the Brewster angle, at which the reflected beam is completely polarized. This angle has a simple relation with respect to the index of refraction n of the dielectric:

$$\tan \alpha_B = n \,. \tag{3.29}$$

However, the reflected intensity is rather small such that a polarizer based on the Brewster-angle reflection is not very effective at the wavelengths considered here. However, the transmitted beam is somewhat polarized, and the transmittance for that polarization state is 100%. A stack of dielectric plates can therefore be used as a polarizer with good transmittance.

3.1.8. *Thin-film polarizers*

Thin-film polarizers are mostly used in the form of cube beam-splitters where the two orthogonally polarized beams emerge at right angles. Polarizing cube beam-splitters are based on a thin film stack on the inside of two glass prisms that are cemented together such that total internal reflection occurs at the Brewster angle within the thin film. This type of polarizer has a limited extinction ratio and wavelength range, but it can be manufactured relatively cheaply even at large apertures (5–10 cm). Thin-film stacks that act as polarizers can also be deposited on oblique plates from which the polarized beam is reflected.

3.1.9. *Comparison of polarizers*

Table 3 summarizes the properties of polarizers that are typically used in astronomy. Crystal-based polarizers are expensive and typically limited in aperture to about 40 mm. Polaroid sheet polarizers are less transparent and do not have as high an extinction ratio, but they are available in sizes up to about 50 cm and are cheap.

type	extinction ratio	transmission (polarized)	wavelength range (nm)	bandpass (nm)	acceptance angle (°)
Glan	$> 10^5$	$> 84\%$	300-2700	full	8
Glan-Thompson	$> 10^6$	$> 92\%$	300-2700	full	15-25
Wollaston	$> 10^6$	$> 92\%$	300-2200	full	20
Polarcor	$> 10^4$	$> 80\%$	633-1550	150	> 20
Polaroid	$150 - 10^4$	$> 75\%$	310-2000	200	> 20
Polarizing cube	> 500	$> 90\%$	400-1600	200-400	10
Wire Grid	> 100	$> 90\%$	$10^3 - 10^6$		> 20

TABLE 3. Comparison of various types of polarizers that are used in astronomy.

3.2. *Fixed linear retarders*

A retarder is an optical element that splits an incoming beam into two components, retards the phase of one of these components, and reunites the components at the exit into a single beam. An ideal retarder does not change the intensity of the light, nor does it change the degree of polarization. Any retarder can be characterized by the two (not identical) Stokes vectors of incoming light that are not changed by the retarder. These two Stokes vectors are sometimes called the *eigenvectors* of the retarder. Depending on whether these Stokes vectors describe linear, circular, or elliptical polarization, the retarder is called a linear, circular, or elliptical retarder. In the following, we will only consider linear retarders because they are by far the most common type of retarder.

3.2.1. *Mueller and Jones matrices for linear retarders*

A linear retarder with its fast axis at $0°$ is characterized by a Jones matrix of the form

$$\mathsf{J}_r\left(\delta\right) = \begin{pmatrix} e^{i\frac{\delta}{2}} & 0 \\ 0 & e^{-i\frac{\delta}{2}} \end{pmatrix}, \qquad (3.30)$$

where δ is the phase shift between the two linear polarization components expressed in radians. Since the absolute phase does not matter, it is possible to write the Jones matrix of a retarder with the retardation in only one matrix element. However, it is advisable to use the 'symmetric' version given above because it avoids introducing an absolute phase that depends on the retardation. Use of the 'asymmetric' version has led to some erroneous theoretical calculations of the instrumental Mueller matrix of telescopes.

The corresponding Mueller matrix is given by

$$\mathsf{M}_r = \begin{pmatrix} 1 & 0 & 0 & 0 \\ 0 & 1 & 0 & 0 \\ 0 & 0 & \cos\delta & -\sin\delta \\ 0 & 0 & \sin\delta & \cos\delta \end{pmatrix}. \qquad (3.31)$$

It is important to realize that the combination of two or more linear retarders in series will, in general, not be equivalent to a linear retarder, but be equivalent to a single elliptical retarder.

3.2.2. *Zero and multiple order linear retarders*

Most retarders are based on birefringent materials that have different indices of refraction for different angles of the incoming linear polarization. Typical birefringent materials that are used are quartz, mica, and polymer films with oriented molecules.

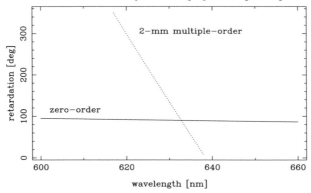

FIGURE 6. Wavelength dependence of a 2-mm multi-order and a true zero-order quartz quarter-wave retarder.

The retardation (delay between ordinary and extraordinary ray) is given by

$$N\lambda = d\,(n_e - n_o) \;,\tag{3.32}$$

where d is the geometrical thickness, λ is the wavelength, and n_e and n_o are the indices of refraction for the extraordinary and the ordinary rays, respectively. N is the retardation expressed in waves. A quarter-wave plate is obtained with $N = m + \frac{1}{4}$ and m being an integer. If $m = 0$, we call it a *true zero-order* retarder. If $m > 0$, we call it a *multi-order* retarder.

Of course, a retardation of 1.25 waves has the same effect on polarization as a retardation of 0.25 waves. However, it is evident that the thicker the retarder becomes (larger d), the faster the retardation changes as a function of wavelength, even if the indices of refraction n_o and n_e would not depend on the wavelength (see Fig. 6).

3.2.3. *Crystal retarders*

Quartz is available in fairly large sizes and can be produced artificially. It is therefore the most commonly used crystal material for high-quality retarders. A true zero-order quarter-wave retarder in the visible is about 15 μm thick. While such true zero-order quartz retarders can now be fabricated on glass substrates, the usual way to obtain a (so-called *compound*) zero-order retarder consists in combining two approximately 1-mm thick plates cut parallel to the optic axis with a difference in thickness that corresponds to the path difference of a true zero-order retarder. The two plates are optically contacted with their fast axes at $90°$ with respect to each other. The retardation of the two plates therefore cancels except for the small path-length difference. The usable spectral range of quartz retarders is from about 180 nm to 2700 nm. For wavelengths below 230 nm, the plates are manufactured from synthetic crystal quartz.

Mica is another material that is often used for commercial retarders. It is cheap and available in large sizes (20 cm by 20 cm). Mica crystals can easily be cleaved into very thin sheets of appropriate thickness to obtain true zero-order retarders. A quarter-wave plate in the visible is about 50 μm thick. Mica is transparent from about 350 nm to 6 μm, but it absorbs even in the visible. Since the required thickness at longer wavelengths, and therefore the absorption, become rather large, mica retarders are normally not used for wavelengths larger than about 1.6 μm.

Other crystals that are used for manufacturing retarders include MgF_2, which transmits over a large wavelength range from the ultraviolet to the mid-infrared and stressed LiF for wavelengths around 100 nm.

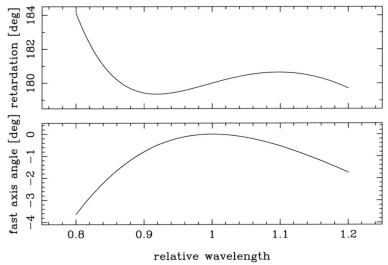

FIGURE 7. Theoretical variation of retardation and fast axis orientation as a function of relative wavelength for a Pancharatnam achromatic half-wave plate.

3.2.4. *Polymer retarders*

Stretched polymers (e.g. polyvinyl alcohol) are also birefringent. The fast axis is perpendicular to the stretch direction. They were invented by Land, West, and Makas around 1946. The thickness of a quarter-wave retarder is about 20 μm in the visible. Polymer retarders are true zero-order retarders that are highly transparent even in the ultraviolet. For certain retardance values, polymer retarders in sizes of up to 40 cm can be obtained.

3.2.5. *Achromatic retarders*

As seen above, retarders are highly wavelength sensitive. By combining two materials with opposite variations of $\delta n = n_e - n_o$ with wavelength, and by choosing appropriate thicknesses, an achromatic retarder can be built whose retardance is correct at two wavelengths. The most used combination of materials is quartz and MgF_2. The useful wavelength range of this type of achromatic retarders is about 50% of the central wavelength.

Another approach uses three identical retarders. Invented by Pancharatnam (1955) for half-wave plates, the outer plates have parallel fast axes, while the fast axis of the inner plate is rotated by about 60°. In contrast to the achromatic wave-plates discussed above, the fast axis direction of the combined retarder depends on the wavelength. Figure 7 shows the variation of the retardation and the fast axis direction as a function of wavelength. Achromatic quarter-wave plates can be constructed in a similar way, although the performance is not as good as for the half-wave retarders.

By combining three identical crystal achromatic retarders with the Pancharatnam approach, superachromatic half-wave plates can be constructed. Again, the fast axis direction depends on the wavelength. However, the angular acceptance angle is very limited because of the thickness of the combined 6 plates (see below). The useful wavelength range for commercially available superachromatic waveplates extends from about 300 nm to 1100 nm.

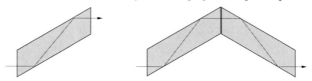

FIGURE 8. Traditional arrangements for quarter-wave (left) and half-wave (right) Fresnel rhombs.

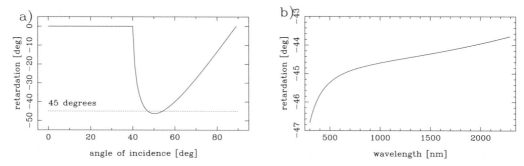

FIGURE 9. Variation of retardance on total internal reflection with a) angle at 632.8 nm and b) wavelength at 55.08° for a BK7 and air interface.

3.2.6. *Fresnel rhombs*

Another approach for obtaining achromatic retarders makes use of the phase shift on total internal reflection on an interface between dielectrica. Reflections on dielectric interfaces do not produce a phase shift (or retardance) except for total internal reflection. Retarders using total internal reflection are typically called *Fresnel* rhombs. When working in the visible, it is normally not possible to achieve a 90° phase shift on a single reflection. However, a single total internal reflection can be arranged such that a retardation of $\lambda/8 = 45°$ is achieved. Two (suitably aligned) reflections therefore provide a retardation of $\lambda/4$, and a half-wave retardation can be achieved with four total internal reflections. Figure 8 shows two possible arrangements for quarter- and half-wave Fresnel rhombs. Historically, this was the first type of retarder that was developed (by Fresnel).

The requirement for total internal reflection on a glass-air interface is

$$n_i \sin \beta > 1 \,, \tag{3.33}$$

where n_i is the index of refraction of the glass and β is the (internal) angle of incidence. The phase shift is given by

$$\tan \delta/2 = -\frac{\cos \beta \sqrt{n_i^2 \sin^2 \beta - 1}}{n_i \sin^2 \beta} \,. \tag{3.34}$$

For BK7, the most common type of optical glass, at a wavelength of 632.8 nm (HeNe laser wavelength) the index of refraction is 1.5151, and at an angle of incidence of 55.08°, the retardation is 45° (or $\lambda/8$). Figure 9a shows the variation of the retardation with the angle of incidence. Below an angle of incidence of 41.3°, no internal reflection occurs. The retardation strongly depends on the angle of incidence. Therefore, Fresnel rhombs have only a small acceptance angle. A deviation of ±0.5° from normal incidence changes the retardation by ±0.5%.

For a material with an index of refraction of 1.5538, an angle of incidence of 45.0° produces exactly a 45° phase retardation. Two right-angle prisms with total internal

type	retardance accuracy (%)	wavelength range (nm)	bandpass (nm)	acceptance angle (°)
quartz	0.4	180-2700	100	3
MgF$_2$	0.4	140-6200	100	3
mica	4	350-1550	100	10
polymer	0.6	400-1800	100	10
Fresnel	2	240-2000	330-1000	2

TABLE 4. Comparison of various types of commercially available zero-order retarders that are used in astronomy. For quartz and MgF$_2$, it is assumed that they are compound zero-order retarders. The accuracy in percent refers to a half-wave plate.

reflection form a quarter-wave plate. When using right-angle prisms for beam deflection, one needs to keep in mind that they act as significant retarders.

The variation of the retardance with wavelength is purely due to the variation of the index of refraction with wavelength, which is generally small. Therefore, Fresnel rhombs are very achromatic. Figure 9b shows the theoretically calculated variation of the retardance for a single total internal reflection for BK7 at 55.08° from 300 nm to 2300 nm. The performance of Fresnel rhombs with respect to the angle of incidence and the variation of the retardance with wavelength can be improved by coating the reflecting surfaces with thin films.

3.2.7. *Comparison of zero-order retarders*

Table 4 summarizes the properties of commercially available zero-order retarders that are typically used in astronomy. Mica and polymer retarders can be produced with fairly large apertures, however, the retardance accuracy tends to be reduced for larger apertures. Both mica and polymer retarders have the advantage of large acceptance angles because they are true zero-order retarders. Crystal-based retarders are limited in aperture size. Fresnel rhombs are the only achromatic retarders consisting of a single piece of optics, but have a very limited acceptance angle.

3.3. *Variable retarders*

For building sensitive polarimeters, it is often desirable to have retarders whose retardance can be varied quickly. This can be achieved either by changing the birefringence (liquid crystals, Kerr and Pockels cells, photoelastic modulators) or by changing the geometrical thickness (e.g. Soleil compensators). Since the latter is necessarily associated with a mechanical motion, it is normally not used for polarimeters and will therefore not be discussed any further.

3.3.1. *Nematic liquid crystal retarders*

Liquid crystals are fluids whose molecules are elongated. At high temperatures, the liquid crystal is isotropic. In the *nematic phase*, the molecules are randomly positioned but aligned essentially in one direction. Some liquid crystals line up parallel or perpendicular to an outside electrical field. For these, the dielectric constant anisotropy is often large, making the liquid crystal very responsive to changes in the applied electric field. The birefringence δn can be very large (larger than typical crystal birefringence). Sheets of liquid crystal can therefore behave like an electronically adjustable optical retarder. The liquid crystal layer is only a few μm thick and represents a true zero-order retarder.

The anisotropy of liquid crystals, and therefore also their birefringence, shows a strong temperature dependence. With zero voltage applied externally, the liquid crystal molecules are parallel to the substrates due to an alignment layer, therefore maximizing the retardation. With an electrical field applied, the liquid crystal molecules tip perpendicular to the substrate causing a reduction in the effective birefringence and hence, the retardation. The alignment layer between the substrate and the liquid crystal prevents the molecules at the surface to rotate freely. This causes a residual retardance of about 30 nm even at high voltages (about 20 V). The retardance changes by about −0.4% per °C. The response time of nematic liquid crystal retarders is proportional to the square of the layer thickness (=total retardation) and of the order of 20 ms.

3.3.2. *Ferro-electric liquid crystal retarders*

The *smectic* liquid crystal phases are characterized by well-defined layers that can slide over one another. The molecules are positionally ordered along one direction. In the smectic C phase, the molecules are tilted away from the layer normal. *Ferroelectric liquid crystals* (FLCs) are the tilted phases of chiral molecules (called smectic C*), which have a permanent polarization, which is why they are called ferroelectric. They respond much more quickly to externally applied fields than nematic liquid crystals and can be used to make fast, bistable electro-optic devices when placed between closely-spaced, electrically conducting glass plates. FLC variable retarders act like retarders with a fixed retardation where the direction of the fast axis can be switched by about 45° (switching angle) by alternating the sign of the applied electrical field. They are true zero-order retarders with switching times on the order of 150 μs. The switching angle is rather temperature sensitive, while the retardance tends to be rather insensitive to temperature variations.

3.3.3. *Photoelastic modulators (PEMs)*

Isotrop and uniform optical materials such as glass become birefringent when strain (e.g. due to compression) is applied in one axis. This is commonly referred to as stress-induced birefringence and the effect is called piezo-optical or photo-elastic effect. For example, a block of a few cm in side length of common BK7 glass can be stressed enough by hand such as to introduce a quarter-wave retardation.

The stress-induced birefringence is proportional to the strain σ. The retardation in waves is therefore given by

$$\delta = \frac{1}{\lambda} C d \sigma \; , \tag{3.35}$$

where C is the stress optical constant, d is the thickness of the variable retarder, and λ is the wavelength. It is possible to construct variable retarders by just adequately compressing optical glass. However, one needs to apply considerable mechanical power to modulate the stress-induced birefringence.

One way to obtain a birefringence modulation with much reduced power is to use a mechanically resonant oscillation since the required mechanical power is proportional to one over the mechanical Q, which is on the order of 10^3 to 10^4 for most glasses.

If a slab of length L is excited at its fundamental mode, a standing acoustic wave with a wavelength of $2L$ is produced. The frequency of the oscillation is given by

$$\omega = \frac{c_s}{2L} \; , \tag{3.36}$$

where c_s is the sound speed in the optical material. For a 57-mm-long fused silica slab, the resonance frequency is 50 kHz.

It is easy to calculate the resulting stress and hence the retardation as a function of

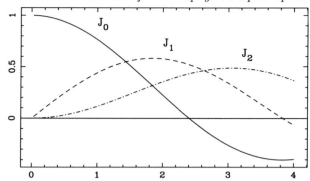

FIGURE 10. Bessel functions of order 0, 1, and 2.

position x and time t. The stress-induced birefringence $\delta(x,t)$ is given by

$$\delta(x,t) = A \, \sin \omega t \, \sin(\frac{\pi x}{L}) \, , \tag{3.37}$$

where A is the amplitude of the oscillation and x extends from 0 to L.

To make the slab oscillate, a quartz crystal with electrodes on its surfaces is forced to oscillate by an externally applied electrical field via the piezo effect. The quartz slab is mechanically coupled to the modulator slab, and the electrical field is driven at the mechanical resonance frequency. The oscillation amplitude A can be regulated with an electronic feedback circuit.

The oscillation is dampened by friction losses within the modulator material. Since the energy loss is inversely proportional to the mechanical Q, and since Q is very large, the energy loss in the modulator is not critical, and the required drive power is small (0.1 to 1 W). A material with a high Q such as fused silica ($Q \approx 10^4$) is desirable. Typical glass has a Q of about 10^3 and is therefore less often used. It is interesting to note that the required drive power does not depend on the length of the slab.

By grouping the $\sin(\frac{\pi x}{L})$ term and the amplitude A into a spatially varying amplitude $A(x)$, the birefringence can be rewritten as

$$\delta(x,t) = A(x) \sin(\omega t) \, . \tag{3.38}$$

The PEM represents a true zero-order retarder. Its Mueller matrix corresponds to a retarder with a time dependent retardation and therefore contains elements with $\sin \delta(x,t)$ and $\cos \delta(x,t)$. By expanding $\sin(\sin(\cdot))$ and $\cos(\sin(\cdot))$ terms into Bessel functions, the Mueller matrix elements become

$$\sin \delta(\mathrm{x,t}) = 2J_1(A(x)) \sin \omega t + \cdots \, , \tag{3.39}$$

$$\cos \delta(x,t) = J_0(A(x)) + 2J_2(A(x)) \cos 2\omega t + \cdots \, , \tag{3.40}$$

$$\tag{3.41}$$

where $J_{0,1,2}$ are the Bessel functions of order 0,1 and 2 (see Fig. 10).

PEMs are stable in operation, show no degrading at high intensity levels and/or UV irradiation, have good optical properties, a large spatial and angular aperture, and require only low voltages at moderate driving powers of less than 1 W. The disadvantages are the sinusoidal modulation (as compared to the more efficient square-wave modulation that can be achieved with liquid crystals) and the very high modulation frequency of 20 to 50 kHz, which requires specialized array detectors (see below).

Modulation Scheme	Advantages	Disadvantages
rotating retarder	high stability	relatively slow modulation
	large wavelength range	beam motion
		needs 8 measurements for all Stokes parameters
liquid crystals	relatively fast modulation	narrow simultaneous wavelength range
	only 4 measurements for all Stokes parameters	limited temporal stability
	no moving parts	damaged by strong UV light
PEM	very fast modulation	narrow simultaneous wavelength range
	high stability	needs special CCD camera
	no moving parts	spatial retardance variation

TABLE 5. Comparison of various temporal modulation schemes employed in astronomical spectropolarimeters.

3.3.4. *Pockels and Kerr cells*

Anisotropy in an otherwise isotropic material can also be introduced by an externally applied electrical field. The material then behaves like a uniaxial crystal. This effect was discovered by John Kerr in 1875 using glass and a high voltage electrical field. This effect also occurs in liquids and some gases. The birefringence induced by the Kerr effect is proportional to the square of the electric field. The latter implies that high voltages are required for Kerr cells.

F.R. Pockels discovered that an electrical field applied along the line of sight and parallel to the crystal optic axis produces birefringence that is proportional to the applied electrical field. Therefore, Pockels cells do not require high voltages. Since crystals are required for Pockels cells, the aperture of commercially available cells is typically limited to about 7 cm. The world's largest Pockels cells of 40 cm by 40 cm clear aperture are currently being developed for the National Ignition Facility, but the technology that is used to bring the electrical current to the crystal faces is not useful for astronomy.

3.4. *Comparison of modulation schemes*

Now that we have reviewed all the ingredients of typical astronomical polarimeters, we can compare different polarimeter concepts employing temporal modulation schemes. Table 5 compares three different approaches. The rotating retarder approach is common in both night-time and solar polarimeters. A large simultaneous wavelength range can be observed simultaneously by using superachromatic waveplates or Fresnel rhombs. While variable retarders avoid many of the problems of the rotating waveplate approach, they are typically limited to a narrow spectral range because of the current absence of achromatic variable retarders.

Achromatic variable retarders would provide a major new addition to astronomical polarimetry because they would have all the advantages of current variable retarders but could be used over a large, simultaneous wavelength range. Achromatic retarders using two different materials are not useful because each of the materials needs to have a retardance of many waves, which cannot be achieved with any of the variable retarders available today. The Pancharatnam approach looks more feasible. However, variable retarders that rely on a change in birefringence (such as nematic liquid crystals and PEMs) will not work because the way the Pancharatnam approach works, which minimizes the overall retardance dependence on the retardance of the individual components. However, three half-wave FLCs in a Pancharatnam configuration should provide an excellent achromatic half-wave plate whose fast axis can be switched by 45°. Such a device has not yet been built to my knowledge.

4. Instrumental errors

It might seem excessive to devote a whole section to instrumental errors in spectropolarimeters. However, astronomical spectropolarimetry is often limited by systematic instrumental errors rather than by statistical errors such as photon and read-out noise, in particular in highly sensitive solar observations.

The following instrumental errors are commonly encountered in high-precision spectropolarimetry:

- Atmospheric seeing and guiding errors
- Instrumental polarization due to
  - Telescope and instrument optics
  - Polarized scattered light in telescope and instrument
  - Spectrograph slit polarization
  - Angle, wavelength, and temperature dependence of retarders and polarizers
  - Crystal aberrations
  - Polarized fringes
- Ghost images
- Variable sky background
- Unpolarized scattered light in atmosphere and optics
- Limited calibration accuracy

In the following, we will not discuss the influence of gratings on polarization because that is a rather complicated issue outside the scope of this chapter. In general, low-order gratings exhibit large variations of the polarization with wavelength. At some wavelengths, they often act as complete polarizers. Maximum transmission is normally obtained with the polarization parallel to the grating lines. High-order echelle gratings typically have only a minor influence on the polarization.

Finally, there will also be some systematic errors due to the data reduction. We will deal with the most important of these error sources in detail in the following sections.

In many cases the instrumental effects due to the telescope and the instrument can be described by Mueller matrices, which have the general form

$$
\begin{pmatrix}
I \to I & Q \to I & U \to I & V \to I \\
I \to Q & Q \to Q & U \to Q & V \to Q \\
I \to U & Q \to U & U \to U & V \to U \\
I \to V & Q \to V & U \to V & V \to V
\end{pmatrix}. \tag{4.42}
$$

The various terms are grouped into three categories:

- $I \to X, X = Q, U, V$: instrumentally induced *polarization*
- $X_1 \to X_{2 \neq 1}$ and $X \to I$: instrumentally introduced *cross-talk*
- $X \to X$: instrumentally introduced *depolarization*

4.1. *Seeing and guiding errors*

Fortunately, air is not birefringent. Therefore, seeing does not produce polarization per se. However, whenever polarization measurements are not carried out simultaneously, seeing can introduce spurious polarization signals because sequentially recorded images will be differently distorted by seeing. The same holds for telescope guiding errors. Therefore, sequential polarization measurements should be modulated at a frequency that is faster than typical seeing frequencies, which are on the order of a few hundred Hz. To completely avoid this issue, measurements should be carried out simultaneously.

4.2. *Polarizing telescopes*

Every telescope introduces some polarization, although the amount may be very small. Rotationally symmetric telescopes are often called *polarization-free* because, theoretically, they would not introduce any net polarization at the very center of the field of view, although they do introduce a very small amount of depolarization. For points away from the center, even a rotationally symmetric telescope will introduce polarization (e.g. Sen & Kakati 1997). Furthermore, seeing destroys the rotational symmetry even in the center of the field of view (Sánchez Almeida 1994). Therefore no telescope should be considered to be completely free of instrumental polarization.

4.2.1. *Stress induced birefringence in glass*

Every piece of glass has some remaining internal stress from the manufacturing process. While these stresses can be minimized by extended annealing periods during the manufacturing, there will always be some remaining stress, which introduces birefringence. As a rule of thumb, one should expect about 5 nm of birefringence for every cm of high-quality glass thickness. In the visible, this will lead to a cross-talk of about 1% between V and Q, U for every cm of glass.

Apart from these static stresses, there is also temperature-induced stress that leads to a time-dependent birefringence. Temperature gradients in glass lead to stresses due to the varying thermal expansion. A careful choice of the type of glass can often alleviate this problem. For instance, fused silica is 12 times better than BK7 as far as temperature-induced birefringence is concerned. Temperature gradients also introduce optical aberrations because of the temperature dependence of the index of refraction. A correlation between birefringence and optical aberrations will lead to different PSFs for different Stokes parameters, which can amount to up to a few percent for diffraction-limited imaging. However, in the following, we will always assume that the observations are averaged over the PSF, i.e. the observations are not diffraction-limited.

4.2.2. *Oblique transmission and reflection*

Oblique reflections off and transmission through optical surfaces such as mirrors introduce polarization and cross-talk between the Stokes parameters. Transmission through dielectric materials such as glass introduces retardation, but no polarization. Oblique reflections on mirrors do not only occur on solar telescopes that have often complicated mirror, but they also occur in night-time telescopes, e.g. in Nasmyth and Coudé foci. The accurate modeling of these reflections is not easy because of oxide and sometimes oil layers on the mirrors and their associated interference effects. Similar issues occur with oblique transmissions through glass surfaces with multi-layer coatings. As an example, Figure 11 shows the variation of the $I \rightarrow Q, U, V$ instrumental polarization as a function of time for the Swedish Vacuum Solar Telescope. Note the large discrepancy between the observed polarization and the theoretical model (Fig. 11a). Only by introducing purely empirical offsets can the theoretical model be fitted reasonably well. This

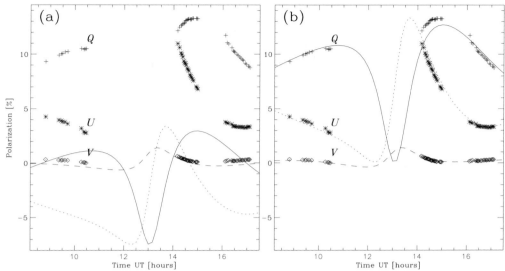

FIGURE 11. Instrumental polarization of the Swedish Vacuum Solar Telescope as a function of time for one particular day at 525 nm (markers). a) Lines show a best-fit theoretical model of the telescope polarization. b) Lines show the same theoretical model plus appropriate offsets to fit the data. (Courtesy Pietro Bernasconi)

illustrates the difficulty of modeling the instrumental effects of telescopes. This why the instrumental polarization and cross-talk are often measured directly and not modeled in a self-consistent way using tabulated optical constants for the involved materials.

The polarization of a stack of thin films on a substrate can be calculated using a matrix theory that is used in thin-film design (e.g. Macleod 1969). For the following, the index of refraction is generally a complex quantity of the form

$$n = \tilde{n} - ik \, , \qquad (4.43)$$

where $\tilde{n}$ is the (real) index of refraction and k is the extinction coefficient. $k = 0$ for dielectrica. Both values are tabulated for common optical materials as a function of wavelength.

The layers are numbered from 1 to L with the first layer being closest to the substrate. Each layer has an associated index of refraction n_j and a geometrical thickness d_j. The substrate and the exterior medium have refractive indices n_s and n_m, respectively.

The complex reflection and transmission coefficients for an angle of incidence of θ_0 are given by

$$r = \frac{\eta_m E_m - H_m}{\eta_m E_m + H_m} \qquad (4.44)$$

and

$$t = \frac{2\eta_m}{\eta_m E_m + H_m} \, , \qquad (4.45)$$

where

$$\begin{pmatrix} E_m \\ H_m \end{pmatrix} = \mathsf{M} \begin{pmatrix} 1 \\ \eta_s \end{pmatrix} \, . \qquad (4.46)$$

The matrix M is the product of all the matrices that describe the various layers, i.e.

$$\mathsf{M} = \mathsf{M}_L \mathsf{M}_{L-1} ... \mathsf{M}_2 \mathsf{M}_1 \, . \qquad (4.47)$$

The 2 by 2 complex matrices M_j are given by

$$\mathsf{M}_j = \begin{pmatrix} \cos\delta_j & \frac{i}{\eta_j}\sin\delta_j \\ i\eta_j\sin\delta_j & \cos\delta_j \end{pmatrix} , \tag{4.48}$$

where

$$\delta_j = \frac{2\pi}{\lambda}n_j d_j \cos\theta_j . \tag{4.49}$$

θ_j can be calculated from Snell's law,

$$n_m \sin\theta_0 = n_j \sin\theta_j . \tag{4.50}$$

For s polarization, η is defined as

$$\eta = n\cos\theta , \tag{4.51}$$

and for p polarization, it is defined as

$$\eta = \frac{n}{\cos\theta} . \tag{4.52}$$

The Mueller matrix for transmission is then given by

$$\frac{1}{2}\begin{pmatrix} (T_s + T_p) & (T_s - T_p) & 0 & 0 \\ (T_s - T_p) & (T_s + T_p) & 0 & 0 \\ 0 & 0 & 2\sqrt{T_p T_s}\cos(\epsilon_p - \epsilon_s) & 2\sqrt{T_p T_s}\sin(\epsilon_p - \epsilon_s) \\ 0 & 0 & -2\sqrt{T_p T_s}\sin(\epsilon_p - \epsilon_s) & 2\sqrt{T_p T_s}\cos(\epsilon_p - \epsilon_s) \end{pmatrix} , \tag{4.53}$$

where

$$T_{s,p} = \frac{\eta_s}{\eta_m}|t_{s,p}|^2 \tag{4.54}$$

and

$$\epsilon_{s,p} = arg(t_{s,p}) . \tag{4.55}$$

To obtain the Mueller matrix for the reflected beam, $T_{s,p}$ is replaced by

$$R_{s,p} = |r_{s,p}|^2 , \tag{4.56}$$

and $\epsilon_{s,p}$ is given by

$$\epsilon_{s,p} = arg(r_{s,p}) . \tag{4.57}$$

Furthermore, the signs of the Mueller matrix elements in the lower right have to be changed to conform with the conventions for a reflected beam. The Mueller matrix for the reflected beam then becomes

$$\frac{1}{2}\begin{pmatrix} (R_s + R_p) & (R_s - R_p) & 0 & 0 \\ (R_s - R_p) & (R_s + R_p) & 0 & 0 \\ 0 & 0 & -2\sqrt{R_p R_s}\cos(\epsilon_p - \epsilon_s) & -2\sqrt{R_p R_s}\sin(\epsilon_p - \epsilon_s) \\ 0 & 0 & 2\sqrt{R_p R_s}\sin(\epsilon_p - \epsilon_s) & -2\sqrt{R_p R_s}\cos(\epsilon_p - \epsilon_s) \end{pmatrix} . \tag{4.58}$$

As an example, let us consider a mirror with $n = 1.2$, $k = 7.5$ (typical for aluminum at 630 nm). A beam incident at $45°$ will have a normalized reflection Mueller matrix of

$$\begin{pmatrix} 1.000 & 0.028 & 0.000 & 0.000 \\ 0.028 & 1.000 & 0.000 & 0.000 \\ 0.000 & 0.000 & -0.983 & -0.180 \\ 0.000 & 0.000 & 0.180 & -0.983 \end{pmatrix} . \tag{4.59}$$

However, if we add a 126-nm thick dielectric layer with an index of refraction of 1.4

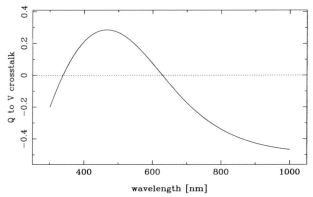

FIGURE 12. Variation of the $V \to Q$ cross-talk as a function of wavelength for an aluminum mirror overcoated with a 126-nm thick dielectrica with an index of refraction of 1.4 and an angle of incidence of $45°$.

on top of the aluminum, we obtain

$$
\begin{pmatrix}
1.000 & -0.009 & 0.000 & 0.000 \\
-0.009 & 1.000 & 0.000 & 0.000 \\
0.000 & 0.000 & -1.000 & 0.000 \\
0.000 & 0.000 & 0.000 & -1.000
\end{pmatrix}.
\tag{4.60}
$$

The cross-talk between Q and V has disappeared. Indeed, mirrors can be constructed that minimize the cross-talk using dielectric thin-film stacks. Nevertheless, while the I to and from Q terms are substantially reduced, they have not been eliminated completely. Furthermore, if we look at the wavelength-dependence of the V to Q cross-talk (see Fig. 12) even when assuming that the optical constants n and k are constant with wavelength, we see that there is a large variation with wavelength and that even the sign of the cross-talk changes. This is due to interference effects in the dielectric layer.

A dielectric layer of the same type with a thickness of only 65 nm leads to the following Mueller matrix

$$
\begin{pmatrix}
1.000 & 0.014 & 0.000 & 0.000 \\
0.014 & 1.000 & 0.000 & 0.000 \\
0.000 & 0.000 & -0.858 & -0.514 \\
0.000 & 0.000 & 0.514 & -0.858
\end{pmatrix},
\tag{4.61}
$$

which maximizes the crosstalk between Q and V. Indeed, it has been noticed that a thin oil film on an aluminum mirror can lead to a big change in the telescope Mueller matrix and that it varies strongly with wavelength. In general, aluminum is covered with a thin layer of aluminum oxide that changes the Mueller matrix with respect to a clean aluminum layer (Sankarasubramanian et al. 1999).

If no thin-film layers are used, the matrix theory approach reduces to the usual Mueller matrices for reflection and transmission on/through dielectric and metallic surfaces, which, for an air-dielectric interface, are given by

$$
\mathsf{M}_r = \frac{1}{2} \left(\frac{\tan \alpha_-}{\sin \alpha_+} \right)^2
\begin{pmatrix}
c_-^2 + c_+^2 & c_-^2 - c_+^2 & 0 & 0 \\
c_-^2 - c_+^2 & c_-^2 + c_+^2 & 0 & 0 \\
0 & 0 & -2c_+ c_- & 0 \\
0 & 0 & 0 & -2c_+ c_-
\end{pmatrix}
\tag{4.62}
$$

and

$$M_t = \frac{1}{2} \frac{\sin 2i \sin 2r}{(\sin\alpha_+ \cos\alpha_-)^2} \begin{pmatrix} c_-^2 + 1 & c_-^2 - 1 & 0 & 0 \\ c_-^2 - 1 & c_-^2 + 1 & 0 & 0 \\ 0 & 0 & 2c_- & 0 \\ 0 & 0 & 0 & 2c_- \end{pmatrix} , \tag{4.63}$$

where

$$\alpha_\pm = i \pm r , \tag{4.64}$$

$$c_\pm = \cos\alpha_\pm , \tag{4.65}$$

and i and r are the angle of the incident light and the refracted light, which are related by Snell's law according to

$$\sin i = n \sin r , \tag{4.66}$$

where n is the index of refraction of the dielectric.

For reflections off metal surfaces, the Mueller matrix for the reflected beam is given by

$$M_r = \frac{1}{2} \begin{pmatrix} \rho_s^2 + \rho_p^2 & \rho_s^2 - \rho_p^2 & 0 & 0 \\ \rho_s^2 - \rho_p^2 & \rho_s^2 + \rho_p^2 & 0 & 0 \\ 0 & 0 & 2\rho_s\rho_p \cos\delta & 2\rho_s\rho_p \sin\delta \\ 0 & 0 & -2\rho_s\rho_p \sin\delta & 2\rho_s\rho_p \cos\delta \end{pmatrix} , \tag{4.67}$$

where the real quantities ρ_s, ρ_p, and $\delta = \phi_s - \phi_p$ are defined by

$$\rho_s e^{i\phi_s} = -\frac{\sin(i-r)}{\sin(i+r)} = \frac{\cos i - n\cos r}{\cos i + n\cos r} \tag{4.68}$$

and

$$\rho_p e^{i\phi_p} = \frac{\tan(i-r)}{\tan(i+r)} = \left(\frac{n\cos r - \cos i}{\cos i + n\cos r}\right)\left(\frac{n\cos r \cos i - \sin^2 i}{n\cos r \cos i + \sin^2 i}\right) \tag{4.69}$$

and i and r are the angles of the incident light and the (complex) angle of the refracted light, which are related by Snell's law according to

$$\sin i = n \sin r , \tag{4.70}$$

where n is the complex index of refraction of the metal. To evaluate the equations above, it is useful to realize that

$$n \cos r = \sqrt{n^2 - \sin^2 i} . \tag{4.71}$$

4.2.3. *Polarized scattered light*

When observing the polarization of a faint source close to a very bright source, scattered, instrumentally polarized light from the bright source can influence the polarimetry of the faint source. The following example and discussion is taken from Keller and Sheeley (1999), who observed the polarization of the solar chromosphere just above the limb of the solar disk using the McMath-Pierce solar telescope.

Figure 13 shows how the apparent polarization of the OI 777 nm triplet varies with geocentric position around the solar limb. Clearly, the effect reverses sign at a position angle near 135° from the geographic north direction. However, the reversal does not occur at the same position angle for all three lines, as if the instrumental effect were being supplemented by different scattering polarizations from each line. Other data sets also showed OI 777.5 nm reversals close to the 45° and 135° locations.

The magnitude of this instrumental effect is on the order of 10^{-3} to 10^{-2}, i.e. it is comparable to any true solar scattering polarization signal. Regular instrumental polarization cannot be the explanation since the Q/I signal due to instrumental polarization

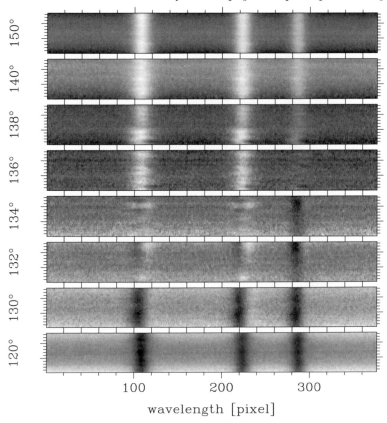

wavelength [pixel]

FIGURE 13. Observed Stokes Q/I signal in the OI 777 nm triplet about $1''$ above the limb as a function of position angle, measured clockwise from geographic north. Lighter-than-average shading denotes polarization along the limb and darker-than-average shading denotes polarization perpendicular to the limb.

has no spectral signature. Empirically, the effect seems to be most pronounced for lines that are in absorption near disk center, but which weaken toward the limb and then go into emission above the limb. Also, telluric lines do not show this effect at all.

When observing above the solar limb, there is a contribution to the signal due to scattered light from the solar disk. While light from the solar limb due to seeing and telescope motion will have the same telescope Mueller matrix as compared to the chromospheric light, there will also be scattered light from all over the solar disk. Since the latter is not following the same path in the optical system as the true chromospheric spectrum (see Fig. 14), it exhibits a different instrumental polarization. Furthermore, the Mueller matrix describing the scattering process is not necessarily equivalent to the regular Mueller matrix for oblique reflection on a metallic coating with a different angle of incidence (e.g. Harvey & Vernold, 1997). In general, the Mueller matrix for scattering has to be calculated by performing a Fourier decomposition of the mirror surface profile that includes the dust on the mirror. The diffraction from each sinusoidal grating (corresponding to one Fourier component) has to be determined. The coherent sum of all these diffracted rays has to be considered when calculating the Mueller matrix for the scattering process.

In the following we construct a model of the influence of polarized, instrumentally

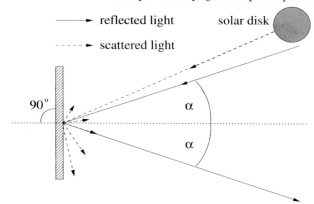

FIGURE 14. Light paths for obliquely reflected (from above the solar limb) and scattered light (from all over the solar disk) on a flat mirror. Note that scattered light does not follow the rule of equal angles of incidence for the incoming and outgoing beams. Part of the scattered light goes in the same direction as the reflected light, thus contributing to the measured signal.

scattered light on precision linear polarization measurements above the limb. For simplicity we will assume that there is only instrumental cross-talk between I and Q, i.e. we assume that there are no U and V signals. The instrumental cross-talk can then be described by

$$\begin{pmatrix} i \\ q \end{pmatrix} = \begin{pmatrix} M_{11} & M_{12} \\ M_{21} & M_{22} \end{pmatrix} \begin{pmatrix} I \\ Q \end{pmatrix} , \qquad (4.72)$$

where lower case letters indicate the measured quantities and upper case letters indicate the true solar signal. With the Mueller matrices M^r and M^s describing the reflected and scattered light, respectively, the observed Stokes q/i signal is given by

$$\frac{q}{i} = \frac{M_{21}^r I_r + M_{21}^s I_s + M_{22}^r Q_r + M_{22}^s Q_s}{M_{11}^r I_r + M_{11}^s I_s + M_{12}^r Q_r + M_{12}^s Q_s} , \qquad (4.73)$$

where I_r, Q_r, I_s, and Q_s are the solar Stokes I and Q signals for the reflected and the scattered beams, respectively.

This equation can be simplified by realizing that the terms $M_{12}^r Q_r$ and $M_{12}^s Q_s$ in the denominator are about three orders of magnitude smaller than the other terms in the denominator. $M_{12}^{r,s}$ as well as $Q^{r,s}$ are all on the order of a few percent or less as compared to $M_{11}^{r,s}$ and $I^{r,s}$, which are of order unity. Furthermore, since the instrumentally scattered light comes from all over the solar disk, we would not expect any net contribution to the true polarization signal, i.e. $Q_s = 0$ is a good assumption. Finally, $M_{11}^{r,s} = M_{22}^r = 1$ is a reasonable assumption that has no significant influence on our analysis. We thus obtain

$$\frac{q}{i} = M_{21}^r + \Delta M \frac{I_s}{I_r + I_s} + \frac{Q_r}{I_r + I_s} , \qquad (4.74)$$

where $\Delta M = M_{21}^s - M_{21}^r$. M_{21}^r is the linear polarization induced by the telescope mirrors and simply adds an offset. ΔM does not change with time as the heliostat rotates because the scattered and reflected light share the same optical path after leaving the heliostat. Finally, I_s is expected to show a spectral dependence similar to the integrated photospheric flux spectrum and should be independent of the exact height above the solar limb because it is due to large-angle (on the order of 0.25 degrees) scattering that should not vary rapidly over small angles of the order of $1''$.

Although the $Q_r/(I_r + I_s)$ term may cause slight variations in q/i, the main influence of instrumentally scattered polarized light is contained in the term $\Delta M I_s/(I_r + I_s)$. For

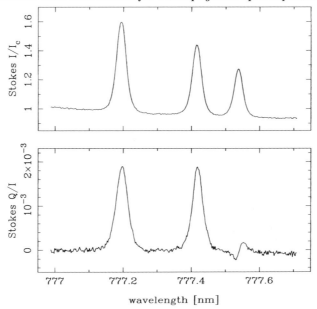

FIGURE 15. True scattering polarization in the OI triplet at 777 nm about $1''$ above the limb. The polarization of the continuum has been set to zero because the observations do not allow a correct measurement of the continuum polarization.

telluric lines, this term is independent of wavelength because they have the same shape for reflected and scattered light, which explains why they do not show the effect. Also, the spectral variation of this term is largest for lines that show a large difference between the photospheric and the chromospheric spectra, which explains why the effect is largest for strong emission lines that are in absorption on the disk.

The dependence of ΔM on position angle β of the polarimeter with respect to the geographic north direction can easily be understood. The combined Mueller matrices for all telescope elements for reflected and scattered light have all reflections in common after the heliostat mirror. Under the assumption of weakly polarizing elements, the multiplication of the Mueller matrices can be approximated by the sum of the Mueller matrices (see Stenflo 1994). The difference between the two Mueller matrices for the whole telescope is therefore almost equivalent to the difference between the Mueller matrices for reflection and scattering at the heliostat alone. These two matrices depend on the sun's declination (which can be assumed to be constant during a day) and on the angle, β, of the positive Stokes Q direction with respect to the geographic north direction. The latter corresponds to a simple rotation of the Stokes coordinate system, hence ΔM is proportional to $\cos 2\beta$.

The simple dependence of ΔM on position angle can be used to determine the true solar signal. If q/i is measured at two or more position angles, we can remove the β-dependence by performing a linear regression with respect to $\cos 2\beta$, and obtain the polarization $Q^r(\lambda)/(I^r(\lambda) + I^s(\lambda))$ up to a constant that corresponds to the continuum polarization. Figure 15 shows this quantity for the OI triplet, and, as expected, indicates much less scattering for the 777.5 nm line than for the other two lines. A similar regression can be made if observations are taken at various distances from the limb.

To avoid problems with instrumentally polarized scattered light, the foremost requirements is to keep the mirrors clean. Careful washing can reduce scattered light drastically.

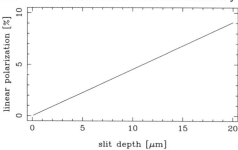
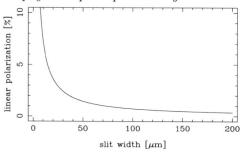

FIGURE 16. Variation of linear polarization induced by a narrow and thin steal slit as a function of slit depth (for a width of 16μm) and width (for a depth of 10μm) at 632.8 nm according to the theory of Slater (1942).

4.2.4. *Polarization due a narrow spectrograph entrance slit*

While the spectrograph entrance slit is really part of an instrument and not the telescope, it is adequate to discuss this effect here. When the width of a slit is comparable to the wavelength of the incident light, the slit acts as a partial polarizer. Indeed, entrance slits of modern spectrographs are often on the order of 10 to 20 times the wavelength, where such effects may become important. Slit polarization was first observed by Fizeau in 1861 and later by Zeeman in 1912 when he measured the polarization of spectral lines. It is obvious that a slit polarizes in the same way that a wire-grid polarizer works. Therefore one might expect that the slit polarization would be much reduced when using dielectric media for the slit. However, various reports in the literature seem to indicate that it does not matter much whether the slit is made of a conducting or a dielectric material (to within a factor of 2). Various models have been developed to calculate this polarization as a function of width, depth, and material properties of the slit.

An overview of various theories has been given by Ismail (1985). The theory by Slater (1942) based on microwave wave-guide theory seems to compare best with actual measurements. It also generally leads to larger values of the polarization than other theories, therefore providing a reliable upper limit to the expected slit polarization.

The expected fractional polarization from Slater's (1942) theory is given by

$$P = \frac{e^{-2a_p z} - e^{-2a_s z}}{e^{-2a_p z} + e^{-2a_s z}} \tag{4.75}$$

where

$$a_s = \frac{2}{b} \sqrt{\frac{\omega \epsilon_0}{2\sigma}} \left[1 - \left(\frac{n\lambda}{2b} \right)^2 \right]^{-\frac{1}{2}} \tag{4.76}$$

and

$$a_p = a_s \left(\frac{n\lambda}{2b} \right)^2 . \tag{4.77}$$

b is the slit width, z is the slit depth, ϵ_0 is the permittivity of free space, ω is the angular frequency of the light and λ is the wavelength of light, σ is the conductivity of the slit material, and n is the wave order. Since $n = 1$ gives the largest amount of polarization, it is sufficient to assume $n = 1$ to estimate an upper limit for the slit polarization. Figure 16 shows theoretical calculations of the induced linear polarization for a steel slit as a function of slit depth and width, which is in reasonable agreement with observations.

These calculations indicate that a narrow slit should be made as thin as possible, at least at the edges to reduce the amount of polarization introduced by the narrow slit.

Very thin slits can be manufactured by evaporating a metal film on a glass substrate or by laser-cutting a very thin metal foil.

4.2.5. *Mitigating instrumental polarization and cross-talk*

There are various approaches to mitigate instrumental polarization and cross-talk. Among the more common approaches are:
- avoiding oblique reflections;
- compensating instrumental polarization and cross-talk with:
  ○ retarders,
  ○ partial polarizers such as tilted glass plates,
  ○ crossing mirrors at 90° (see above),
  ○ use two or four mirror arrangements to compensate for a single oblique reflection off a mirror;
- measure and take into account in data reduction.

4.3. *Angle-dependence of polarizers and retarders*

4.3.1. *Angle-dependence of polarizers*

While sheet polarizers and similarly constructed polarizers are basically insensitive to the angle of incidence, crystal polarizers have a limited angle of incidence under which they work. For calcite-based polarizers, this is on the order of 10°, which is due to the critical angles for total internal reflection at the inner surfaces being different for ordinary and extraordinary rays. The transmitted light is polarized when the incidence angle is below this limit. Otherwise rays are either both reflected or transmitted, depending on the construction of the polarizer.

Beams through crystals behave optically differently from glass because the index of refraction depends on the angle of the ray with respect to the crystal optic axis. The most common crystal aberrations are defocus and astigmatism. Recent versions of some optical ray-tracing software (e.g. ZEMAX) can perform the basic crystal calculations.

As an example, we consider a calcite beam splitter. Obviously, ordinary and extraordinary rays travel different distances (see Fig. 5). Therefore, there will be a focus difference between the two beams. To avoid this problem, the calcite block can be split into two pieces, which are crossed at 90°. This is called a *Savart plate*, which has a splitting that is reduced by factor of $\sqrt{2}$, but the defocus has disappeared. However, the two beams will show astigmatism in opposite directions, which is due to the crystal aberrations, something that is not observed in isotropic materials.

Another approach consist in using the two calcite blocks in the same orientation, but add a half-wave plate between them to exchange the ordinary and extraordinary beams, which is called a *modified* Savart plate. The crystal astigmatism is now the same in the two beams. By adding a cylindrical lens, the astigmatism can be compensated for. As a rule of thumb for calcite, the astigmatic focus difference amounts to about 5% of the calcite thickness.

4.3.2. *Angle-dependence of retarders*

Retarders can have strong variations of retardation with angle of incidence. For a rotation around an axis that is parallel to the optic axis, the angle between the beam and the optic axis of the crystal does not change. Therefore, both e and o beams will travel at the same speed, but through more material. For small angles of incidence θ, Snell's law tells us that the angle of the beam within the retarder has an angle of θ/n, where n is the geometrical mean of the two indices of refraction, i.e. $n = \sqrt{n_e n_o}$. The

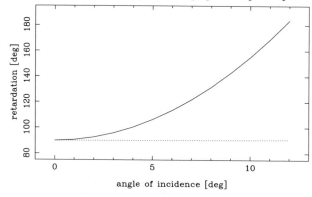

FIGURE 17. Variation of retardance with angle of incidence for 2-mm thick compound zero-order (solid line) and true zero-order (dotted line) quartz quarter-wave plates at 632.8 nm.

increase of the path length due to the inclined beam is therefore given by $d/\cos(\theta/n)$ such that the relative path difference becomes

$$(N\lambda)\,(\theta) = \frac{d}{\cos\theta/n}\,(n_e - n_o)\ . \tag{4.78}$$

With θ measured in radians and using the approximation $\cos x \approx 1 - x^2/2$, and using the expression for $N\lambda$ at normal incidence, we obtain the (approximate) angular variation of the retardation as

$$\delta\,(N\lambda)\,(\theta) = +\frac{N\theta^2}{2n^2}\ . \tag{4.79}$$

For a 2-mm thick compound zero-order quartz quarter-wave plate, a rotation of $10°$ will make it almost into a half-wave plate (see Fig. 17), while a true zero-order retarder shows barely any change in retardation.

For a rotation perpendicular to the optic axis, one derives the same equation, but with the opposite sign. For larger angles the theory needs be corrected in the sense that the e and o beams in the retarder have different angles and therefore also exit the beam at two slightly different locations.

It is evident that the angle-dependent variation is linear in N and quadratic in the angle θ. Also, a compound zero-order retarder made from to thick retarders of the same material has the same angle dependence as a single multi-order retarder of the same thickness because one plate is rotated around an axis parallel to the optic axis and the other plate around an axis perpendicular to the optic axis. While some catalogs and even text books claim that compound zero-order and true-zero order waveplates show the same angle dependence, this is not correct. If one has a fast beam, it is important to use true zero-order retarders.

Kerr cells, liquid crystals, and PEMs all behave like true zero-order retarders. However, for PEMs, the retardance decreases with increasing distance from the center. Hence, it is possible to place a PEM into an optical beam in such a way that the angular and spatial variation compensate each other to a large degree. It is therefore possible to use PEMs with very fast beams.

Pockels cells have only a few degrees acceptance angle because of strong crystal effects.

4.4. *Temperature dependence of retarders*

Retarders can be particularly temperature sensitive, in particular multiple order wave-plates. The optical path difference variation between ordinary and extraordinary rays, to first order, is given by

$$\delta d \left(n_e - n_o\right) + d \left(\delta n_e - \delta n_o\right) , \tag{4.80}$$

where δ indicates variations with temperature. Using the coefficient of thermal expansion $\alpha = \delta d / d$, we obtain

$$\delta N = N \left(\alpha + \frac{\delta n_e - \delta n_o}{n_e - n_o}\right) \tag{4.81}$$

For quartz, the quantity in parentheses amount to -1.0×10^{-4} per degree Kelvin at a wavelength of 632.8 nm. For a 2 mm thick multi-order quarter-wave retarder, this translates into a retardation variation of about 1 degree per degree Kelvin. Compound zero-order waveplates have the same temperature dependence as true zero-order re-tarders because the two plates compensate each other to a large degree. Achromatic zero-order retarders made from different materials can have larger effects due to temper-ature changes because the two materials have different coefficients of thermal expansion and temperature variations of the index of refraction.

4.5. *Polarized interference fringes*

When the retardation of a variable retarder is varied, the optical path length within the retarder also changes. Multiple reflections between the surfaces of the retarder lead to spectral fringes, whose pattern changes when the retardation is changed (because of the change in optical path length). A polarized spectrum (difference between two measurements with different retardations) will show a fringe pattern (see Fig. 18). This effect has been described by Oakberg (1995) for photoelastic modulators. Similar fringe patterns have also been observed in liquid crystal retarders and fixed retarders. The latter is due to the fact that the expression for the retardance of a birefringent plate given above is only a first-order approximation. Multiple reflections at the surfaces must be taken into account to obtain the higher-order correction terms. When including the higher-order terms, fringes in the retardation occur (e.g. Clarke & Grainger 1971). These can be reduced by applying an appropriate anti-reflection coating to the birefringent material.

Tilting, wedging, and coating the retarders reduces the fringe amplitude considerably. However, at the 1×10^{-5} level, polarized fringes are almost always present. Often the fringes can be removed during the data reduction by an appropriate filtering in the Fourier domain.

4.6. *Detector-induced errors*

Any real detector system has non-linearities, which are mostly due to the analog read-out electronics. When trying to look for very small polarization signals (e.g. 1×10^{-5} of the intensity) on top of a small instrumental polarization signal of 1%, non-linearities become important, as is shown in the following.

Let the measured signal S be a quadratic function of the incoming intensity I

$$S = aI^2 + bI + c , \tag{4.82}$$

which is the most simple form for a non-linear behavior. b represents the (arbitrary) gain, c any remaining influence of bias or dark current that has not been correctly removed, and a models the non-linearity. The constant term also models effects of stray-light in a filter or a spectrograph.

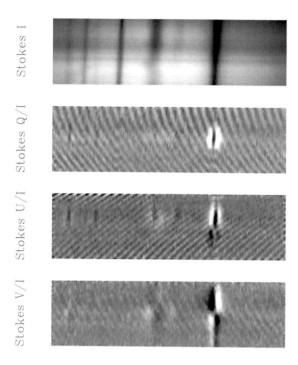

Stokes I

Stokes Q/I

Stokes U/I

Stokes V/I

FIGURE 18. Polarized interference fringes in highly sensitive vector measurements of a sunspot spectrum around the HeI 1083.0 nm line observed with ZIMPOL I (see below).

For unpolarized incoming light, the two opposite polarization states measured by the polarimeter are

$$I^+ = I + \delta I \ , \quad I^- = I - \delta I \ , \tag{4.83}$$

where δI is the instrumental polarization signal.

The measured amount of polarization is determined from

$$P_m = \frac{S^+ - S^-}{S^+ + S^-} \ , \tag{4.84}$$

which in the ideal case of $a = 0$, $c = 0$ corresponds to the instrumentally introduced polarization

$$P = \frac{bI^+ - bI^-}{bI^+ + bI^-} = \delta \ . \tag{4.85}$$

In the case of non-linearities and an offset error, the apparent, measured polarization signal becomes

$$P_m = \frac{2a\delta I^2 + b\delta I}{aI^2 + a\delta^2 I^2 + bI + c} \ . \tag{4.86}$$

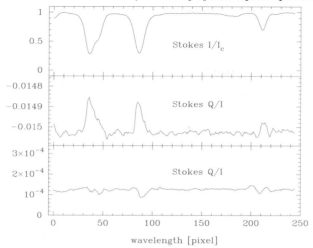

FIGURE 19. Artificial polarization signal created by the coupling of non-linearities in the CCD readout electronics and instrumental polarization for a solar spectrum observed at disk center. The top panel shows the Stokes I profile normalized with the continuum intensity I_c. The center panel shows Stokes Q/I in the case of -1.5% linear polarization induced by oblique reflections in the telescope. The lower panel shows the same linear polarization after compensating the linear polarization induced by the telescope. The remaining structure seen in the lower panel is only due to magnetic fields in the quiet sun.

As expected, in the absence of instrumental polarization, i.e. $\delta = 0$, no polarization would be measured.

In the following we will only keep terms up to second order in a, δ, and their cross-products. First, we assume that $c = 0$. The measured polarization then becomes

$$P_m = \delta \left(1 + \frac{a}{b}I\right) . \tag{4.87}$$

The observed polarization has therefore an additional component that is essentially proportional to aI, the coefficient of the non-linear term times the intensity. If the detected signal corresponds to a spectrum, the measured polarization is not constant anymore, but has a Stokes I-like additive component.

To measure very small polarization signals in the presence of non-linearities in the detector system, it is necessary to minimize the instrumental polarization. After normalizing Stokes I with the continuum intensity I_c, variations of Stokes I are of order unity. A typical value for the non-linear term is $a/b = 0.01$. To achieve a sensitivity of 1×10^{-5}, the instrumental polarization δ must be smaller than 1×10^{-3}. This is particularly hard to achieve for linear polarization where a typical value for existing solar telescopes is $\delta = 0.05$. Therefore, a typical magnitude of the coupling of instrumental polarization with non-linearities in the detector system is 5×10^{-4}.

Figure 19 compares two linear polarization measurements of the same spectral region with and without compensation of the instrumental polarization. The ZIMPOL I camera used for those measurements did show a non-linearity of about $a/b = 1.4\%$. The simple model developed above explains the sign as well as the magnitude of the effect.

The other case is $a = 0$. The measured polarization then becomes

$$P_m = \delta \left(1 - \frac{c}{bI}\right) . \tag{4.88}$$

Since I shows structure, e.g. spectral lines, the observed polarization will show signatures proportional to $1/I$. Any offset such as bias, dark current, and stray-light must be removed to a high degree. However, due to minuscule changes in the observing conditions (e.g. variation of the bias with time), there is often an offset error of approximately 1×10^{-3}. Therefore, this effect is about one order of magnitude smaller than the influence of non-linearities under realistic circumstances. It disappears completely if there is no instrumental polarization.

5. Examples of astronomical spectropolarimeters

5.1. *Introduction*

A large variety of astronomical spectropolarimeters have been designed and built. It is outside of the scope of this chapter to give an overview of all the various approaches and instruments. Instead, I selected a few examples with which I have personal experience. Other polarimeters that are used for night-time applications have been discussed by Tinbergen (1996). Solar polarimeters that have produced significant scientific results but are not discussed any further here include the Advanced Stokes Polarimeter (Lites 1996) based on a rotating retarder and the Tenerife Infrared Polarimeter (Martínez Pillet et al. 1999), which is based on FLCs.

5.2. *Sensitive stellar polarimeter*

A sensitive method for stellar circular polarimetry was developed by Semel et al. (1993). It is based on a rotating quarter-wave plate and a double-calcite beam-splitter, which produce two beams corresponding to opposite circularly polarized light. The quarter-wave plate can be rotated to $+45°$ and $-45°$ with respect to the polarization axes of the beam-splitter. Both beams are recorded simultaneously. The quarter-wave plate is then rotated by $90°$ and another image is exposed. The four measurements of the same object are then combined to obtain an estimate of the Stokes V/I ratio that is largely free of effects from seeing and gain variations between different detector areas if the polarization signal is small. This approach has been used in stellar polarimetry with great success by Donati et al. (1990, 1999). It can be applied to any polarized Stokes parameter.

This approach also works very well for solar applications where the spectrum in the first and the second exposures are different. Consider the measured intensities in the two beams in the first exposure after subtraction of the dark current (for the case of Stokes V)

$$S_1^l = g_l \alpha_1 (I_1 + V_1), \quad S_1^r = g_r \alpha_1 (I_1 - V_1). \tag{5.89}$$

The subscript 1 indicates the first exposure, the subscripts l and r indicate the left and the right beams of the polarizing beam-splitter. S describes the measured signal, g the gain in a particular beam, and α the average transmission of the atmosphere and the instrument for a given exposure.

In the second exposure, after the retarder has been rotated, the measured signals are given by

$$S_2^l = g_l \alpha_2 (I_2 - V_2), \quad S_2^r = g_r \alpha_2 (I_2 + V_2). \tag{5.90}$$

Note that the incoming intensity in the second exposure may be completely different from the first exposure. Such changes may be due to seeing as well as instrumental changes. This also includes a shift of the two beams between exposures due to beam-wobble induced by rotation of a wave plate.

The following combination of these four measured intensities removes the effect of

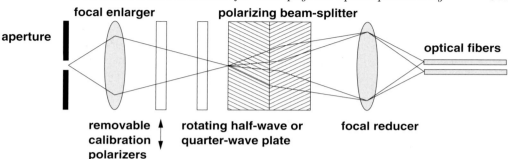

FIGURE 20. Optical layout of the MuSiCoS polarimeter by Donati et al. (1999). An entrance aperture provides a fixed field of view independent of seeing. A focal enlarger provides a slow beam through the polarimeter to avoid large angles of inclination on the polarization optics. Circular and linear polarizers can be inserted to calibrate the polarimeter. A rotating quarter-wave or half-wave plate provides the beam exchange. A polarizing beam-splitter splits the beam into two opposite polarization states that are reimaged onto the ends of two fibers that lead to the echelle spectrograph.

transmission changes and differential gain variations of different detector areas:

$$\frac{1}{4}\left(\frac{S_1^l}{S_2^l}\frac{S_2^r}{S_1^r} - 1\right) = \frac{1}{2}\frac{I_2 V_1 + I_1 V_2}{I_1 I_2 - I_2 V_1 - I_1 V_2 + V_1 V_2} \ . \tag{5.91}$$

In the case of $V \ll I$, this is equivalent to

$$\frac{1}{2}\left(\frac{V_1}{I_1} + \frac{V_2}{I_2}\right) . \tag{5.92}$$

Therefore we obtain the average V/I signal of the two exposures. No spurious polarization signals are introduced. If V is comparable to I, the method can be extended to higher orders (Bianda et al. 1998).

Figure 20 shows the layout and the operating principle behind the MuSiCoS spectrograph developed by Donati et al. (1999). Due to problems with polarized fringes in the superachromatic retarders, the instrument currently uses a crystalline achromatic quarter-wave plate for circular polarization measurements and a rotation of the whole instrument for the linear polarimetry. It is planned to upgrade the system with Fresnel rhombs that are very achromatic but do not show significant polarized fringes at the spectral resolution of the MuSiCoS spectrograph. By combining many spectral lines from the echelle spectra, it is possible to achieve sensitivities on the order of $1 \cdot 10^{-4}$ with high spectral resolution. Results from this polarimeter are shown in the chapter by Mathys.

5.3. *Zurich Imaging Stokes Polarimeters*

5.3.1. *Introduction*

Photoelastic modulators are the preferred variable retarders for polarization modulation above 1 kHz, but their high modulation frequencies of 20 to 50 kHz are incompatible with the read-out rate of standard array detectors. A new instrument concept developed by Povel et al. (1990) reconciled the incompatibility between the slow CCD image sensor and the fast photoelastic modulators by using the CCD directly as a demodulator.

Here we describe two instruments that are based on this approach. ZIMPOL (**Z**urich **I**maging **Pol**arimeter) I has been used for almost 10 years now, while the second generation instrument, ZIMPOL II has just started delivering the first scientific results. A more detailed description of the overall concepts can be found in Povel (1995).

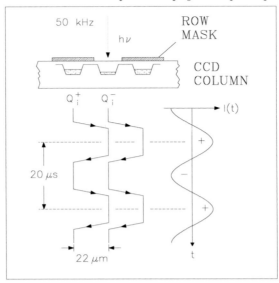

FIGURE 21. Principle of the ZIMPOL I CCD demodulator. A cross section through a part of the CCD sensor is shown at the top. The motion of the charge packets in the CCD are shown as a function of the modulation indicated in the right.

5.3.2. *CCD array as a fast demodulator*

The frame rate of regular CCD array detectors with a useful number of pixels is limited to about 1 kHz. For the detection of fast intensity modulations with frequencies of 50 to 100 kHz typical of PEMs, regular CCD array detectors cannot be used.

The charge shifting and storing capabilities of a CCD array detector, however, may be employed to operate the detector as a synchronous demodulator. The array is divided into photo-sensitive rows and storage rows that are shielded from light by a mask as shown in Fig. 21. The photo charges Q^+ generated in the photo-sensitive rows during the first modulation half-cycle are shifted into the storage row at the transition to the second modulation half-cycle. The photo charges Q^- generated during the second modulation half-cycle are shifted into the other storage row at the transition to the first modulation half-cycle. Q^+ and Q^- are simultaneously shifted from left to right and vice versa. While this procedure is repeated over many modulation cycles, Q^+ and Q^- are integrated alternately and synchronously to the modulation. Whenever the desired amount of charges have been accumulated, the charges are read out. From the digitized charge signals the normalized difference $P = (Q^+ - Q^-)/(Q^+ + Q^-)$ is calculated after correction for dark current and electronic offsets. P is proportional to the amplitude of the modulated signal. Note that pixel-to-pixel gain variations do not affect P. To avoid a reduction of the measured modulation amplitude by effects due to the finite transfer time of the charges from photo-sensitive to masked rows and vice versa, the transfer time must be about a factor of 100 faster than the modulation frequency. Furthermore, the charge transfer efficiency in both directions needs to be very high. Three-phase buried channel frame transfer CCD arrays designed for video applications meet these requirements.

A similar, yet much slower method was used by Stockman (1982) for differential imaging in astronomical applications.

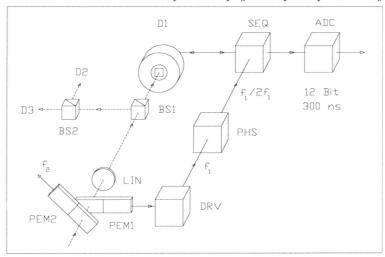

FIGURE 22. Setup for recording all four Stokes parameters with two photoelastic modulators. The orientation of the components is given with respect to Stokes Q. PEM1 ($0°$), PEM2 ($45°$): photoelastic modulators, LIN ($22.5°$): Glan linear polarizer, BS1, BS2: beam splitter cubes that direct the modulated light to detectors D1, D2, and D3.

5.3.3. *Polarimeter with PEM*

If a PEM is oriented under $45°$ with respect to Stokes Q and a linear polarizer follows at $-45°$ with respect to the PEM, the output intensity is given by

$$I'(t) = \frac{1}{2}\left(I + Q\cos\delta(t) + V\sin\delta(t)\right) , \qquad (5.93)$$

where the retardation $\delta(t)$ has been given above. After the expansion in Bessel functions, Q is modulated at a frequency of $2\omega t$, while V is modulated at ωt. These two intensity modulations can be recovered separately by two different CCD cameras.

5.3.4. *ZIMPOL I*

The polarization modulator of ZIMPOL I consists of two photoelastic modulators and a linear polarizer (see Fig. 22). This configuration has been discussed in detail by Stenflo (1984, 1991). The combination of PEMs and the linear polarizer will henceforth been called modulator package. The modulators are commercially available devices that oscillate at a frequency of about 42 kHz.

A light beam with Stokes vector (I, Q, U, V) constant in time entering the modulator package produces intensity variations according to

$$I'(t) = \frac{1}{2}\left(I + Q\sqrt{2}J_2(A)\cos 2\omega_1 t + U\sqrt{2}J_2(A)\cos 2\omega_2 t + V\sqrt{2}J_1(A)\sin\omega_1 t\right). \quad (5.94)$$

The oscillation frequencies of the two PEMs are given by ω_1 and ω_2, and J_1 and J_2 are the Bessel functions of order 1 and 2. The amplitude of both PEMs, A, is chosen such that $J_0(A) = 0$. For simultaneous measurements of all four Stokes parameters the light needs to be distributed to three separate detectors demodulating at frequencies ω_1, $2\omega_1$, and $2\omega_2$. This can be accomplished by two beam splitter cubes (see Fig. 22). More information on the design of ZIMPOL I can be found in Keller et al. (1992) and Povel et al. (1994).

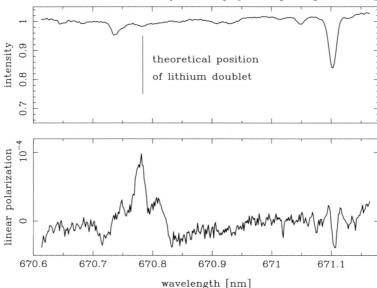

FIGURE 23. The lithium doublet at 670.9 nm is very difficult to measure in the solar intensity spectrum. The weak absorption line seen in intensity is due to CN. However, the lithium lines are resonance lines and therefore exhibit linear scattering polarization close to the solar limb. Since there are no other strongly polarizing lines close by, the lithium lines stand out in the polarization spectrum.

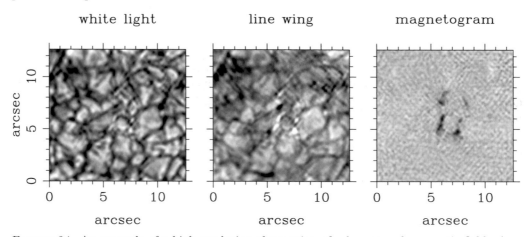

FIGURE 24. An example of a high-resolution observation of solar network magnetic fields close to disk center in CaI 610.3 nm using ZIMPOL I and speckle polarimetry to remove the effects of seeing. The magnetic fields are concentrated in very small areas and tend to be bright in both the white-light and the line-wing images.

5.3.5. *Scientific results from ZIMPOL I*

ZIMPOL I has provided a wealth of scientific results on solar polarization thanks to its high sensitivity that reaches down to a polarization of a few times 10^{-6}. Among many other things, it provided the first direct measurement of the field strength in solar intra-network fields (Keller et al. 1994) and opened the window to the *second solar spectrum* (see Stenflo in this volume). Figure 23 shows an example of the second solar spectrum. Thanks to its fast read-out rate and high modulation frequency, ZIMPOL I has been

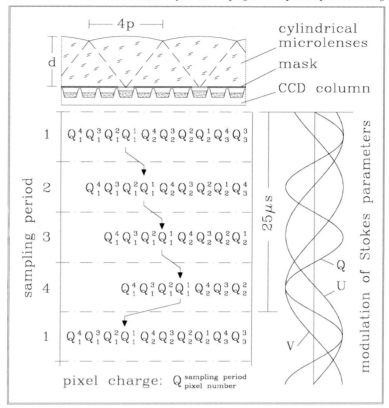

FIGURE 25. Principle of the ZIMPOL II demodulation scheme. The upper part shows a cross-section of the micro-lens array in a plane perpendicular to the pixel rows. p is the pixel size ($22.5~\mu$m). d is 1 mm. The right part shows the modulation of the Stokes parameters Q, U, and V. The left part shows the charge shifting scheme with respect to the modulation as envisioned in the conceptual design.

the premier instrument for image reconstruction techniques such as speckle imaging (see Fig. 24) and behind adaptive optics systems.

5.3.6. *ZIMPOL* II

A drawback of ZIMPOL I is its low efficiency due to the mask (a factor of 2) and the restriction of a single CCD demodulator to one frequency (a factor of 3). This efficiency loss of a factor of 6 is overcome by ZIMPOL II (see Stenflo et al. 1992), which was designed to use a cylindrical micro-lens array, two phase-coupled, synchronized PEMs with the same frequency, and an achromatic quarter-wave plate. Stokes V is modulated at $\sin(\omega t - \pi/4)$, Stokes Q at $\cos 2(\omega t - \pi/4)$, and Stokes U at $\sin(\omega t + \pi/4)$. Four sampling intervals in one modulation cycle are needed to record all four modulation states with a single, specially masked CCD (see Fig. 25). The light is focused on the CCD with a micro-lens array such that out of four pixels in one column only one is used for light detection while the remaining three are used for temporary charge storage. The four charge packets corresponding to four independent linear combinations of the four polarization states are sequentially shifted between the pixel rows in phase with the four-state modulation. After the CCD is read out, the four Stokes parameters can be calculated from the digitized pixel charges in the four interlaced images. In this

FIGURE 26. Artists impression of SOLIS on top of the Kitt Peak Vacuum Telescope building. The instrument with the large entrance window is the Vector-Spectromagnetograph.

way a precision imaging polarimeter recording simultaneously all four Stokes parameters may be realized with a single CCD sensor. In contrast, the three-CCD system used in ZIMPOL I requires accurate mechanical alignment and careful post-processing to combine the images from the three CCD cameras.

While possible, it proved hard to accurately align the micro-lenses on the CCDs (Gale et al. 1997). Furthermore, stray-light within the micro-lens-CCD assembly led to significant problems during the data reduction. Finally, the coupled PEMs have not been realized so far, which has prevented ZIMPOL II from measuring all four Stokes parameters simultaneously. However, using a modulator package based on FLCs will provide the required modulation waveforms (Gandorfer 1999). First scientific results from ZIMPOL II include an atlas of the second solar spectrum (Gandorfer 2000).

5.4. *SOLIS Vector-Spectromagnetograph*

5.4.1. *Introduction*

SOLIS, the Synoptic Optical Long-term Investigations of the Sun, will provide unique, modern observations of the Sun on a continuous basis for several decades. SOLIS consists of three instruments (a 50-cm vector spectromagnetograph, a 14-cm full-disk imager, and an 8-mm sun-as-a-star spectrometer) attached to a single equatorial mount. The mount is currently installed for testing at a preliminary site in Tucson. In its final configuration, SOLIS will be located on top of the Kitt Peak Vacuum Telescope (KPVT, see Fig. 26) and become operational by the end of 2001. More information can be found at http://www.nso.noao.edu/solis.

The SOLIS Vector Spectromagnetograph (VSM) operates in four different observing modes at three different wavelengths. The corresponding data products in the following list are given in parentheses.

(*a*) Photospheric full-disk vector-magnetograms using the FeI 630.15 and FeI 630.25 nm lines (field strength, azimuth, inclination, flux, Doppler velocity, continuum intensity);

Parameter	Specification
angular element	1″.0 by 1″.0
angular coverage	2048″ by 2048″
format	2048″ by 1″
geometric accuracy	<0″.5 rms after remapping
instrumental mtf	measurable to ±0.01
total mtf	<0.1 at frequencies greater than Nyquist
motion in RA	±0.25° for flat-fielding
scan rate	0.2-5.0 s/″
timing accuracy	better than 1 ms
spectral lines	FeI 630.1515, FeI 630.2507 nm, CaII 854.2089 nm, HeI 1083.0 nm
spectral resolution	200,000
wavelength ranges	630.1015-630.3007 nm, 854.2±0.1 nm, 1083.0±0.5 nm
polarimetry	630.2 nm: I,Q,U,V; 854.2 nm: I,V; 1083.0 nm: I
sensitivity	0.0002 per pixel in 0.5 s
relative accuracy	0.001
image stabilization	>40 Hz to improve spatial resolution
seeing monitoring	for information only
cloud detection	at user-specified level
cloud interruptions	for user-determined period

TABLE 6. Technical specifications for the SOLIS Vector-Spectromagnetograph (VSM)

(*b*) Chromospheric full-disk magnetograms using the CaII 854.2 nm line (line-of-sight magnetic flux, Doppler velocity, line core intensity);

(*c*) Full-disk HeI 1083.0 nm line characteristics (equivalent width, continuum intensity, Doppler velocity, line depth, line asymmetry, Doppler width, Si line width, Si line depth, Si Doppler velocity);

(*d*) Photospheric full-disk longitudinal magnetograms using the FeI 630.15 and 630.25 nm lines (line-of-sight magnetic flux).

Data products (a), (b), and (c) will be produced three times a day, while data product (d) will be produced once a day. Users may observe the same or subsets of these products such as vector-magnetograms of active regions at high temporal cadence or HeI 1083.0 nm equivalent width measurements of selected areas.

The CaII 854.2 nm and the HeI 1083.0 nm lines were chosen to provide a continuing record of the current data set from the KPVT. To measure vector magnetic fields outside sunspots in the visible part of the spectrum, it is indispensable to observe at least two spectral lines with different Landé factors. The FeI 630.15 and 630.25 nm lines were chosen because they are very suitable to measure vector magnetic fields in quiet as well as active regions of the photosphere. The Advanced Stokes Polarimeter (ASP), an instrument that delivers precise vector field measurements (Lites 1996), has also used the FeI lines at 630.2 nm.

The most important improvement over the currently produced data sets from the KPVT are the precise vector polarimetry, which allows the derivation of the true magnetic field vector as compared to the current longitudinal flux measurements.

5.4.2. *Specifications*

Table 6 shows the specifications for the VSM according to which the instrument has been designed and is now under construction.

FIGURE 27. The optical layout of the SOLIS Vector-Spectromagnetograph. The entrance window is on the upper right, and the folded Littrow spectrograph is on the lower left side of the drawing.

5.4.3. *Design*

The design of the VSM faced many challenges. In particular, the following requirements have made the design effort demanding:

- compact instrument no longer than 2.5 m
- athermal optical design to keep instrument stable at varying ambient temperatures
- high guiding accuracy of better than $0\!''\!5$ rms
- low instrumental polarization of less than $1 \cdot 10^{-3}$
- large wavelength range from 630 nm to 1090 nm with constant magnification
- high spectral resolution of about 200,000
- highest possible throughput
- high energy densities of up to 0.2 MW/m^2
- high data rates of over 300 MByte/s

An overview of the optical layout is shown in Fig. 27. The mechanical layout is shown in Fig. 28.

A 50-cm effective aperture telescope is sufficient to achieve the required polarization sensitivity. To match the CCD pixel size of 16 μm per pixel ($=1''$) an f/6.6 beam is needed. The telescope has a Ritchey-Chrétien (RC) configuration with a two-lens field corrector to provide adequate image quality over the whole field of view, minimal geometric distortion, equal image size for all wavelengths, and a roughly telecentric beam to minimize field of view effects in the polarization modulators and the spectrograph.

The 6-mm thick entrance window reduces contamination of the optics and allows us to fill the whole instrument with helium to minimize internal seeing. The window is slightly wedged and anti-reflection coated to reduce fringes. Birefringence in the fused silica window is reduced by making it thin and oversized.

FIGURE 28. The mechanical layout of the SOLIS Vector-Spectromagnetograph. The entrance window is on the lower left, and the folded Littrow spectrograph is on the upper right side of the drawing.

The primary mirror is made of ULE to minimize temperature-induced aberrations. It will be coated with over-coated silver like all the other reflective optics to maximize the throughput. The secondary mirror is made of a silicon single crystal, which provides high thermal conductivity. Regular glass-based mirrors with their low thermal conductivity would lead to large thermal gradients within the mirror substrate and therefore to intolerable optical aberrations. A similar secondary mirror has been used in the Flare Genesis telescope (Bernasconi et al. 1999). Fans circulate the helium in such a way that the secondary mirror is cooled, which is mounted on a fast tip/tilt stage to compensate for image motion.

The requirements of low instrumental polarization and high guiding accuracy have led to an innovative design of the entrance slit area. Four linear arrays are cofocal and parallel to the entrance slit and provide an accurate guiding signal to the secondary mirror tip/tilt system and the mount. This avoids differential guiding errors and instrumental polarization and cross-talk due to a beam-splitter. The location of the limbs on the linear arrays will provide the solar disk position, the sharpness of the limb, and the sky conditions.

The fast beam will lead to an energy density of about 0.2 MW/m^2 in the focal plane. The energy that does not go to the guider arrays or through the entrance slit will be absorbed by an actively cooled plate made out of a copper-silicon carbide composite that is used for high-energy laser optics.

The 16-μm wide, thin entrance slit covers an area of 2048″ by 1″ and is followed by the modulator packages. The entrance slit is curved to compensate for a 26-pixel spectral line curvature. The spectrograph is a Littrow arrangement with a doublet lens. The Richardson Grating Laboratory Echelle grating has 79 grooves/mm, a 63.5° blaze angle, a 204 by 408 mm ruled area, and an efficiency of about 50%, which is almost independent of the polarization state at 630.2 nm. Purely static mechanical structures will not be able to keep the grating stable enough. Therefore, the grating has actuators in two axes that are controlled by error signals generated by the CCD cameras in the final focal plane.

Since we were not able to find CCD cameras that would cover the full length of the slit with 2048 pixels, there is a beam-splitter in the spectrograph focal plane that splits the spectrum perpendicular to the spectral lines into two equal parts. The polarizing beam-splitters in front of the cameras also require a mask in the focal plane that limits the area that falls onto the detectors. Otherwise, the two polarization states would overlap. The focal mask and beam-splitter are put into the exit focal plane, which is reimaged with an Offner system that provides achromatic one-to-one reimaging with good field properties.

To achieve a high signal-to-noise ratio (SNR) in a short time, the CCD cameras must be able to read out at a high frame rate. In addition, the modulation frequency, and therefore the read-out frequency, must be faster than typical correlation times of seeing. The required SNR led to a frame rate of 300 Hz, which is also a good rate to minimize the influence of seeing. Of course, the CCDs should have the highest possible quantum efficiency and a read-out noise that is significantly below the photon shot noise. The VSM will use two 300 frame/s 1024 by 256 pixel backside-illuminated frame transfer CCD cameras from PixelVision. The cameras will take unshuttered 3.3 ms exposures of the two orthogonally polarized spectra, which are transferred to a storage area in less than 0.15 ms. Readout will occur during the next 3.3 ms exposure with 16 parallel channels at 14 bits each.

Each of the 32 channels from the two CCD cameras delivers its data via fiber optical links to an image acquisition system, which will add up frames in buffers according to the states of the polarization modulators. At least 8 frames need to be summed to reduce the initial 300 MByte/s to less than 40 MByte/s. At this reduced data rate, the accumulated data can be sent directly to the data handling system.

The polarization measurement optics consist of a polarization calibration section, a polarization modulator section, and polarizing beam-splitters.

The two polarization calibration packages (one for 630.2 nm and one for 854.2 nm) have interference filters to reduce the overall light level except for the observed band pass. Without those filters, the polarizers would be damaged due to the high flux level. The filters are followed by a dichroic polarizer and a quarter-wave plate.

For 630.2 nm, the polarizer and the quarter-wave plate can be positioned independently at various angles with respect to the entrance slit and with respect to each other. This allows us to estimate the Mueller matrices of the corrector lenses, the modulator packages, and determine small deviations of the characteristics of the calibration polarizer and the waveplate. This procedure is modeled after the successful ASP approach (Skumanich et al. 1997). For 854.2 nm, the polarizer and the quarter-wave plate are combined to provide circularly polarized light.

Each of the three modulator packages consists of one or two ferroelectric liquid crystal (FLC) modulators from Displaytech, each being a half-wave retarder at 630.2 nm, and fixed polymer retarders from Meadowlark. An FLC corresponds to a true zero-order

Stokes parameters	wavelength [nm]	construction
I, Q, U, V	630.2	interference filter $\lambda/2$ FLC at $-22.5°$ $\lambda/4$ retarder at $0°$ $\lambda/2$ FLC at $0°$ $\lambda/4$ retarder at $-22.5°$ polarizing beam-splitter at $0°$
I, V	630.2	interference filter $\lambda/4$ retarder at $0°$ $\lambda/2$ FLC at $0°$ polarizing beam-splitter at $0°$
I, V	854.2	interference filter $\lambda/6$ (at 854.2 nm) retarder at $-90°$ $\lambda/2$ (at 630.2 nm) FLC at $-20.6°$ polarizing beam-splitter at $0°$

TABLE 7. Design of the three modulator packages for the SOLIS Vector-Spectromagnetograph.

half-wave plate whose fast axis can be switched between $0°$ and $45°$ within about 40 μs. There are no polarizers in the modulator packages.

The modulation scheme for vector polarimetry is an improved version of the schemes proposed by Gandorfer (1999) and Rabin (private communication). The SOLIS VSM scheme has the advantage that both Stokes Q and U have the same noise characteristics and have slightly better SNR than Stokes V. In addition, it does not require any eighth-wave plates. Furthermore, the scheme is less sensitive to temperature fluctuations and more efficient than the scheme proposed by Gandorfer.

Table 7 outlines the constructions of the three modulator packages. The positive Q direction is defined at $22.5°$ with respect to the spectrograph slit and the polarizing beam-splitter.

The polarizing beam splitters are located just in front of the detectors to produce two orthogonal, linearly polarized spectral images. The images will be offset in the spectral direction with two calcite plates and a half-wave plate in between so that both ordinary and extraordinary beams travel equal optical path lengths and experience the same amount of crystal astigmatism, which is compensated for by a weak cylinder lens. The spectrograph and the associated optics are built in such a way as to minimize the instrumental polarization between the modulators and the polarizing beam splitters.

The advantage of this approach is that there are no moving parts for the polarization analysis, that the switching of the polarization states can occur rapidly, and that both polarization states are detected simultaneously after having passed through the same optics (except for the polarizing beam-splitter and the camera window).

5.4.4. *Instrumental polarization and cross-talk*

The only optical elements whose polarization properties are not calibrated regularly are the entrance window and the primary and secondary mirrors. All the other optical elements are located after the polarization calibration optics. Therefore, the combined properties of those elements can be calibrated regularly. However, we still try to minimize the polarization introduced by those elements because the coupling of instrumental polarization and non-linearities in the CCD camera read-out electronics can lead to effects that are hard to calibrate.

The static birefringence of the window is due to remaining stress from the annealing process. Measurements of the mounted window showed values of less than 3 nm. The expected crosstalk is nevertheless unacceptable. We will therefore measure the window Muller matrix with large sheet polarizers and take it into account when reducing the data.

5.4.5. *Data Reduction*

The first level of data reduction produces spectra of Stokes parameters that need to be reduced to physical quantities such as field strength, azimuth, and inclination. A quick look analysis, which provides rapid derivation of approximate values for the physical quantities will be available within 10 minutes after the observations have been taken. The fully analyzed data, which provides accurate determinations of the physical quantities, will be available within 24 hours of the observations.

Jack Harvey, Peter Povel, and Jan Stenflo taught me most of what I know about polarimetry. I thank Javier Trujillo Bueno, Fernando Moreno Insertis, Lourdes González, and Nieves Villoslada for an excellently organized winter school and the audience for many inspiring discussion. I also thank my wife Karen and my son Philip for their support during the preparation of the lectures and the manuscript. The National Solar Observatory is one of the National Optical Astronomy Observatories, which are operated by the Association of Universities for Research in Astronomy, Inc. (AURA) under cooperative agreement with the National Science Foundation.

REFERENCES

ALBERT, A. 1972 *Regression and the Moore-Penrose Pseudoinverse.* Academic.

AZZAM, R. M. A. & BASHARA, N. M. 1987 *Ellipsometry and Polarized Light.* North-Holland.

BERNASCONI, P., RUST, D., MURPHY, G., & EATON H. 1999 High Resolution Polarimetry With a Balloon-Borne Telescope: The Flare Genesis Experiment. In *High Resolution Solar Physics: Theory, Observations, and Techniques* (ed. T. R. Rimmele, K. S. Balasubramaniam, & R. R. Radick). ASP Conf. Ser., vol. 183, pp. 279–287. ASP.

BIANDA, M., SOLANKI, S. K., & STENFLO, J. O. 1998 Hanle depolarisation in the solar chromosphere. *Astron. Astrophys.* **331**, 760–770.

BILLINGS, B. H. 1990 *Selected Papers on Polarization.* SPIE Optical Engineering Press.

CHIPMAN, R. A. 1995 Mechanics of polarization ray tracing. *Opt. Eng.* **34**, 1636–1645.

CLARKE, D. & GRAINGER, J. F. 1971 *Polarized Light and Optical Measurements.* Pergamon Press.

COLLETT, E. 1993 *Polarized light: fundamentals and applications.* Marcel Decker.

DEL TORO INIESTA, J. C. & COLLADOS, M. 2000 Optimum modulation and demodulation matrices for solar polarimetry. *Applied Optics* **39**, 1637–1642.

DONATI, J.-F., SEMEL, M., REES, D. E., TAYLOR, K., & ROBINSON, R. D. 1990 Detection of a magnetic region on HR 1099. *Astron. Astrophys.* **232**, L1–L4.

DONATI, J.-F., CATALA, C., WADE, G. A., GALLOU, G., DELAIGUE, G., RABOU, P. 1999 A dedicated polarimeter for the MuSiCoS échelle spectrograph. *Astron. Astrophys. Suppl. Ser.* **134**, 149–159.

GALE, M. T., PEDERSEN, J., SCHUETZ, H., POVEL, H., GANDORFER, A., STEINER, P., & BERNASCONI, P. 1997 Active alignment of replicated microlens arrays on a charge-coupled device imager. *Opt. Eng.* **36**, 1510–1517.

GANDORFER, A. 1999 Ferroelectric retarders as an alternative to piezoelastic modulators for use in solar Stokes vector polarimetry. *Opt. Eng.* **38**, 1402–1408.

GANDORFER, A. 2000 *The second solar spectrum.* vdf.

GEHRELS, T. 1974 *Planets, Stars and Nebulae studied with photopolarimetry.* The University of Arizona Press.

HARVEY, J. E. & VERNOLD, C. L. 1997 Transfer function characterization of scattering surfaces revisited. *SPIE Proc.* **3141**, 113–127.

ISMAIL, M. A. 1986 Transmission of light by slit designed for astronomical spectrograph. *Astropys. and Space Science* **22**, 1–32.

KELLER, C. U. & SHEELEY, N. R. 1999 Scattering polarization in the chromosphere. In *Proc. of the 2nd Solar Polarization Workshop* (ed. K. N. Nagendra & J. O. Stenflo). pp. 17–30. Kluwer.

KELLER, C. U., AEBERSOLD, F., EGGER, U., POVEL, H. P., STEINER, P., & STENFLO, J. O. 1992 Zurich Imaging Stokes Polarimeter – Design Review. *LEST Technical Report* **53**. University of Oslo.

KELLER, C. U., DEUBNER, F. L., EGGER, U., FLECK, B., & POVEL, H. P. 1994 On the strength of solar intra-network fields. *Astron. Astrophys.* **286**, 626–634.

KLIGER, D. S., LEWIS J. W. & RANDALL, C. E. 1990 *Polarized Light in Optics and Spectroscopy.* Academic Press.

LEROY, J.-L. 2000 *Polarization of Light and Astronomical Observations.* Gordon and Breach.

LITES, B. W. 1996 Performance Characteristics of the Advanced Stokes Polarimeter. *Solar Phys.* **163**, 223–230.

LIVINGSTON, W. C. 1993 *Selected Papers on Instrumentation in Astronomy.* SPIE.

MACLEOD, H. A. 1969 *Thin-Film Optical Filters.* Elsevier.

MARTÍNEZ PILLET, V., COLLADOS, M., BELLOT RUBIO, L. R., RODRÍGUEZ HIDALGO, I., RUIZ COBO, B., SOLTAU D. 1999 TIP: The Tenerife Infrared Polarimeter AG Abstract Services **15**.

OAKBERG, T. C. 1995 Modulated interference effects: use of photoelastic modulators with lasers. *Opt. Eng.* **34**, 1545–1550.

PANCHARATNAM, S. 1955 Achromatic combinations of birefringent plates. Part II. An achromatic quarter-wave plate. Proc. Indian Acad. Sci. **A41**, 137–144.

POVEL, H. P. 1995 Imaging Stokes polarimetry with piezoelastic modulators and charge-coupled-device image sensors. *Opt. Eng.* **34**, 1870–1878.

POVEL, H. P., AEBERSOLD, H., & STENFLO, J. O. 1990 Charge Coupled Device Image Sensor as a Demodulator in a 2D Polarimeter with piezo-elastic Modulators. *Applied Optics* **29**, 1186–1190.

POVEL, H. P., KELLER, C. U., & YADIGAROGLU, I.-A. 1994 Two-dimensional Polarimeter with a Charge-Coupled-Device Image Sensor and a Piezo-elastic Modulator. *Applied Optics* **33**, 4254–4260.

RAMACHANDRAN, G. N. & RAMASESHAN, S. 1962 Crystal Optics. In *Encyclopedia of Physics* (ed. S. Flügge) vol. XXXV/1.

SÁNCHEZ ALMEIDA, J. 1994 Instrumental polarization in the focal plane of telescopes. 2: Effects induced by seeing. *Astron. Astrophys.* **292**, 713–721.

SÁNCHEZ ALMEIDA, J. & MARTÍNEZ PILLET, V. 1992 Instrumental polarization in the focal plane of telescopes. *Astron. Astrophys.* **260**, 543–555.

SANKARASUBRAMANIAN, K., SAMSON, J. P. A., & VENKATAKRISHNAN, P. 1999 Measurement of instrumental polarisation of the Kodaikanal tunnel tower telescope. In *: Proc. of the 2nd Solar Polarization Workshop* (ed. K. N. Nagendra & J. O. Stenflo). pp. 313–320. Kluwer.

SEMEL, M., DONATI, J.-F., & REES, D. E. 1993 Zeeman-Doppler imaging of active stars. 3: Instrumental and technical considerations. *Astron. Astrophys.* **278**, 231–237.

SEN, A. K. & KAKATI, M. 1997 Instrumental polarization caused by telescope optics during wide field imaging. *Astron. Atsrophys. Suppl. Ser.* **126**, 113–119.

SHURCLIFF, W. A. 1962 *Polarized light-production and use.* Harvard University Press

SKUMANICH, A., LITES, B. W., MARTÍNEZ PILLET, V., & SEAGRAVES, P. 1997 The Calibration of the Advanced Stokes Polarimeter. *Astrophys. J. Suppl.* **110**, 357–380.

SLATER, J. C. 1942 *Microwave Transmission*. McGraw-Hill.

STENFLO, J. O. 1984 Solar magnetic and velocity-field measurements: new instrumental concepts. *Applied Optics* **23**, 1267–1278.

STENFLO, J. O. 1991 Optimization of the LEST polarization modulation system. *LEST Technical Report* **44**. University of Oslo.

STENFLO J. O. 1994 *Solar Magnetic Fields: Polarized Radiation Diagnostics*. Kluwer.

STENFLO, J. O., KELLER, C. U., & POVEL, H. P. 1992 Demodulation of all Four Stokes Parameters With a Single CCD. ZIMPOL II - Conceptual Design. *LEST Technical Report* **54**. University of Oslo.

STOCKMAN, H. S. 1982 Differential imaging using charge-coupled image device (CCD) imagers with on-chip charge storage. *SPIE Proc.* **331**, 76–80.

SWINDELL, W. 1975 *Polarized Light*. Dowden, Hutchinson & Ross.

TINBERGEN, J. 1996 *Astronomical polarimetry*. Cambridge University Press

WAHLSTROHM, E. E. 1960 *Optical Crystallography*. Wiley.